■ ■ ■ ■ ■ ■ ■ ■ ■

SPICE
for Power Electronics
and Electric Power

Muhammad H. Rashid

Ph.D., Fellow IEE
Professor of Electrical Engineering
Purdue University at Fort Wayne

PRENTICE HALL, Englewood Cliffs, New Jersey 07632

Library of Congress Cataloging-in-Publication Data

Rashid, M. H.
    SPICE for power electronics and electric power / Muhammad H.
Rashid.
        p.   cm.
    Includes bibliographical references and index.
    ISBN 0-13-030420-4
    1. Power electronics—Data processing.   2. Electronic circuit
design—Data processing.   3. Electric circuit analysis—Data
processing.   4. SPICE (Computer file)   I. Title.
TK7881.15.R38   1993
621.31'7—dc20   92-40235
                                                              CIP

Publisher: Alan Apt
Production Editor: Mona Pompili
Cover Designer: Wanda Lubelska Design
Copy Editor: Barbara Zeiders
Prepress Buyer: Linda Behrens
Manufacturing Buyer: Dave Dickey
Editorial Assistant: Shirley McGuire

 ©1993 by Prentice-Hall, Inc.
A Paramount Communications Company
Englewood Cliffs, New Jersey 07632

Printed in the United States of America

10 9 8 7 6 5 4 3

ISBN 0-13-030420-4

Prentice-Hall International (UK) Limited, London
Prentice-Hall of Australia Pty. Limited, Sydney
Prentice-Hall Canada, Inc., Toronto
Prentice-Hall Hispanoamericana, S.A., Mexico
Prentice-Hall of India Private Limited, New Delhi
Prentice-Hall of Japan, Inc., Tokyo
Simon & Schuster Asia Pte. Ltd, Singapore
Editora Prentice-Hall do Brasil, Ltda., Rio de Janeiro

TRADEMARK INFORMATION

PSpice and Probe are registered trademarks of
    MicroSim Corporation.
IBM-PC is a registered trademark of International
    Business Machines Corporation.
Macintosh II is a registered trademark of Apple
    Computer Inc.
WordStar and WordStar 2000 are the registered
    trademarks of Micropro International
    Corporation.
Word is a registered trademark of Microsoft
    Corporation.
Program Editor is a registered trademark of
    WordPerfect Corporation.

To my parents, my wife Fatema,
and
my children, Fa-eza, Farzana, and Hasan

Preface

Power electronics is normally offered as a technical elective. It is an application oriented and interdisciplinary course, which requires a background in mathematics, electrical circuits, control system, analog and digital electronics, microprocessor, electric power, and electrical machines.

The understanding of the operation of a power electronics circuit requires a clear knowledge of the transient behavior of current and voltage waveforms for each and every circuit element at every instant of time.

These features make power electronics a difficult course for students to understand, and for professors to teach. A laboratory helps in understanding power electronics and its control interfacing circuits. Development of a power electronics laboratory is relatively expensive compared to other courses in EE curriculum. The power electronics is playing a key role in industrial power control.

The Engineering Accreditation Commission of the Accreditation Board for Engineering and Technology (EAC/ABET) requirements specify the computer integration and design content in the EE-curriculum. To be competitive, a power electronics course should integrate a design content of approximately 50% and an extensive use of computer-aided analysis.

The student version of PSpice, which is available free to students, is ideal for class-room use and for assignments requiring computer-aided simulation and analysis. Without any additional resources and lecture time, PSpice can also be integrated into power electronics.

Probe is a graphics post-processor in PSpice, and is very useful in plotting the results of simulation, especially with the capability of arithmetic operation it can be used to plot impedance, power, etc. Once the students get experience in simulating on PSpice, they really appreciate the advantages of .Probe command. *Probe* is an option on PSpice, but it comes with the student version. Running *Probe* does not require a math co-processor. The students can also get the normal printer output or printer plotting. The prints and plots are very helpful for students in relating their theoretical understanding and making judgment on the merits of a circuit and its characteristics.

Probe is like a theoretical oscilloscope with the special features to perform arithmetic operations, and it can be used as a laboratory bench to view the waveforms of currents, voltages, power, power factor, etc. The Fourier analysis gives the total harmonic distortion (THD) of any waveform.

The capability of *Probe* along with other features to represent data in Table, Value, Function, Polynomial, Laplace, Param, Step makes PSpice a versatile simulation tool for power electronics and electric power courses. Students can design power electronics circuits, use the PSpice simulator to verify the design, and make necessary design modifications. In the absence of a dedicated power electronics laboratory, the laboratory assignments could be design problems, which are to be simulated and verified by PSpice.

This book is based on the author's experience in integrating 50% design content, and SPICE on a power electronics course of 3 credit hours. The students were assigned design problems and asked to use PSpice to verify their designs by plotting and/or printing the output waveform(s), and to confirm the ratings of devices and components by plotting the instantaneous voltage, current and power. The objective of this book is to integrate the SPICE simulator to a power electronics course at the junior level or senior level with a minimum amount of time and effort. This book assumes no prior knowledge about the SPICE simulator, and introduces the applications of various SPICE commands through numerous examples of power electronics circuits.

This book can be divided into nine parts: (1) introduction to SPICE simulation—chapters 1, 2 and 3; (2) source and element modeling—chapters 4 and 5; (3) SPICE commands—chapter 6; (4) Rectifiers—Chapters 7 and 8; (5) AC voltage controllers—Chapter 9; (6) DC choppers—chapters 10 and 11; (7) Inverters—chapters 12 and 13; (8) Control Applications—chapters 14 and 15, and (9) difficulties—chapter 16. Chapters 7 to 13 use simple models for power semiconductor switches, leaving the complex models for special projects and assignments. Chapter 15 uses the simple circuit models of DC motors and AC inductor motors to predict their control characteristics. Two reference tables are included to aid in choosing a device, component, and/or command.

This book is intended to demonstrate the techniques for power conversions and the quality of the output waveforms, rather than the accurate modeling of power semiconductor devices. This approach has the advantage that the students can compare the results with those obtained in a class-room environment with simple switch models of devices.

This book can be used as a textbook on SPICE for students specializing in power electronics, and power systems. It can also be a supplement to any standard textbook on power electronics, and power systems. The following sequence is recommended:

1. Supplement to a basic power systems (or electrical machine) course with three hours of lectures (or equivalent Lab hours) and self-study assignments

from chapters 1 to 6. Starting from chapter 2, the students should work with PCs.

2. Continue as a supplement to a Power Electronics course with two hours of lectures (or equivalent Lab hours) and self-study assignments from chapters 7 to 15.

Without any prior experience on SPICE and integrating SPICE at the power electronics level, two hours of lectures (or equivalent Lab hours) are recommended on chapters 1 to 6. Chapters 7 to 15 could be left for self-study assignments. From the author's experience in the class, it has been observed that after two lectures of 50 minutes duration, all students could solve assignments independently without any difficulty. The class could progress in a normal manner with one assignment per week on power electronics circuits simulation and analysis with SPICE.

The book has sections on suggested laboratory experiments and design problems on power electronics. The complete laboratory guidelines for each experiment are presented. Thus, the book can also be used as a laboratory manual for power electronics. The design problems can be used as assignments for a design-oriented simulation laboratory.

Although the materials of this book have been developed for engineering students, but the book is also strongly recommended for EET-students specializing in power electronics, and power systems.

Muhammad H. Rashid
Fort Wayne, Indiana

Acknowledgments

I would like to thank the following reviewers for their comments and suggestions:

Frederick C. Brockhurst—Rose-Hulman Institute of Technology
A. P. Saki Meliopoulos—Georgia Institute of Technology
Peter Lauritzen—University of Washington
Saburo Matsusaki—TDK Corporation, Japan

It has been a great pleasure working with the editorial staff—Alan Apt, Sondra Chavez, and Shirley McGuire. Finally, I would thank my family for their love, patience, and understanding.

PSpice SOFTWARE

The PSpice student version software is available from Prentice-Hall, Inc. To order software, please see the form which is included at the end of this book.

- PSpice student version disks (two 5¼″) IBM PC compatible (73476-4)
- PSpice student version disks (two 3½″) IBM PC compatible (73475-6)
- PSpice student version disks (three 3½″) MAC II compatible (73474-7)

It is free to instructors who want to duplicate it for classroom use. Instructors can obtain the student version software directly from:

MicroSim Corporation
20 Fairbanks, Irvine, CA 92718, USA
Tel: (800) 245-3022 (Toll Free)
     (714) 770-3022
Fax: (714) 455-0554

Any comments and suggestions regarding this book are welcomed and should be addressed to the author.

Dr. Muhammad H. Rashid
Professor of Electrical Engineering
Indiana University—Purdue University at Fort Wayne
Fort Wayne, IN 46805-1499, USA

■■■■■■■■■

Contents

Chapter 1

■■■■■■■■

Introduction

1-1 INTRODUCTION

Electronic circuit design requires accurate methods of evaluating circuit performance. Because of the enormous complexity of modern integrated circuits, computer-aided circuit analysis is essential and can provide information about circuit performance that is almost impossible to obtain with laboratory prototype measurements. Computer-aided analysis makes possible the following procedures:

1. Evaluation of the effects of variations in such elements as resistors, transistors, and transformers
2. Assessment of performance improvements or degradations
3. Evaluation of the effects of noise and signal distortion without the need for expensive measuring instruments
4. Sensitivity analysis to determine the permissible bounds due to tolerances on every element value or parameter of active elements
5. Fourier analysis without expensive wave analyzers
6. Evaluation of the effects of nonlinear elements on circuit performance
7. Optimization of the design of electronic circuits in terms of circuit parameters

SPICE (simulation program with integrated circuit emphasis) is a general-purpose circuit program that simulates electronic circuits. SPICE can perform analyses on various aspects of electronic circuits: the operating (or quiescent) points of transistors, time-domain response, small-signal frequency response, and so on. SPICE contains models for common circuit elements, active as well as passive, and it is capable of simulating most electronic circuits. It is a versatile program and is widely used in both industry and academic institutions.

Until recently, SPICE was available only on mainframe computers. In addition to the cost of the computer system, such a machine can be inconvenient for classroom use. In 1984, MicroSim introduced the PSpice simulator, which is similar to the Berkeley SPICE and runs on an IBM-PC or compatible, and is available free of cost to students for classroom use. PSpice thus widens the scope for the integration of computer-aided circuit analysis into electronic circuits courses at the undergraduate level. Other versions of PSpice, which will run on the Macintosh II, 486-based processor, VAX, SUN, NEC, and other computers, are also available.

1-2 DESCRIPTIONS OF SPICE

PSpice is a member of the SPICE family of circuit simulators, all of which originate from the SPICE2 circuit simulator, whose development spans a period of about 30 years. During the mid-1960s, the program ECAP was developed at IBM [1]. Later, ECAP served as the starting point for the development of the program CANCER at the University of California (UC)–Berkeley in the late 1960s. Using CANCER as the basis, SPICE was developed at Berkeley in the early 1970s. During the mid-1970s, SPICE2, which is an improved version of SPICE, was developed at UC–Berkeley. The algorithms of SPICE2 are robust, powerful, and general in nature, and SPICE2 has become an industry-standard tool for circuit simulation. SPICE3, a variation of SPICE2, is designed especially to support computer-aided design (CAD) research programs at UC–Berkeley. As the development of SPICE2 was supported using public funds, this software is in public domain, which means that it may be used freely by all U.S. citizens.

SPICE2, referred to simply as SPICE, has become an industry standard. The input syntax for SPICE is a free-format style that does not require that data be entered in fixed column locations. SPICE assumes reasonable default values for unspecified circuit parameters. In addition, it performs a considerable amount of error checking to ensure that a circuit has been entered correctly.

PSpice, which uses the same algorithms as SPICE2, is equally useful for simulating all types of circuits in a wide range of applications. A circuit is described by statements stored in a file called the *circuit file*. The circuit file is read by the SPICE simulator. Each statement is self-contained and independent of every other statement and does not interact with other statements. SPICE (or PSpice) statements are easy to learn and to use.

1-3 TYPES OF SPICE

The commercially supported versions of SPICE2 can be divided into two types: mainframe versions and PC-based versions. Their methods of computation may differ, but their features are almost identical. However, some may include such additions as a pre-processor or shell program to manage input and provide interactive control, as well as a post-processor used to refine the normal SPICE output. A person who is used to one SPICE version (e.g., PSpice) should be able to work with other versions.

Mainframe versions are:

HSPICE (from Meta-Software), which is designed for integrated-circuit design with special device models

RAD-SPICE (from Meta-Software), which simulates circuits subjected to ionizing radiation

IG-SPICE (from A.B. Associates), which is designed for "interactive" circuit simulation with graphics output

I-SPICE (from NCSS Time Sharing), which is designed for "interactive" circuit simulation with graphics output

Precise (from Electronic Engineering Software)

PSpice (from MicroSim)

AccuSim (from Mentor Graphics)

Spectre (from Cadence Design)

SPICE-Plus (from Valid Logic)

The PC versions include the following:

AllSpice (from Acotech)

IS-SPICE (from Intusoft)

Z-SPICE (from Z-Tech)

SPICE-Plus (from Analog Design Tools)

DSPICE (from Daisy Systems)

PSpice (from MicroSim)

1-4 LIMITATIONS OF PSpice

As a circuit simulator, PSpice has the following limitations:

1. The PC-based student version of PSpice is restricted to circuits with 10 transistors only. However, the professional (or production) version can simulate a circuit with up to 200 bipolar transistors (or 150 MOSFETs).

2. The program is not interactive; that is, the circuit cannot be analyzed for various component values without editing the program statements.

3. PSpice does not support an iterative method of solution. If the elements of a circuit are specified, the output can be predicted. On the other hand, if the output is specified, PSpice cannot be used to synthesize the circuit elements.

4. The input impedance cannot be determined directly without running the graphic post-processor, Probe. Although the student version does not re-

quire a floating-point co-processor for running Probe, the professional version does require such a co-processor.

5. To run the PC version requires 512 kilobytes of memory (RAM).

6. Distortion analysis is not available.

7. The output impedance of a circuit cannot be printed or plotted directly.

8. The student version will run *with* or *without* a floating-point co-processor (8087, 80287, 80387, or 80487). If a co-processor is present, the program will run at full speed. Otherwise, it will run 5 to 15 times slower. The professional version requires a co-processor.

SUGGESTED READING

1. R. W. Jensen and M. D. Liberman, *IBM Electronic Circuit Analysis Program and Applications*. Englewood Cliffs, N.J.: Prentice Hall, 1968.

2. R. W. Jensen and L. P. McNamee, *Handbook of Circuit Analysis Languages and Techniques*. Englewood Cliffs, N.J.: Prentice Hall, 1976.

3. *PSpice Manual*. Irvine, Calif.: MicroSim Corporation, 1992.

Chapter 2

■■■■■■■■■

Circuit descriptions

2-1 INTRODUCTION

PSpice is a general-purpose circuit program that can be applied to simulate electronic and electrical circuits. A circuit must be specified in terms of element names, element values, nodes, variable parameters, and sources. The input to the circuit shown in Fig. 2-1(a) is a pulse voltage, as shown in Fig. 2-1(b). The circuit is to be simulated for calculating and plotting the transient response from 0 to 400 μs with an increment of 1 μs. The Fourier series coefficients and total harmonic distortion (THD) are to be printed. We discuss (1) how to describe this circuit to PSpice, (2) how to specify the type of analysis to be performed, and (3) how to define the output variables required. Description and analysis of a circuit require that the following be specified:

Nodes
Element values
Circuit elements
Element models
Sources
Output variables
Types of analysis
PSpice output commands
Format of circuit files
Format of output files

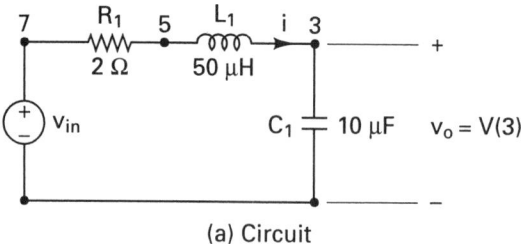

(a) Circuit

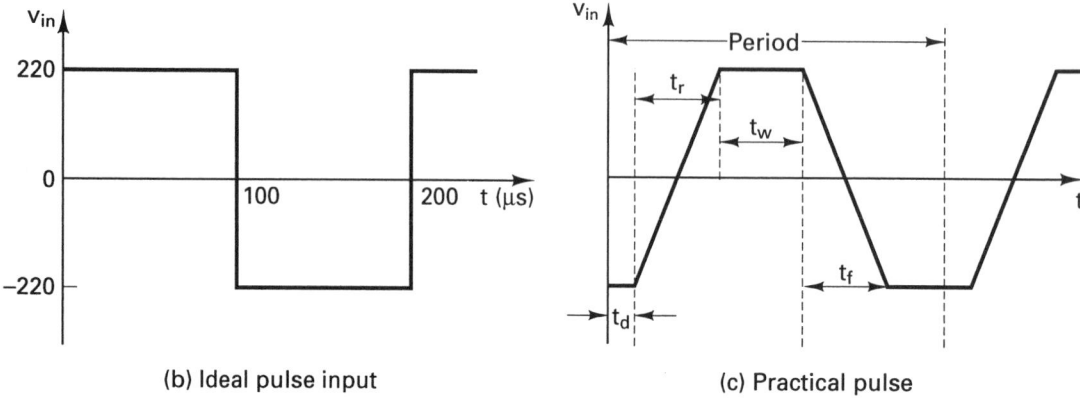

(b) Ideal pulse input

(c) Practical pulse

Figure 2-1 *RLC* circuit with pulse input.

2-2 NODES

Nodes numbers are assigned to the circuit of Fig. 2-1. Elements are connected between nodes. The node numbers to which an element is connected are specified after the name of the element. Node numbers must be integers from 0 to 9999 but need not be sequential. Node 0 is predefined as the ground. All nodes must be connected to at least two elements and should therefore appear at least twice. All nodes must have a dc path to the ground node. This condition, which is not met in all circuits, is normally satisfied by connecting very large resistors (see Section 16-10).

2-3 ELEMENT VALUES

The value of a circuit element is written after the nodes to which the element is connected. The values are written in standard floating-point notation with optional scale and units suffixes. Some values without suffixes that are allowable by PSpice are

5   5.   5.0   5E+3   5.0E+3   5.E3

There are two types of suffixes: the scale suffix and the units suffix. The scale suffix multiplies the number that they follow. Scale suffixes recognized by PSpice are

F	1E−15
P	1E−12
N	1E−9
U	1E−6
MIL	25.4E−6
M	1E−3
K	1E3
MEG	1E6
G	1E9
T	1E12

The units suffixes that are normally used are

V	volt
A	ampere
HZ	hertz
OHM	ohm
H	henry
F	farad
DEG	degree

The first suffix is always the scale suffix; the units suffix follows the scale suffix. In the absence of a scale suffix, the first suffix may be a units suffix, provided that it is not the symbol of scale suffixes. The units suffixes are always ignored by PSpice. If the value of an inductor is 15 μH, it is written as 15U or 15UH. In the absence of scale and units suffixes, the units of voltage, current, frequency, inductance, capacitance, and angle are understood by default to be volts, amperes, hertz, henrys, farads, and degrees, respectively. PSpice ignores any units suffix, and the following values are equivalent:

25E−3   25.0E−3   25M   25MA   25MV   25MOHM   25MH

Notes

1. The scale suffixes are all uppercase letters.
2. M means "milli," not "mega." 2 MΩ is written as 2MEG or 2MEGOHM.

Circuit elements are identified by name. A name must start with a letter symbol corresponding to the element, but after that it can contain either letters or numbers. Names can be up to eight characters long. Table 2-1 shows the first letter of elements and sources. For example, the name of a capacitor must start with a C.

TABLE 2-1 SYMBOLS OF CIRCUIT ELEMENTS AND SOURCES

First letter	Circuit elements and sources
B	GaAs MES field-effect transistor
C	Capacitor
D	Diode
E	Voltage-controlled voltage source
F	Current-controlled current source
G	Voltage-controlled current source
H	Current-controlled voltage source
I	Independent current source
J	Junction field-effect transistor
K	Mutual inductors (transformer)
L	Inductor
M	MOS field-effect transistor
Q	Bipolar junction transistor
R	Resistor
S	Voltage-controlled switch[a]
T	Transmission line
V	Independent voltage source
W	Current-controlled switch[a]

[a] Not available in SPICE2 but available in SPICE3.

The format for describing passive elements is

⟨element name⟩ ⟨positive node⟩ ⟨negative node⟩ ⟨value⟩

where the current is assumed to flow from the positive node, N+, to the negative node, N−. The formats for passive elements are described in Chapters 4 and 5. The passive elements of Fig. 2-1 are described as follows:

The statement that R_1 has a value of 2 Ω and is connected between nodes 7 and 5 is

```
R1   7   5   2
```

The statement that L_1 has a value of 50 μH and is connected between nodes 5 and 3 is

```
L1   5   3   50UH
```

The statement that C_1 has a value of 10 μF and is connected between nodes 3 and 0 is

```
C1   3   0   10UF
```

2-5 ELEMENT MODELS

The values of some circuit elements are dependent on other parameters, such as the initial condition of an inductor, the capacitance as a function of its voltage, and the resistance as a function of temperature. Models may be used to assign values to the various parameters of circuit elements. The techniques for specifying models of sources, passive elements, and active elements are described in Chapters 4, 5, and 6, respectively.

We shall represent the source voltage by a pulse, which has a model of the form

```
PULSE  (−VS +VS TD TR TF PW PER)
```

where −VS, +VS = negative and positive values of the pulse, respectively
TD, TR, TF = delay time, rise time, and fall time, respectively
PW = width of the pulse
PER = period of the pulse

In practice, it is not possible to generate a pulse with a zero value for the delay time and the fall time. Thus TR and TF should have a small but finite value. A practical pulse is shown in Fig. 2-1(c). Let us assume that TD = 0 and TR = TF = 1 ns. The model for the input voltage of Fig. 2-1(b) becomes

```
PULSE  (−220V   220V   0   1NS   1NS   100US   200US)
```

2-6 SOURCES

Voltage (or current) sources can be either dependent or independent. Also listed in Table 2-1 are the letter symbols for the types of source. An independent voltage (or current) source can be dc, sinusoidal, pulse, exponential, polynomial, piecewise linear, or single-frequency frequency modulation. Models for describing source parameters are described in Chapter 4.

The format for a source is

⟨source name⟩   ⟨positive node⟩   ⟨negative node⟩   ⟨source model⟩

where the current is assumed to flow into the source from positive node N+ to negative node N−. The order of nodes N+ and N− is critical. Assuming that node 7 has a higher potential than node 0, the statement for the input source v_{in}

connected between nodes 7 and 0 is

```
VIN  7  0  PULSE (—220V  220V  0  1NS  1NS  100US  200US)
```

2-7 OUTPUT VARIABLES

PSpice has some unique features in printing or plotting output voltages or currents. The various types of output variables permitted by PSpice are discussed in Chapter 3. The voltage of node 3 with respect to node 0 is specified by V(3,0) or V(3). The voltage of node 7 with respect to node 0 is specified by V(7,0) or V(7).

2-8 TYPES OF ANALYSIS

PSpice allows various types of analysis. Each type is invoked by including its command statement. For example, a statement beginning with a .DC command will cause a dc sweep to be done. The types of analysis and their corresponding . (dot) commands are:

Dc analysis:
   Dc sweep of an input voltage–current source, a model parameter, or temperature (.DC)
   Linearized device model parameterization (.OP)
   Dc operating point (.OP)
   Small-signal transfer function (Thévenin's equivalent) (.TF)
   Small-signal sensitivities (.SENS)

Transient analysis:
   Time-domain response (.TRAN)
   Fourier analysis (.FOUR)

Ac analysis:
   Small-signal frequency response (.AC)
   Noise analysis (.NOISE)

It should be noted that the . (dot) is an integral part of a command. The various dot commands are discussed in detail in Chapter 6.
   The format for performing transient response is

```
TRAN  TSTEP  TSTOP
```

where TSTEP is the time increment and TSTOP is the final (stop) time. Therefore, the statement for the transient response from 0 to 400 μs with a 1-μs increment is

```
.TRAN   1US   400US
```

PSpice performs a Fourier analysis from the results of the transient analysis. The command for the Fourier analysis of voltage V(N) is

```
.FOUR   FREQ   VN
```

where FREQ is the fundamental frequency. Thus the duration of the transient analysis must be at least one period long, PERIOD = 1/FREQ. The command for the Fourier analysis of voltage V(3) with PERIOD = 200 μs and FREQ = 1/200 μs = 5 kHz is

```
.FOUR   5KHZ   V(3)
```

2-9 PSpice OUTPUT COMMANDS

The most common forms of output are print tables and plots. The transient response (.TRAN), dc sweep (.DC), frequency response (.AC), and noise (.NOISE) analysis can produce output in the form of print tables and plots. The command for output in the form of tables is .PRINT, that for output plots is .PLOT, and that for graphical output is .PROBE.

The statement for the plots of V(3) and V(7) from the results of transient analysis is

```
.PLOT   TRAN   V(3)   V(7)
```

The statement for the tables of V(3) and V(7) from the results of transient analysis is

```
.PRINT   TRAN   V(3)   V(7)
```

The outputs of .PRINT and .PLOT commands are stored in an output file created automatically by PSpice.

Probe is the *graphics post-processor* of PSpice, and the statement for the .PROBE command is

```
.PROBE
```

which causes the results of simulation to be available in graphical outputs on both the display and the hard copy. After executing the .PROBE command, Probe will put up a menu on the screen to obtain graphical output. It is very easy to use Probe. With the .PROBE command, there is no need for the .PLOT command. .PLOT generates the plot on the output file, while .PROBE gives graphical output on the monitor screen, which can be dumped directly into a plotter and/or printer. The output commands are discussed in Section 6-3.

2-10 FORMAT OF CIRCUIT FILES

A circuit file that is read by PSpice may be divided into five parts: (1) the title, which describes the type of circuit or any comments; (2) the circuit description, which defines the circuit elements and the set of model parameters; (3) the analysis description, which defines the type of analysis; (4) the output description, which defines the way the output is to be presented; and (5) the end of program. The format for a circuit file is as follows,

> Title
> Circuit description
> Analysis description
> Output description
> End-of-file statement (.END)

Notes

1. The first line is the title line, and it may contain any type of text.
2. The last line must be the .END command.
3. The order of the intervening lines is not important and does not affect the results of simulations.
4. If a PSpice statement is longer than one line, the statement can continue to the next line. A continuation line is identified by a plus sign (+) in the first column of the next line. The continuation lines must follow one another in the proper order.
5. A comment line may be included anywhere and is preceded by an asterisk (*).
6. The number of blanks between items is not significant (except for the title line). The tabs and commas are equivalent to blanks. For example, " " and " " and "," and " , " are all equivalent.
7. PSpice statements or comments can be in either upper- or lowercase letters.
8. SPICE2 statements must be in uppercase letters. *It is advisable to type the PSpice statements in uppercase so that the same circuit file can be run on SPICE2 also.*
9. If you are not sure of a command or statement, use that command or statement to run the circuit file to see what happens. SPICE is user-friendly software that provides an error message on the output file.
10. In electrical circuits, subscripts are normally assigned to symbols for voltages, currents, and circuit elements. However, in SPICE, the symbols are

represented without subscripts. For example, v_s, i_s, L_1, C_1, and R_1 are represented by VS, IS, L1, C1, and R1, respectively. As a result, the SPICE circuit description of voltages, currents, and circuit elements is often different from that of the circuit symbols.

2-11 FORMAT OF OUTPUT FILES

The results of simulation by PSpice are stored in an output file. It is possible to control the type and amount of output by various commands. If there is any error in the circuit file, PSpice will display a message on the screen indicating that there is an error and will suggest looking at the output file for details. The output falls into four types:

1. A description of the circuit itself that includes the net list, the device list, the model parameter list, and so on.
2. Direct output from some of the analyses without the .PLOT and .PRINT commands. This includes the output from the .OP, .TF, .SENS, .NOISE, and .FOUR analyses.
3. Prints and plots resulting from .PRINT and .PLOT commands, including output from the .DC, .AC, and .TRAN analyses.
4. Run statistics, which include various types of summary information about the entire run, including times required by various analyses and the amount of memory used.

Example 2-1

The *RLC* circuit of Fig. 2-1(a) is to be simulated on PSpice to calculate and plot the transient response from 0 to 400 μs with an increment of 1 μs. The capacitor voltage V(3) and the current through R_1, I(R1), are to be plotted. The Fourier series coefficients and total harmonic distortion (THD) are to be printed. The circuit file's name is EX2-1.CIR, and the outputs are to be stored in the file EX2-1.OUT. The .PROBE command will make the results available both in display and on hard copy.

Solution The circuit file contains the following statements.

■ ■

Example 2-1   Pulse response of an *RLC* circuit

SOURCE ■
```
* The format for a pulse source is
* PULSE (-VS +VS TD TR TF PW PER)
* Refer to Chapter 4 for modeling sources.
* vin is connected between nodes 7 and 0, assuming that node 7
* is at a higher potential than node 0.
* For -220 to +220 V, a delay of td = 0, a rise time of tr = 1 ns,
* a fall time of tf = 1 ns, a pulse width = 100 µs, and
* a period = 200 µs, the source is described by
VIN   7   0   PULSE (-220V  220V  0  1NS  1NS  100US  200US)
```

■ ■ * R_1, with a value of 2 Ω, is connected between nodes 7 and 5.
* Assuming that current flows into R_1 from node 7 to node 5 and that
* the voltage of node 7 with respect to node 5, V(7,5), is positive,
* R_1 is described by
R1   7   5   2
* L_1, with a value of 50 μH, is connected between nodes 5 and 3.
* Assuming that current flows into L_1 from node 5 to node 3 and that
* the voltage of node 5 with respect to node 3, V(5,3), is positive,
* L_1 is described by
L1   5   3   50UH
* C_1, with a value of 10 μF, is connected between nodes 3 and 0.
* Assuming that current flows into C_1 from node 3 to node 0 and that
* the voltage of node 3 with respect to node 0, V(3), is positive,
* C_1 is described by
C1   3   0   10UF

■ ■ ■ * Transient analysis is invoked by the .TRAN command, whose simple
format is
* .TRAN TSTEP TSTOP
* Refer to Chapter 6 for dot commands. For transient analysis
* from 0 to 400 μs with an increment of 1 μs, the statement is
.TRAN   1US   400US
* Prints the results of transient analysis for V(R1), V(L1), and V(C1):
.PRINT TRAN   V(R1)   V(L1)   V(C1)
* Plots the results of transient analysis for V(3) and I(R1):
.PLOT   TRAN   V(3)   I(R1)
* Refer to Chapter 6 for dot commands. For Fourier analysis
* of V(3) at a fundamental frequency of 5 kHz, the statement is
.FOUR   5KHZ   V(3)
* Graphic output can be obtained simply by invoking the .PROBE command
* (refer to Chapter 6 for dot commands):
.PROBE
* The end of program is invoked by the .END command:
.END
■ ■

If the PSpice programs are loaded in a fixed (hard) disk and the circuit file is stored in a floppy diskette on drive A:, the general command to run the circuit file is

PSPICE A: ⟨input file⟩   A: ⟨output file⟩

For an input file EX2-1.CIR and the output file EX2-1.OUT, the command is

PSPICE A:EX2-1.CIR   A:EX2-1.OUT

If the output file name is omitted, the results are stored by default on an output file that has the same name as the input file and is in the same drive, but with an extension of .OUT. It is a good practice to have .CIR and .OUT extensions on circuit files so that the circuit file and the corresponding output file can be identified. Thus the command can simply be

PSPICE A:EX2-1.CIR

The results of the transient response that are obtained on the display by the .PROBE command are shown in Fig. 2-2. The results of the .PRINT statement can be obtained by printing the contents of the output file EX2-1.OUT. From the output file EX2-1.OUT, the results of the Fourier analysis are as follows:

```
FOURIER COMPONENTS OF TRANSIENT RESPONSE V(3)
DC COMPONENT = −1.726830E−01
```

HARMONIC NO	FREQUENCY (HZ)	FOURIER COMPONENT	NORMALIZED COMPONENT	PHASE (DEG)	NORMALIZED PHASE (DEG)
1	5.000E+03	3.502E+02	1.000E+00	−5.171E+01	0.000E+00
2	1.000E+04	2.718E−01	7.762E−04	−1.795E+02	−1.278E+02
3	1.500E+04	2.245E+01	6.410E−02	−1.517E+02	−1.000E+02
4	2.000E+04	8.506E−02	2.429E−04	1.708E+02	2.225E+02
5	2.500E+04	4.411E+00	1.259E−02	−1.601E+02	−1.084E+02
6	3.000E+04	5.024E−02	1.435E−04	1.753E+02	2.270E+02
7	3.500E+04	1.683E+00	4.806E−03	−1.595E+02	−1.078E+02
8	4.000E+04	3.617E−02	1.033E−04	1.791E+02	2.309E+02
9	4.500E+04	8.936E−01	2.552E−03	−1.608E+02	−1.090E+02

```
TOTAL HARMONIC DISTORTION = 6.555345E+00 PERCENT
```

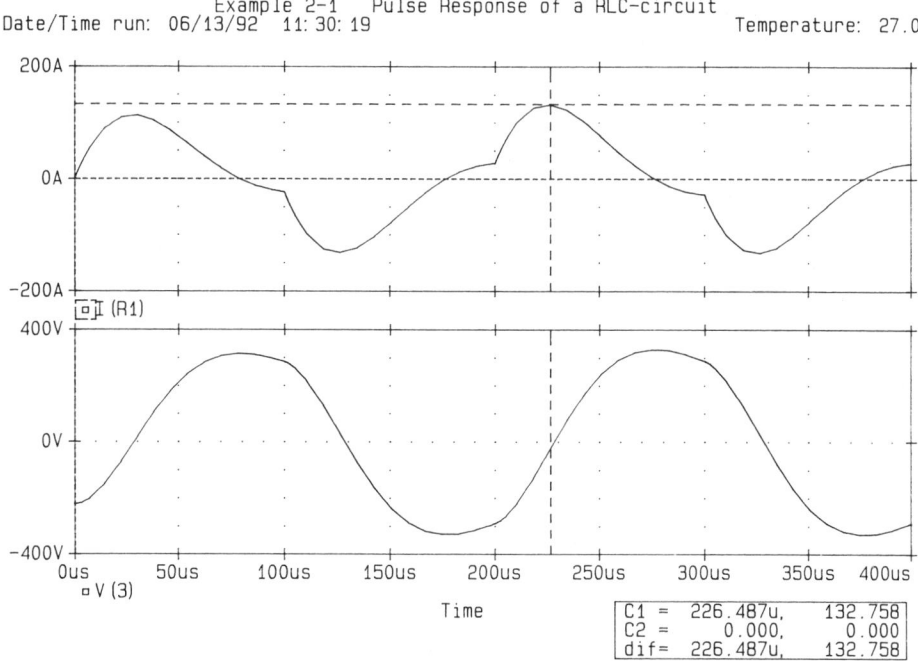

Figure 2-2   Pulse response for Example 2-1.

Example 2-2

Three *RLC* circuits with $R = 2 \ \Omega$, $1 \ \Omega$, and $8 \ \Omega$ are shown in Fig. 2-3(a). The inputs are identical step voltages, as shown in Fig. 2-3(b). Use PSpice to calculate and plot

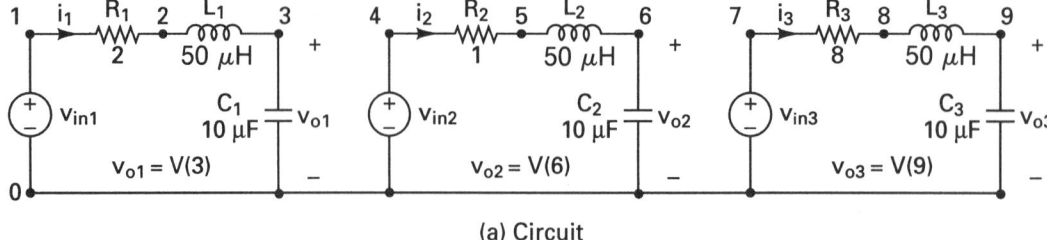

(a) Circuit

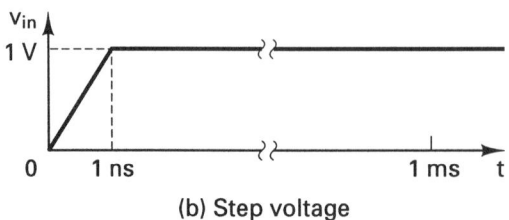

(b) Step voltage

Figure 2-3 *RLC* circuits with step-pulse input voltages.

the transient response from 0 to 400 μs with an increment of 1 μs. The capacitor voltages are the outputs V(3), V(6), and V(9), which are to be plotted. The circuit is to be stored in file EX2-1.CIR, and the outputs are to be stored in file EX2-2.OUT. The results should also be available for display and hard copy by the .PROBE command.

Solution The description of the circuit file is similar to that of Example 2-1, except that the input is a step voltage instead of a pulse voltage. The circuit may be regarded as three *RLC* circuits having three separate inputs. The step signal can be represented by piecewise linear source and it is described in general by

```
PWL (T1 V1 T2 V2 ... TN VN)
```

where VN is the voltage at time TN. Assuming a rise time of 1 ns, the step voltage of Fig. 2-3(b) can be described by

```
PWL (0  0  1NS  1V  1MS  1V)
```

The list of the circuit file is as follows:

■ ■

Example 2-2   Step response of series *RLC* circuits

SOURCE	■ VI1	1	0	PWL (0	0	1NS	1V	1MS	1V)	; Step of 1 V
	VI2	4	0	PWL (0	0	1NS	1V	1MS	1V)	; Step of 1 V
	VI3	7	0	PWL (0	0	1NS	1V	1MS	1V)	; Step of 1 V
CIRCUIT	■ ■ R1	1	2	2						
	L1	2	3	50UH						
	C1	3	0	10UF						
	R2	4	5	1						
	L2	5	6	50UH						
	C2	6	0	10UF						

```
                R3    7    8    8
                L3    8    9    50UH
                C3    9    0    10UF
ANALYSIS  ▪ ▪ ▪ .PLOT    TRAN    V(3)    V(6)    V(9)    ; Plot output voltages
                .TRAN  1US    400US                      ; Transient analysis
                .PROBE                                   ; Graphics post-processor
          .END
          ▪▪▪▪▪▪▪▪▪▪▪▪▪▪▪▪▪▪▪▪▪▪▪▪▪▪▪▪▪▪▪▪▪▪▪▪▪▪▪▪▪▪▪▪▪▪▪▪▪▪▪▪▪
```

The results of the transient analysis that are obtained on the display by the .PROBE command are shown in Fig. 2-4. The results of the .PRINT statement can be obtained by printing the contents of the output file EX2-2.OUT.

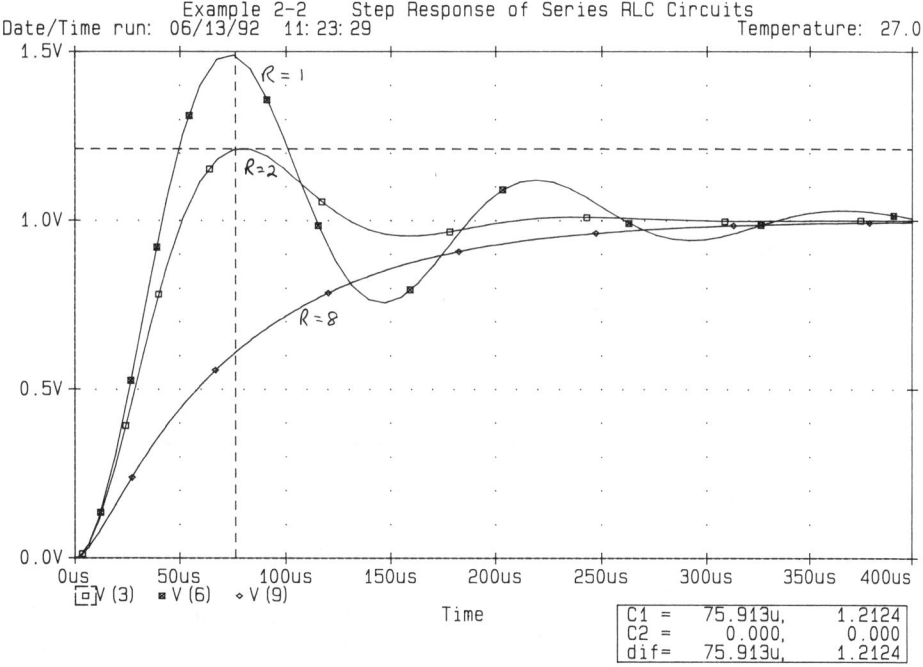

Figure 2-4   Step response of *RLC* circuits for Example 2-2.

Example 2-3

Repeat Example 2-1 if the input voltage is a sine wave of $v_{in} = 10 \sin(2\pi \times 5000t)$.
Solution   The model for a simple sinusoidal source is

```
SIN (VO    VA    FREQ)
```

where VO = offset voltage, V
      VA = peak voltage, V
    FREQ = frequency, Hz

For a sinusoidal voltage $v_{in} = 10 \sin(2\pi \times 5000t)$, the model is

```
SIN (0    10V    5KHZ)
```

The circuit file contains the following statements.

■■

Example 2-3   *RLC* circuit with sinusoidal input voltage

SOURCE   ■ * The format for a simple sinusoidal source is
      * SIN (VO   VA   FREQ)
      * Refer to Chapter 4 for modeling sources.
      * v_{in} is connected between nodes 7 and 0, assuming that node 7
      * is at a higher potential than node 0.
      * With a peak voltage of V_A = 10 V, a frequency = 5 kHz, and
      * an offset value of V_O = 0, the source is described by
      VIN   7   0   SIN (0   10V   5KHZ)   ; Input voltage

CIRCUIT   ■ ■ R1   7   5   2
        L1   5   3   50UH
        C1   3   0   10UF

ANALYSIS   ■ ■ ■ .PLOT   TRAN   V(3)   V(7)   ; Plot output voltages
        .TRAN   1US   500US   ; Transient analysis
        .PROBE   ; Graphics post-processor
.END
■■

The results of the transient response that are obtained on the display by the .PROBE command are shown in Fig. 2-5. The results of the .PRINT statement can be obtained by printing the contents of the output file EX2-3.OUT.

Example 2-4

For the circuit of Fig. 2-3(a), the frequency response is to be calculated and printed over the frequency range 100 Hz to 100 kHz with a decade increment and 100 points per decade. The peak magnitude and phase angle of the voltage across the capacitors are to be plotted on the output file. The results should also be available for display and hard copy by the .PROBE command.

Solution   The circuit file is similar to that of Example 2-2, except that the statements for the type of analysis and output are different. The frequency response analysis is invoked by the .AC command, whose format is

```
.AC    DEC    NP    FSTART    FSTOP
```

where DEC = sweep by decade
      NP = number of points per decade
  FSTART = starting frequency
   FSTOP = ending (or stop) frequency

For NP = 100, FSTART = 100 Hz, and FSTOP = 100 kHz, the statement is

```
.AC    DEC    100    100    100KHZ
```

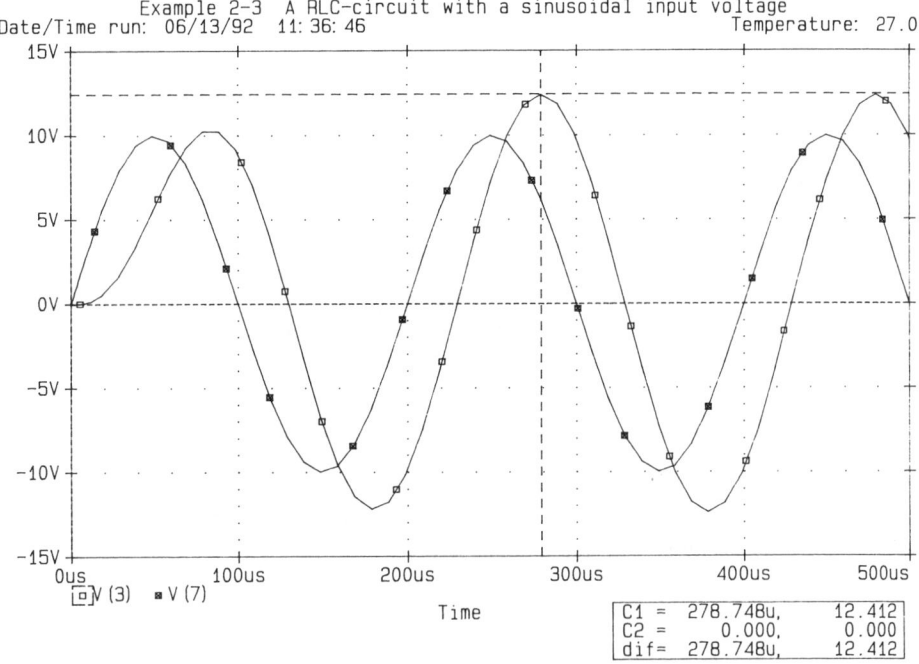

C1 =	278.748u,	12.412
C2 =	0.000,	0.000
dif=	278.748u,	12.412

□V (3)   ■V (7)   Time

Figure 2-5   Transient response for Example 2-3.

The magnitude and phase of voltage V(3) are specified as VM(3) and VP(3). The statement to plot is

```
. PLOT    AC    VM(3)    VP(3)
```

The input voltage is ac and the frequency is variable. We can consider a voltage source with a peak magnitude of 1 V. The statement for an independent voltage source is

```
VIN    7    0    AC    1V
```

The circuit file contains the following statements.

■ ■

Example 2-4   Frequency response of *RLC* circuits

SOURCE ■ * v_{in} is an independent voltage source whose frequency is varied
          * by PSpice during the frequency response analysis.

```
          VI1   1   0   AC   1V ; Ac voltage of 1 V
          VI2   4   0   AC   1V ; Ac voltage of 1 V
          VI3   7   0   AC   1V ; Ac voltage of 1 V
```

CIRCUIT ■ ■ R1   1   2   2

```
          L1   2   3   50UH
          C1   3   0   10UF
```

```
R2      4    5    1
L2      5    6    50UH
C2      6    0    10UF
R3      7    8    8
L3      8    9    50UH
C3      9    0    10UF
```

ANALYSIS ■ ■ ■ *   The frequency response analysis is invoked by the .AC command, whose
      *   format is
      *   .AC   DEC   NP   FSTART   FSTOP
      * Refer to Chapter 6 for dot commands.
      .AC   DEC   100   100HZ   100KHZ
      *   Plots the results of .AC analysis for the magnitude and phase of V(3):
      .PLOT   AC   VM(3)   VP(3)
      .PROBE
.END

The results of the frequency response obtained on the display by the .PROBE
command are shown in Fig. 2-6. The results of the .PLOT statement can be obtained
by printing the contents of the output file EX2-4.OUT. The .AC and .TRAN commands could be added to the same circuit file to perform two analyses.

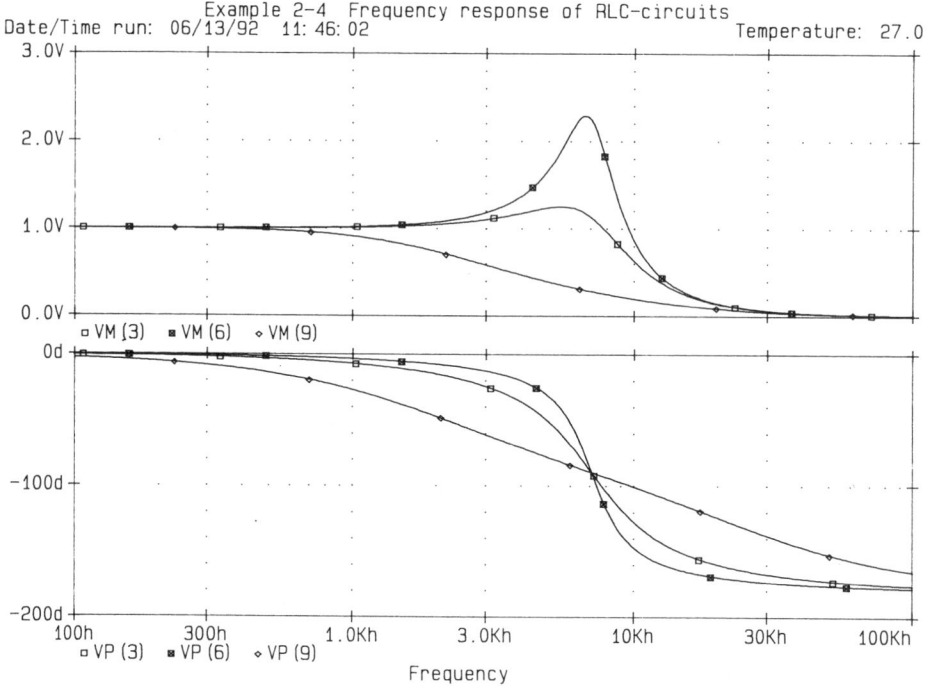

Figure 2-6   Frequency responses of *RLC* circuit for Example 2-4.

SUGGESTED READING

1. M. H. Rashid, *SPICE for Circuits and Electronics Using PSpice.* Englewood Cliffs, N.J.: Prentice Hall, 1990.

2. Paul W. Tuinenga, *SPICE: A Guide to Circuit Simulation and Analysis Using PSPICE.* Englewood Cliffs, N.J.: Prentice Hall, 1992.

PROBLEMS

2-1. The *RLC* circuit of Fig. P2-1 is to be simulated to calculate and plot the transient response from 0 to 2 ms with an increment of 10 μs. The voltage across resistor R is the output. The input and output voltages are to be plotted on an output file. The results should also be available for display and hard copy by the .PROBE command.

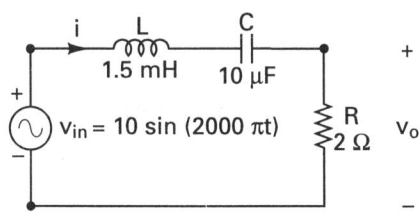

Figure P2-1

2-2. Repeat Problem 2-1 for the circuit of Fig. P2-2, where the output is taken across capacitor C.

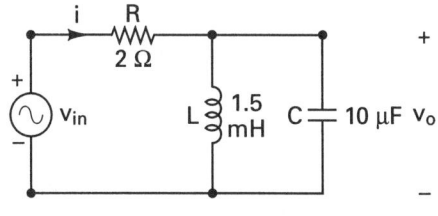

Figure P2-2

2-3. Repeat Problem 2-1 for the circuit of Fig. P2-3, where the output is the current i_s through the circuit.

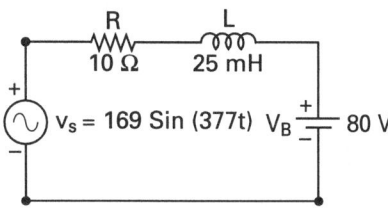

Figure P2-3

2-4. Repeat Problem 2-1 if the input is a step input, as shown in Fig. 2-3(b).

2-5. Repeat Problem 2-2 if the input is a step input, as shown in Fig. 2-3(b).

2-6. The *RLC* circuit of Fig. P2-6(a) is to be simulated to calculate and plot the transient response from 0 to 2 ms with an increment of 5 μs. The input is a step current, as shown in Fig. P2-6(b). The voltage across resistor R is the output. The input and output voltages are to be plotted on an output file. The results should also be available for display and hard copy by the .PROBE command.

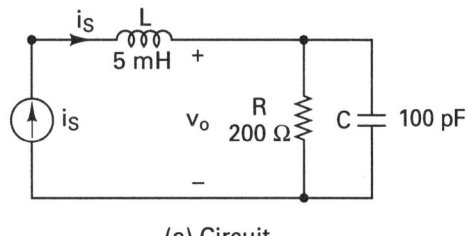

(a) Circuit

Figure P2-6

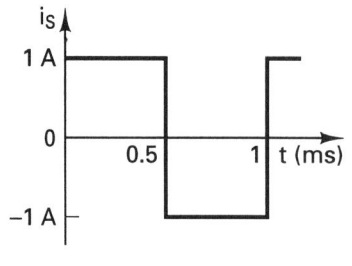

(b) Input current

Figure P2-6 *(Continued)*

2-7. Repeat Problem 2-6 for the circuit of Fig. P2-7, where the output is taken across capacitor *C*.

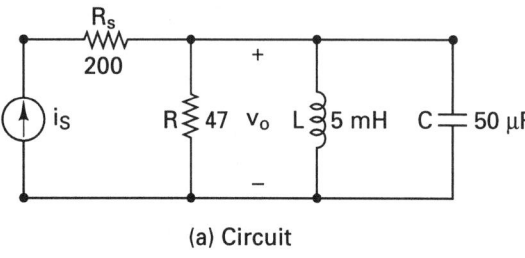

(a) Circuit

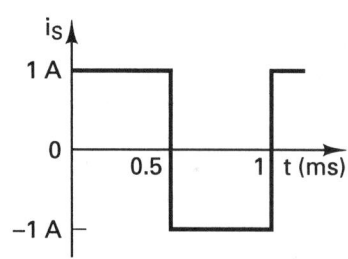

(b) Input current

Figure P2-7

2-8. The circuit of Fig. P2-2 is to be simulated to calculate and print the frequency response over the frequency range 10 Hz to 100 kHz with a decade increment and 10 points per decade. The peak magnitude and phase angle of the voltage across the resistor are to be printed on the output file. The results should also be available for display and hard copy by the .PROBE command.

2-9. Repeat Problem 2-8 for the circuit of Fig. P2-6.

2-10. Repeat Problem 2-8 for the circuit of Fig. P2-7.

2-11. Repeat Problem 2-8 for the circuit of Fig. P2-11.

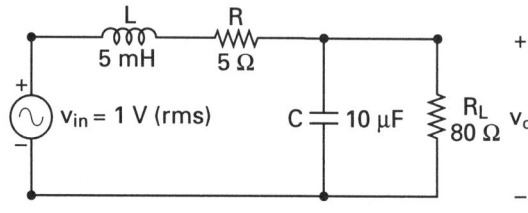

Figure P2-11

Chapter 3

∎∎∎∎∎∎∎∎

Defining output variables

3-1 INTRODUCTION

PSpice has some unique features for printing or plotting output voltages or currents by .PRINT and .PLOT statements. The .PRINT and .PLOT statements, which may have up to eight output variables, are discussed in Chapter 6. The output variables that are allowed in .PRINT and .PLOT statements depend on the type of analysis:

Dc sweep and transient analysis
Ac analysis
Noise analysis

3-2 DC SWEEP AND TRANSIENT ANALYSIS

Dc sweep and transient analysis use the same type of output variables. The output variables can be divided into two types: voltage output and current output. An output variable can be assigned the symbol of a device (or element) or the terminal symbol of a device to identify whether the output is the voltage across the device or the current through the device (or element). Table 3-1 shows the symbols for two-terminal elements. Table 3-2 shows the symbols and terminal symbols for three- and four-terminal devices.

3-2.1 Voltage Output

The output voltages for dc sweep and transient analysis can be obtained by the following statements:

TABLE 3-1 SYMBOLS FOR TWO-TERMINAL ELEMENTS

First letter	Element
C	Capacitor
D	Diode
E	Voltage-controlled voltage source
F	Current-controlled current source
G	Voltage-controlled current source
H	Current-controlled voltage source
I	Independent current source
L	Inductor
R	Resistor
V	Independent voltage source

TABLE 3-2 SYMBOLS AND TERMINAL SYMBOLS FOR THREE- OR FOUR-TERMINAL DEVICES

First letter	Device	Terminals
B	GaAs MESFET	D (drain)
		G (gate)
		S (source)
J	JFET	D (drain)
		G (gate)
		S (source)
M	MOSFET	D (drain)
		G (gate)
		S (source)
		B (bulk, substrate)
Q	BJT	C (collector)
		B (base)
		E (emitter)
		S (substrate)

V($\langle$node$\rangle$)	Voltage at $\langle$node$\rangle$ with respect to ground
V(N1,N2)	Voltage at node N_1 with respect to node N_2
V($\langle$name$\rangle$)	Voltage across two-terminal device, $\langle$name$\rangle$
Vx($\langle$name$\rangle$)	Voltage at terminal x of three-terminal device, $\langle$name$\rangle$
Vxy($\langle$name$\rangle$)	Voltage across terminals x and y of three-terminal device, $\langle$name$\rangle$
Vz($\langle$name$\rangle$)	Voltage at port z of transmission line, $\langle$name$\rangle$

PSpice VARIABLES	MEANING
V(5)	Voltage at node 5 with respect to ground
V(4,2)	Voltage of node 4 with respect to node 2
V(R1)	Voltage of resistor R_1, where the first node (as defined

	in the circuit file) is positive with respect to the second node
V(L1)	Voltage of inductor L_1, where the first node (as defined in the circuit file) is positive with respect to the second node
V(C1)	Voltage of capacitor C_1, where the first node (as defined in the circuit file) is positive with respect to the second node
V(D1)	Voltage across diode D_1, where the anode positive is positive with respect to the cathode
VC(Q3)	Voltage at the collector of transistor Q_3, with respect to ground
VDS(M6)	Drain–source voltage of MOSFET M_6
VB(T1)	Voltage at port B of transmission line T_1

Note. SPICE and some versions of PSpice do not permit measuring voltage across a resistor, an inductor, and a capacitor [e.g., V(R1), V(L1), and V(C1)]. This type of statement is applicable only to outputs by .PLOT and .PRINT commands.

3-2.2 Current Output

The output currents for dc sweep and transient analysis can be obtained by the following statements:

I(⟨name⟩)	Current through ⟨name⟩
Ix(⟨name⟩)	Current into terminal x of ⟨name⟩
Iz(⟨name⟩)	Current at port z of transmission line, ⟨name⟩

PSpice VARIABLES	MEANING
I(VS)	Current flowing into dc source V_s
I(R5)	Current flowing into resistor R_5, where the current is assumed to flow from the first node (as defined in the circuit file) through R_5 to the second node
I(D1)	Current into diode D_1
IC(Q4)	Current into the collector of transistor Q_4
IG(J1)	Current into gate of JFET J_1
ID(M5)	Current into drain of MOSFET M_5
IA(T1)	Current at port A of transmission line T_1

Note. SPICE and some versions of PSpice do not permit measuring the current through a resistor [e.g., I(R5)]. The easiest way is to add a dummy voltage

---

source of 0 V (say, VX = 0 V) and to measure the current through that source [e.g., I(VX)].

Example 3-1

A dc circuit with a bipolar transistor is shown in Fig. 3-1. Write the various currents and voltages in forms that are allowed by PSpice. The dc sources of 0 V are introduced to measure currents I_1 and I_2.

Solution

SYMBOL	PSpice VARIABLE	MEANING
I_B	IB(Q1)	Base current of transistor Q_1
I_C	IC(Q1)	Collector current of transistor Q_1
I_E	IE(Q1)	Emitter current of transistor Q_1
I_S	I(VCC)	Current through voltage source V_{cc}
I_1	I(VX)	Current through voltage source V_x
I_2	I(VY)	Current through voltage source V_y
V_B	VB(Q1)	Voltage at the base of transistor Q_1
V_C	VC(Q1)	Voltage at the collector of transistor Q_1
V_E	VE(Q1)	Voltage at the emitter of transistor Q_1
V_{CE}	VCE(Q1)	Collector–emitter voltage of transistor Q_1
V_{BE}	VBE(Q1)	Base–emitter voltage of transistor Q_1

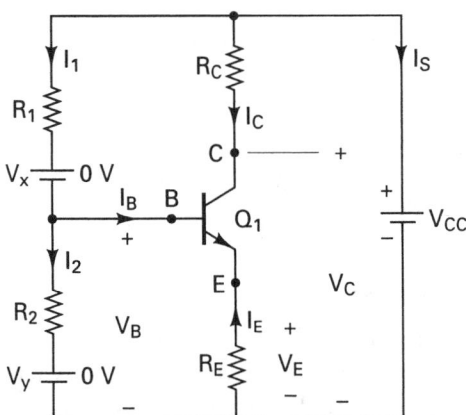

Figure 3-1 Dc circuit with a bipolar transistor.

Example 3-2

An *RLC* circuit with a step input is shown in Fig. 3-2. Write the various currents and voltages in forms that are allowed by PSpice.

Solution

SYMBOL	PSpice VARIABLE	MEANING
i_R	I(R)	Current through resistor R
i_L	I(L)	Current through inductor L

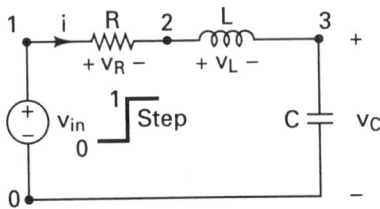

Figure 3-2 *RLC* circuit with a step input.

i_C	I(C)	Current through capacitor C
i_{in}	I(VIN)	Current flowing into voltage source v_{in}
v_3	V(3)	Voltage of node 3 with respect to ground
v_{23}	V(2,3)	Voltage of node 2 with respect to node 3
v_{12}	V(1,2)	Voltage of node 1 with respect to node 2
v_R	V(R)	Voltage of resistor R where the first node (as defined in the circuit file) is positive with respect to the second node
v_L	V(L)	Voltage of inductor L where the first node (as defined in the circuit file) is positive with respect to the second node
v_C	V(C)	Voltage of inductor L where the first node (as defined in the circuit file) is positive with respect to the second node

Note. SPICE and some versions of PSpice do not permit measuring voltage across a resistor, an inductor, and a capacitor [e.g., V(R1), V(L1), and V(C1)]. This type of statement is applicable only to outputs by .PLOT and .PRINT commands.

3-3 AC ANALYSIS

In ac analysis, the output variables are sinusoidal quantities and are represented by complex numbers. An output variable can have magnitude, magnitude in decibels, phase, group delay, real part, and imaginary part. The output variables listed in Sections 3-2.1 and 3-2.2 are augmented by adding a suffix as follows,

SUFFIX	MEANING
(none)	Peak magnitude
M	Peak magnitude
DB	Peak magnitude in decibels
P	Phase in radians
G	Group delay ($-\delta_{phase}/\delta_{frequency}$)
R	Real part
I	Imaginary part

3-3.1 Voltage Output

The statements for ac analysis are similar to those for dc sweep and transient analysis, provided that the suffixes are added as illustrated below.

PSpice VARIABLE	MEANING
VM(5)	Magnitude of voltage at node 5 with respect to ground
VM(4,2)	Magnitude of voltage at node 4 with respect to node 2
VDB(R1)	Decibel magnitude of voltage across resistor R_1, where the first node (as defined in the circuit file) is assumed to be positive with respect to second node
VP(D1)	Phase of anode voltage of diode D_1 with respect to cathode
VCM(Q3)	Magnitude of the collector voltage of transistor Q_3 with respect to ground
VDSP(M6)	Phase of the drain–source voltage of MOSFET M_6
VBP(T1)	Phase of voltage at port B of transmission line T_1
VR(2,3)	Real part of voltage at node 2 with respect to node 3
VI(2,3)	Imaginary part of voltage at node 2 with respect to node 3

3-3.2 Current Output

The statements for ac analysis are similar to those for dc sweep and transient responses. However, only the currents through the elements listed in Table 3-3 are available. For all other elements, a zero-valued voltage source must be placed in series with the device (or device terminal) of interest. Then a print or plot statement should be used to determine the current through this voltage source.

TABLE 3-3 CURRENT THROUGH ELEMENTS FOR AC ANALYSIS

First letter	Element
C	Capacitor
I	Independent current source
L	Inductor
R	Resistor
T	Transmission line
V	Independent voltage source

PSpice VARIABLE	MEANING
IM(R5)	Magnitude of current through resistor R_5
IR(R5)	Real part of current through resistor R_5
II(R5)	Imaginary part of current through resistor R_5

IM(VIN)	Magnitude of current through source v_{in}
IR(VIN)	Real part of current through source v_{in}
II(VIN)	Imaginary part of current through source v_{in}
IAG(T1)	Group delay of current at port A of transmission line T_1

Example 3-3

The frequency response is performed for the *RLC* circuit of Fig. 3-3. Write the various voltages and currents in forms that are allowed by PSpice. The dummy voltage source of 0 V is introduced to measure current, I_L.

Solution

SYMBOL	PSpice VARIABLE	MEANING
V_2	VM(2)	Peak magnitude of voltage at node 2
$\underline{/V_2}$	VP(2)	Phase angle of voltage at node 2
V_{12}	VM(1,2)	Peak magnitude of voltage between nodes 1 and 2
$\underline{/V_{12}}$	VP(1,2)	Phase angle of voltage between nodes 1 and 2
I_R	IM(VX)	Magnitude of current through voltage source V_x
$\underline{/I_R}$	IP(VX)	Phase angle of current through voltage source V_x
I_L	IM(L1)	Magnitude of current through inductor L_1
$\underline{/_L}$	IP(L1)	Phase angle of current through inductor L_1
I_C	IM(C1)	Magnitude of current through capacitor C_1
$\underline{/I_C}$	IP(C1)	Phase angle of current through capacitor C_1

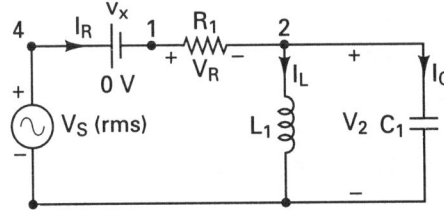

Figure 3-3 *RLC* circuit for Example 3-3.

3-4 NOISE ANALYSIS

For the noise analysis, the output variables are predefined as follows:

OUTPUT VARIABLE	MEANING
ONOISE	Total rms summed noise at output node
INOISE	ONOISE equivalent at the input node
DB(ONOISE)	ONOISE in decibels
DB(INOISE)	INOISE in decibels

Noise Output Statement

`.PRINT NOISE INOISE ONOISE`

Note. The noise output from only one device cannot be obtained by a .PRINT or .PLOT command. However, the print interval on the .NOISE statement can be used to output this information. The .NOISE command is discussed in Section 6-8.

SUMMARY

The PSpice variables can be summarized as follows:

V(⟨node⟩)	Voltage at ⟨node⟩ with respect to ground
V(N1, N2)	Voltage at node N_1 with respect to node N_2
V(⟨name⟩)	Voltage across two-terminal device, ⟨name⟩
Vx(⟨name⟩)	Voltage at terminal x of device, ⟨name⟩
Vxy(⟨name⟩)	Voltage at terminal x with respect to terminal y for device, ⟨name⟩
Vz(⟨name⟩)	Voltage at port z of transmission line, ⟨name⟩
I(⟨name⟩)	Current through device, ⟨name⟩
Ix(⟨name⟩)	Current into terminal x of device, ⟨name⟩
Iz(⟨name⟩)	Current at port z of transmission line, ⟨name⟩
(none)	Magnitude
M	Magnitude
DB	Magnitude in decibels
P	Phase in radians
G	Group delay ($-\delta_{phase}/\delta_{frequency}$)
R	Real part
I	Imaginary part

Chapter 4

■■■■■■■■■

Voltage and current sources

4-1 INTRODUCTION

PSpice allows generating dependent (or independent) voltage and current sources. An independent source can be time variant. A nonlinear source can also be simulated by a polynomial. In this chapter we explain the techniques for generating sources. The PSpice statements for various sources require:

Source modeling
Independent sources
Dependent sources
Behavioral device modeling

4-2 SOURCES MODELING

The independent voltage and current sources that can be modeled by PSpice are:

Pulse
Piecewise linear
Sinusoidal
Exponential
Single-frequency frequency modulation

4-2.1 Pulse Source

The waveform and parameters of a pulse waveform are shown in Fig. 4-1. The symbol of a pulse source is PULSE and the general form is (see Table 4-1)

PULSE (V1 V2 TD TR TF PW PER)

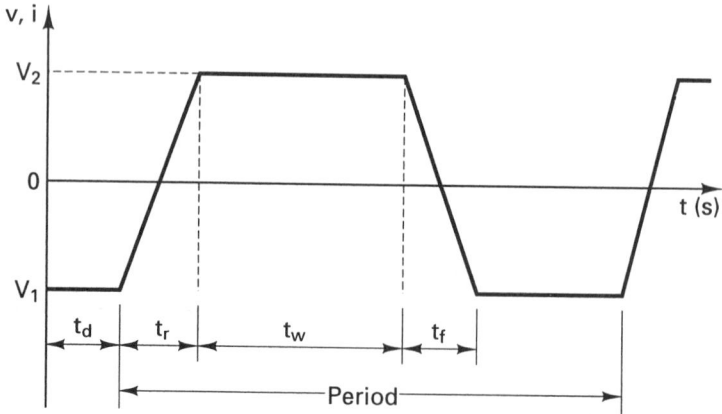

Figure 4-1 Pulse waveform.

TABLE 4-1 MODEL PARAMETERS OF
PULSE SOURCES

Name	Meaning	Unit	Default
V1	Initial voltage	V	None
V2	Pulsed voltage	V	None
TD	Delay time	s	0
TR	Rise time	s	TSTEP
TF	Fall time	s	TSTEP
PW	Pulse width	s	TSTOP
PER	Period	s	TSTOP

V1 and V2 *must* be specified by the user. TSTEP and TSTOP in Table 4-1 are the incrementing time and stop time, respectively, during transient (.TRAN) analysis.

Typical Statements

For $V_1 = -1$, $V_2 = 1$ V, $t_d = 2$ ns, $t_r = 2$ ns, $t_f = 2$ ns, pulse width = 50 ns, and period = 100 ns, the model statement is

PULSE (−1   1   2NS   2NS   2NS   50NS   100NS)

With $V_1 = 0$, $V_2 = 1$, the model becomes

PULSE (0   1   2NS   2NS   2NS   50NS   100NS)

With $V_1 = 0$, $V_2 = -1$, the model becomes

PULSE (0   −1   2NS   2NS   2NS   50NS   100NS)

4-2.2 Piecewise Linear Source

A point in a waveform can be described by time T_i and its value V_i. Every pair of values (T_i, V_i) specifies the source value V_i at time T_i. The voltage at a time between the intermediate points is determined by PSpice by using linear interpolation. The symbol of a piecewise linear source is PWL and the general form is (see Table 4-2)

```
PWL (T1  V1  T2  V2 ... TN  VN)
```

TABLE 4-2 MODEL PARAMETERS OF PWL SOURCES

Name	Meaning	Unit	Default
T_i	Time at a point	s	None
V_i	Voltage at a point	v	None

Typical Statement

The model statement for the typical waveform of Fig. 4-2 is

```
PWL (0  3  10US  3V  15US  6V  40US  6V  45US  2V  60US  0V)
```

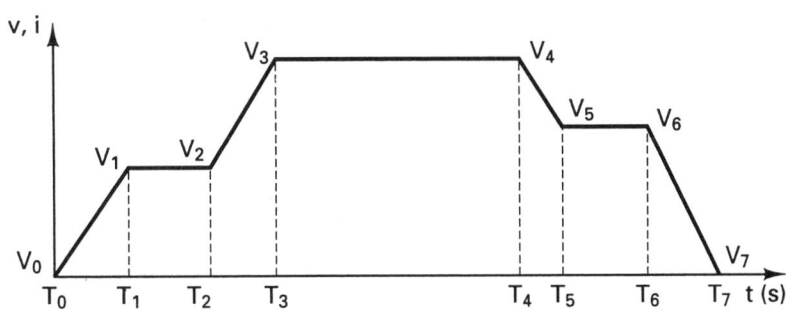

Figure 4-2 Piecewise linear waveform.

4-2.3 Sinusoidal Source

The symbol of a sinusoidal source is SIN and the general form is (see Table 4-3)

```
SIN (VO  VA  FREQ  TD  ALP  THETA)
```

VO and VA *must* be specified by the user. TSTOP in Table 4-3 is the stop time during transient (.TRAN) analysis. The waveform stays at 0 for a time of TD and then the voltage becomes an exponentially damped sine wave. An exponentially

TABLE 4-3 MODEL PARAMETERS OF SIN SOURCES

Name	Meaning	Unit	Default
VO	Offset voltage	V	None
VA	Peak voltage	V	None
FREQ	Frequency	Hz	1/TSTOP
TD	Delay time	s	0
ALPHA	Damping factor	1/s	0
THETA	Phase delay	degrees	0

damped sine wave is described by

$$V = V_O + V_A e^{-\alpha(t - t_d)} \sin[(2\pi f(t - t_d) - \theta]$$

and this is shown in Fig. 4-3.

Typical Statements

```
SIN (0   1V   10KHZ   10US   1E5)
SIN (1   5V   10KHZ   01E5   30DEG)
SIN (0   2V   10KHZ   0 030DEG)
SIN (0   2V   10KHZ)
```

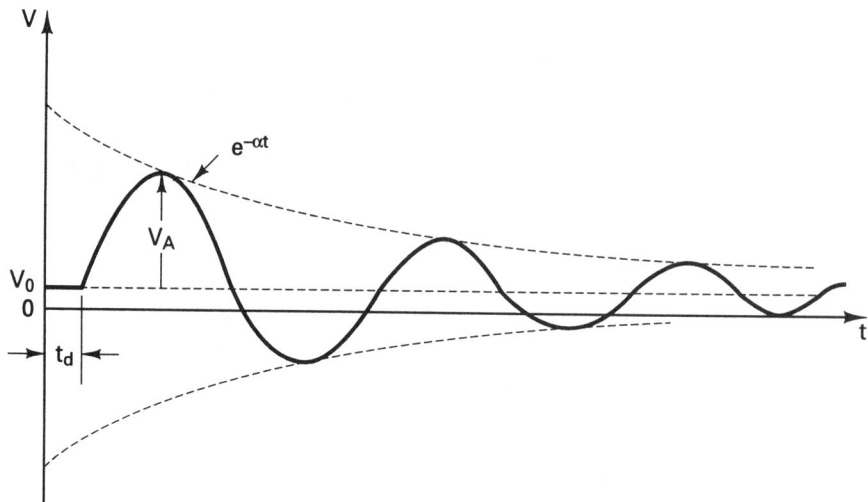

Figure 4-3 Damped sinusoidal waveform.

4-2.4 Exponential Source

The waveform and parameters of an exponential waveform are shown in Fig. 4-4. The symbol of exponential sources is EXP and the general form is (see Table 4-4)

```
EXP (V1  V2  TRD  TRC  TFD   TFC)
```

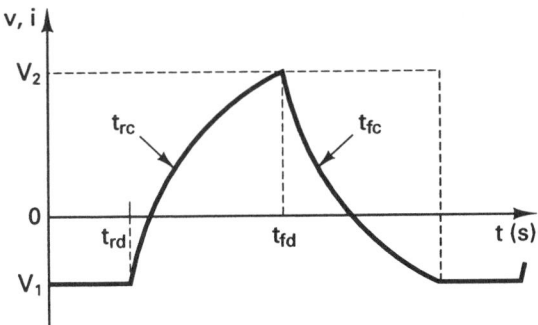

Figure 4-4 Exponential wave-form.

TABLE 4-4 MODEL PARAMETERS OF
EXP SOURCES

Name	Meaning	Unit	Default
V1	Initial voltage	V	None
V2	Pulsed voltage	V	None
TRD	Rise delay time	s	0
TRC	Rise-time constant	s	TSTEP
TFD	Fall delay time	s	TRD + TSTEP
TFC	Fall-time constant	s	TSTEP

V1 and V2 *must* be specified by the user. TSTEP in Table 4-4 is the incrementing time during transient (.TRAN) analysis. In an EXP waveform, the voltage remains V1 for the first TRD seconds. Then the voltage rises exponentially from V1 to V2 with a rise-time constant of TRC. After a time of TFD, the voltage falls exponentially from V2 to V1 with a fall-time constant of TFC.

Note. The values of EXP waveform as well as the values of other time-dependent waveforms at intermediate time points are determined by PSpice by means of linear interpolation.

Typical Statements

For V1 = 0, V2 = 1 V, TRD = 2NS, TRC = 20NS, TFD = 60NS, and TFC = 30NS, the model statement is

```
EXP    (0    1    2NS    20NS    60NS    30NS)
```

With TRD = 0, the statement becomes

```
EXP    (0    1    0    20NS    60NS    30NS)
```

With V1 = −1 V and V2 = 2 V, it is

```
EXP    (−1    2    2NS    20NS    60NS    30NS)
```

4-2.5 Single-Frequency Frequency Modulation Source

The symbol of a source with single-frequency frequency modulation is SFFM, and the general form is (see Table 4-5)

SFFM (VO   VA   FC   MOD   FS)

TABLE 4-5 MODEL PARAMETERS OF
SFFM SOURCES

Name	Meaning	Unit	Default
VO	Offset voltage	V	None
VA	Amplitude of voltage	V	None
FC	Carrier frequency	Hz	1/TSTOP
MOD	Modulation index		0
FS	Signal frequency	Hz	1/TSTOP

VO and VA *must* be specified by the user. TSTOP is the stop time during transient (.TRAN) analysis. The waveform is of the form

$$V = V_O + V_A \sin[(2\pi F_C t) + M \sin(2\pi F_S t)]$$

Typical Statements

For $V_O = 0$, $V_A = 1$ V, $F_C = 30$ MHz, MOD = 5, and $F_S = 5$ kHz, the model statement is

SFFM (0   1V   30MHZ   5   5KHZ)

With $V_O = 1$ mV and $V_A = 2$ V, the model becomes

SFFM (1MV   2V   30MHZ   5   5KHZ)

4-3 INDEPENDENT SOURCES

The independent sources can be time invariant and time variant. They can be currents or voltages, as shown in Fig. 4-5. The following notations are used only to explain the general format of a statement and do not appear in the PSpice statement.

(text)	Text within parentheses is a comment
[item]	Optional item
[item]*	Zero or more of optional item
⟨item⟩	Item required
⟨item⟩*	Zero or more of item required

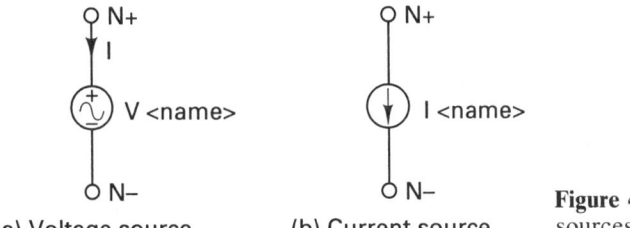

(a) Voltage source (b) Current source

Figure 4-5 Voltage and current sources.

4-3.1 Independent Voltage Source

The symbol of an independent voltage source is V and the general form is

```
V⟨name⟩     N+      N−
+           [dc ⟨value⟩]
+           [ac ⟨(magnitude) value⟩ ⟨(phase) value⟩]
+           [(transient specifications)]
```

Note. The first column with a + (plus) signifies continuation of the PSpice statement. After the + sign, the statement can continue in any column.

The (transient specifications) must be one of the following sources:

PULSE (⟨parameters⟩)	For a pulse waveform
PWL (⟨parameters⟩)	For a piecewise linear waveform
SIN (⟨parameters⟩)	For a sinusoidal waveform
EXP (⟨parameters⟩)	For an exponential waveform
SFFM (⟨parameters⟩)	For a frequency-modulated waveform

N+ is the positive node and N− is the negative node, as shown in Fig. 4-5(a). Positive current flows from node N+, through the voltage source, to the negative node N−. The voltage source need not be grounded. For the dc, ac, and transient values, the default value is zero. None or all of dc, ac, and transient values may be specified. The ⟨(phase) value⟩ is in degrees.

The source is set to the dc value in dc analysis. It is set to ac value in ac analysis. If the ⟨(phase) value⟩ in ac analysis is omitted, the default is 0. The time-dependent source (e.g., PULSE, EXP, SIN, etc.) is assigned for transient analysis. A voltage source may be used as an *ammeter* in PSpice by inserting a zero-valued voltage source into the circuit for the purpose of measuring current. Since a zero-valued source behaves as a short circuit, there will be no effect on circuit operation.

Typical Statements

```
V1       15   0    6V                          ; By default, dc specification of 6 V
V2       15   0    DC    6V                     ; Dc specification of 6 V
VAC      5    6    AC    1V                      ; Ac specification of 1 V with 0° delay
VACP     5    6    AC    1V    45DEG  ; Ac specification of 1 V with 45° delay
VPULSE   10   0    PULSE (0   1   2NS   2NS  2NS  50NS   100NS) ; Transient pulse
VIN      25   22   DC    2  AC  1  30  SIN (0   2V   10KHZ)
```

Note. VIN assumes 2 V for dc analysis, 1 V with a delay angle of 30° for ac analysis, and a sine wave of 2 V at 10 kHz for transient analysis. This allows source specifications for different analyses in the same statement.

4-3.2 Independent Current Source

The symbol of an independent current source is I and the general form is

```
I⟨name⟩   N+      N−
+         [dc ⟨value⟩]
+         [ac ⟨(magnitude) value⟩ ⟨(phase) value⟩]
+         [⟨transient specifications⟩]
```

The ⟨transient specifications⟩ must be one of the following sources:

PULSE (⟨parameters⟩)	For a pulse waveform
PWL (⟨parameters⟩)	For a piecewise linear waveform
SIN (⟨parameters⟩)	For a sinusoidal waveform
EXP (⟨parameters⟩)	For an exponential waveform
SFFM (⟨parameters⟩)	For a frequency-modulated waveform

N+ is the positive node and N− is the negative node, as shown in Fig. 4-5(b). Positive current flows from node N+, through the current source, to the negative node N−. The current source need not be grounded. The source specifications are similar to those of an independent voltage source.

Typical Statements

```
I1      15    0    2.5MA                     ; By default, dc specification of 2.5 mA
I2      15    0    DC    2.5MA               ; Dc specification of 2.5 mA
IAC     5     6    AC    1A                  ; Ac specification of 1 A with 0° delay
IACP    5     6    AC    1A      45DEG       ; Ac specification of 1 V with 45° delay
IPULSE  10    0    PULSE (0   1A  2NS  2NS  2NS  50NS  100NS) ; Transient pulse
IIN     25    22   DC    2A   AC   1A  30DEG  SIN (0  2A  10KHZ)
```

Note. IIN assumes 2 A for dc analysis, 1 A with a delay angle of 30° for ac analysis, and a sine wave of 2 A at 10 kHz for transient analysis. This allows source specifications for different analyses in the same statement.

4-4 DEPENDENT SOURCES

There are five types of dependent sources:

Polynomial source
Voltage-controlled voltage source
Current-controlled current source

Voltage-controlled current source

Current-controlled voltage source

4-4.1 Polynomial Source

Let us call the three controlling variables A, B, and C, and the output source, Y. Figure 4-6 shows a source Y that is controlled by A, B, and C. The output source Y takes the form

$$Y = f(A, B, C, \ldots)$$

where Y can be a voltage or current, and A, B, and C can be a voltage or current or any combination. The symbol of a polynomial or nonlinear source is POLY(n), where n is the number of dimensions of the polynomial. The default value of n is 1. The dimensions depend on the number of controlling sources. The general form is

```
POLY(n) ⟨(controlling) nodes⟩ ⟨(coefficient) values⟩
```

The output sources or the controlling sources can be voltages or currents. For voltage-controlled sources, the number of controlling nodes must be twice the number of dimensions. For current-controlled sources, the number of controlling sources must be equal to the number of dimensions. The number of dimensions and the number of coefficients are arbitrary.

For a polynomial of $n = 1$ with A as the only controlling variable, the source function takes the form

$$Y = P_0 + P_1A + P_2A^2 + P_3A^3 + P_4A^4 + \cdots + P_nA^n$$

where $P_0, P_1, \ldots, P_n$ are the coefficient values, and this is written in PSpice as

```
POLY  NC1+   NC1−   P₀   P₁  P₂   P₃   P₄   P₅ ... Pₙ
```

where NC1+ and NC1− are the positive and negative nodes, respectively, of controlling source A.

For a polynomial of $n = 2$ with A and B as the controlling sources, the source

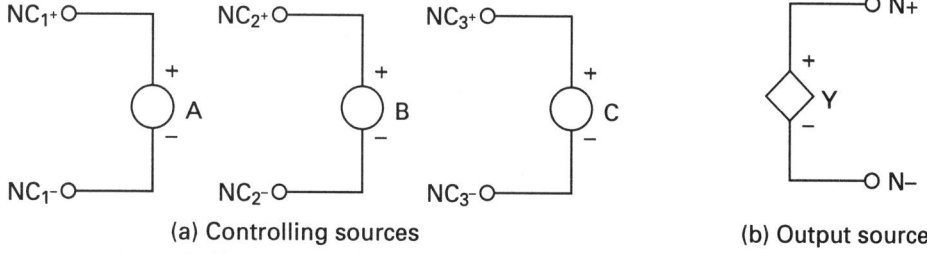

(a) Controlling sources (b) Output source

Figure 4-6 Polynomial source.

function Y takes the form

$$Y = P_0 + P_1A + P_2B + P_3A^2 + P_4AB + P_5B^2 + P_6A^3$$
$$+ P_7A^2B + P_8AB^2 + P_9B^3 + \cdots$$

and this is described in PSpice as

```
POLY(2)  NC1+  NC1-  NC2+  NC2-   P₀   P₁   P₂   P₃   P₄   P₅ ... Pₙ
```

where NC1+, NC2+ and NC1−, NC2− are the positive and negative nodes, respectively, of the controlling sources.

For a polynomial of $n = 3$ with A, B, and C as the controlling sources, the source function Y takes the form

$$Y = P_0 + P_1A + P_2B + P_3C + P_4A^2 + P_5AB + P_6AC + P_7B^2 + P_8BC + P_9C^2$$
$$+ P_{10}A^3 + P_{11}A^2B + P_{12}A^2C + P_{13}AB^2 + P_{14}ABC + P_{15}AC^2 + P_{16}B^3$$
$$+ P_{17}B^2C + P_{18}BC^2 + P_{19}C^3 + P_{20}A^4 + \cdots$$

and this is written in PSpice as

```
POLY(3)  NC1+ NC1- NC2+ NC2- NC3+ NC3-  P₀   P₁   P₂   P₃   P₄   P₅  ...  Pₙ
```

where NC1+, NC2+, NC3+ and NC1−, NC2−, NC3− are the positive and negative nodes, respectively, of the controlling sources.

Typical Model Statements

For $Y = 2V(10)$, the model is

```
POLY     10   0   2.0
```

For $Y = V(5) + 2[V(5)]^2 + 3[V(5)]^3 + 4[V(5)]^4$, the model is

```
POLY      5   0   0.0   1.0   2.0   3.0   4.0
```

For $Y = 0.5 + V(3) + 2V(5) + 3[V(3)]^2 + 4V(3)V(5)$, the model is

```
POLY(2)   3   0   5   0   0.5   1.0   2.0   3.0   4.0
```

For $Y = V(3) + 2V(5) + 3V(10) + 4[V(3)]^2$, the model is

```
POLY(3)   3   0   5   0   10   0   0.0   1.0   2.0   3.0   4.0
```

If $I(VN)$ is the controlling current through voltage source VN, and $Y = I(VN) + 2[I(VN)]^2 + 3[I(VN)]^3 + 4[I(VN)]^4$, the model is

```
POLY     VN   0.0   1.0   2.0   3.0   4.0
```

If $I(VN)$ and $I(VX)$ are the controlling currents and $Y = I(VN) + 2I(VX) +$

$3[I(VN)]^2 + 4I(VN)I(VX)$, the model is

```
POLY(2)  VN   VX   0.0   1.0   2.0   3.0   4.0
```

Note. If the source is of one dimension and only one coefficient is specified, as in the first example, $Y = 2V(10)$, PSpice assumes that $P_0 = 0$ and the value specified is P_1. That is, $Y = 2$ A.

4-4.2 Voltage-Controlled Voltage Source

The dependent sources are shown in Fig. 4-7. The symbol of the voltage-controlled voltage source shown in Fig. 4-7(a) is E, and it takes the linear form

```
E⟨name⟩  N+   N-   NC+   NC-    ⟨(voltage gain) value⟩
```

N+ and N− are the positive and negative output nodes, respectively, and NC+ and NC− are the positive and negative nodes, respectively, of the controlling voltage. The nonlinear form is

```
E⟨name⟩  N+   N-   [POLY (polynomial specifications)]
+     [VALUE (expression)]   [TABLE (expression)]
+     [LAPLACE (expression)]   [FREQ (expression)]
```

The POLY description is described in Section 4-4.1. The number of controlling nodes in POLY is twice the number of dimensions. A particular node may appear more than once and the output and controlling nodes could be the same.

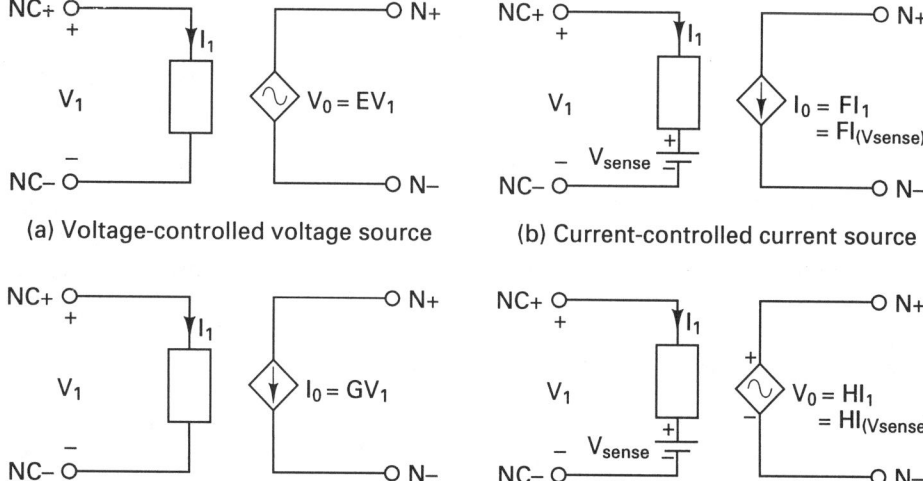

(a) Voltage-controlled voltage source   (b) Current-controlled current source

(c) Voltage-controlled current source   (d) Current-controlled voltage source

Figure 4-7   Dependent sources.

The VALUE, TABLE, LAPLACE, and FREQ descriptions of sources are available only with the analog behavioral modeling option of PSpice and are discussed in Section 4-5.

Typical Statements

```
EAB    1   2   4   6   1.0   ; Voltage gain of 1
EVOLT  4   7  20  22   2E5   ; Voltage gain of 2E5
```

ENONLIN, which is connected between nodes 25 and 40, is controlled by V(3) and V(5). Its value is given by the polynomial $Y = V(3) + 1.5V(5) + 1.2[V(3)]^2 + 1.7V(3)V(5)$, and the model becomes

```
ENONLIN  25  40  POLY(2)  3  0  5  0  0.0  1.0  1.5  1.2  1.7 ; POLY source
```

E2, which is connected between nodes 10 and 12, is controlled by V(5), and its value is given by the polynomial $Y = V(5) + 1.5[V(5)]^2 + 1.2[V(5)]^3 + 1.7[V(5)]^4$, and the model becomes

```
E2    10  12  POLY  5  0  0.0  1.0  1.5  1.2  1.7  ; POLY source
```

4-4.3 Current-Controlled Current Source

The symbol of the current-controlled current source shown in Fig. 4-7(b) is F, and it takes the linear form

```
F⟨name⟩  N+  N−  VN  ⟨(current gain) value⟩
```

N+ and N− are the positive and negative nodes, respectively, of the current source. VN is a voltage source through which the controlling current flows. The controlling current is assumed to flow from the positive node of VN, through the voltage source VN, to the negative node of VN. The current through the controlling voltage source, I(VN), determines the output current.

   The voltage source VN that monitors the controlling current must be an independent voltage source and it can have *zero* or finite value. If the current through a resistor controls the source, a dummy voltage source of 0 V should be connected in series with the resistor to monitor the controlling current.

   The nonlinear form is

```
F⟨name⟩  N+  N−  [POLY (polynomial specifications)]
```

The POLY source is described in Section 4-4.1. The number of controlling current sources for the POLY must be equal to the number of dimensions.

Typical Statements

```
FAB    1   2  VIN  10    ; Current gain of 10
FAMP  13   4  VCC  50    ; Current gain of 50
```

FNONLIN, which is connected between nodes 25 and 40, is controlled by the current through voltage source VN. Its value is given by the polynomial I = I(VN) + 1.5[I(VN)]^2 + 1.2[I(VN)]^3 + 1.7[I(VN)]^4, and the PSpice model becomes

```
FNONLIN  25  40  POLY  VN  0.0  1.0  1.5  1.2  1.7
```

4-4.4 Voltage-Controlled Current Source

The symbol of the voltage-controlled current source shown in Fig. 4-7(c) is G, and it takes the linear form

```
G⟨name⟩  N+  N−  NC+  NC−  ⟨(transconductance) value⟩
```

N+ and N− are the positive and negative output nodes, respectively, and NC+ and NC− are the positive and negative nodes, respectively, of the controlling voltage.

The nonlinear form is

```
G⟨name⟩  N+   N−   [POLY (polynomial specifications)]
+      [VALUE (expression)]     [TABLE (expression)]
+      [LAPLACE (expression)]  [FREQ (expression)]
```

The POLY description is described in Section 4-4.1. The VALUE, TABLE, LAPLACE, and FREQ descriptions of sources are available only with the analog behavioral modeling option of PSpice. These are discussed in Section 4-5.

Typical Statements

```
GAB     1  2   4   6  1.0   ; Transconductance of 1
GVOLT   4  7  20  22  2E5   ; Transconductance of 2E5
```

GNONLIN, which is connected between nodes 25 and 40, is controlled by V(3) and V(5). Its value is given by the polynomial Y = V(3) + 1.5V(5) + 1.2[V(3)]^2 + 1.7V(3)V(5), and the model becomes

```
GNONLIN  25  40  POLY(2) 3  0  5  0  0.0  1.0  1.5  1.2  1.7 ; POLY source
```

G2, which is connected between nodes 10 and 12, is controlled by V(5), and its value is given by the polynomial Y = V(5) + 1.5[V(5)]^2 + 1.2[V(5)]^3 + 1.7[V(5)]^4, and the model becomes

```
G2  10  12  POLY  5  0  0.0  1.0  1.5  1.2  1.7  ; POLY source
```

A voltage-controlled current source can be applied to simulate conductance if the controlling nodes are the same as the output nodes. This is shown in Fig. 4-8(a). For example, the PSpice statement

```
GRES  4  6  4  6  0.1     ; transconductance of 0.1
```

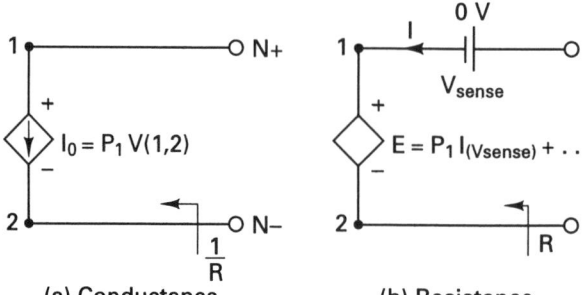

Figure 4-8 Conductance and re-sistance.

(a) Conductance

(b) Resistance

is a linear conductance of 0.1 seimens (Ω^{-1} or mhos) with a resistance of $1/0.1 = 10\ \Omega$. The PSpice statement

```
GMHO  1  2  POLY  1  2  0.0  1.5M  1.7M  ;  POLY source
```

represents a nonlinear conductance (Ω^{-1}) of the polynomial form

$$I = 1.5 \times 1^{-3}V(1,2) + 1.7 \times 10^{-3}[V(1,2)]^2$$

4-4.5 Current-Controlled Voltage Source

The symbol of the current-controlled voltage source shown in Fig. 4-7(d) is H, and it takes the linear form

```
H⟨name⟩  N+  N-  VN  ⟨(transresistance) value⟩
```

N+ and N− are the positive and negative nodes, respectively, of the voltage source. VN is a voltage source through which the controlling current flows, and its specifications are similar to those for a current-controlled current source.
The nonlinear form is

```
H⟨name⟩  N+  N-  [POLY (polynomial specifications)]
```

The POLY source is described in Section 4-4.1. The number of controlling current sources for the POLY must be equal to the number of dimensions.

Typical Statements

```
HAB    1   2  VIN  10
HAMP  13   4  VCC  50
```

HNONLIN, which is connected between nodes 25 and 40, is controlled by I(VN). Its value is given by the polynomial $V = I(VN) + 1.5[I(VN)]^2 + 1.2[I(VN)]^3 + 1.7[I(VN)]^4$, and the model becomes

```
HNONLIN  25  40  POLY  VN  0.0  1.0  1.5  1.2  1.7   ;  POLY source
```

A voltage-controlled current source can be applied to simulate conductance

---

if the controlling current is the same as the current through the voltage between the output nodes. This is shown in Fig. 4-8(b). For example, the PSpice statement

```
HRES   4   6   VN   0.1    ; Transresistance of 0.1
```

is a linear resistance of 10 Ω.
The PSpice statement

```
HOHM   1   2   POLY   VN   0.0   1.5M   1.7M  ; POLY source
```

represents a nonlinear resistance in ohms of the polynomial form

```
H = 1.5 × 1⁻³I(VN) + 1.7 × 10⁻³[I(VN)]²
```

4-5 BEHAVIORAL DEVICE MODELING

PSpice allows characterization of devices in terms of the relation between their input(s) and output(s). This relation is instantaneous. At each moment in time, there is an output for each value of the input. This representation, known as *behavioral modeling*, is available only with the analog behavioral modeling option of PSpice. This option is implemented as a set of extensions to two of the controlled sources: E and G. The behavioral modeling allows specifications in the form

```
VALUE
TABLE
LAPLACE
FREQ
```

4-5.1 Value

The VALUE extension to the controlled sources allows an instantaneous transfer function to be written as a mathematical expression in standard notation. The general forms are

```
E⟨name⟩   N+   N−   VALUE = {⟨expression⟩}
G⟨name⟩   N+   N−   VALUE = {⟨expression⟩}
```

The ⟨expression⟩ itself is enclosed in braces ({ }). It can contain the arithmetical operators ("+", "−", "*", and "/") along with parentheses and the following functions:

FUNCTION	MEANING
ABS(x)	$\|x\|$ (absolute value)
SQRT(x)	$\sqrt{x}$

FUNCTION	MEANING
EXP(x)	e^x
LOG(x)	$\ln(x)$ (log of base e)
LOG10(x)	$\text{Log}(x)$ (log of base 10)
PWR(x,y)	$|x|^y$
PWRS(x,y)	$+|x|^y$ (if $x > 0$), $-|x|^y$ (if $x < 0$)
SIN(x)	$\sin(x)$ (x in radians)
COS(x)	$\cos(x)$ (x in radians)
TAN(x)	$\tan(x)$ (x in radians)
ARCTAN(x)	$\tan^{-1}(x)$ (result in radians)

Typical Statements

```
ESQROOT  2  3   VALUE = {4V*SQRT(V(5))}         ; Square roots
EPWR     1  2   VALUE = {V(4.3)*I(VSENSE)}      ; Product of v and i
ELOG     3  0   VALUE = {10V*LOG(I(VS)/10mA)} ; Log of current ratio
GVCO     4  5   VALUE = {15MA*SIN(6.28*10kHz*TIME*(10V*V(7)))}
GRATIO   3  6   VALUE = {V(8,2)/V(9)}           ; Voltage ratio
```

The VALUE can be used to simulate linear and nonlinear resistances (or conductances) if appropriate functions are used. A resistance is a current-controlled voltage source. For example, the statement

```
ERES  2  3  VALUE = {I(VSENSE)*5K}
```

is a linear resistance with a value of 5 kΩ. VSENSE, which is connected in series with ERES, is needed to measure the current through ERES.

A conductance is a voltage-controlled current source. For example, the statement

```
GCOND  2  3  VALUE = {V(2,3)*1M}
```

is a linear conductance with a value of 1 mΩ^{-1}. The controlling nodes are the same as the output nodes.

Notes

1. VALUE *should* be followed by a space.

2. ⟨expression⟩ *must* fit on one line.

4-5.2 Table

The TABLE extension to the controlled sources allows an instantaneous transfer function to be described by a table. This form is well suited for use with, for example, measured data. The general forms are:

```
E⟨name⟩  N+  N−   TABLE { ⟨expression⟩ } =
+                 ⟨ ⟨(input) value⟩, ⟨(output) value⟩ ⟩*
G⟨name⟩  N+  N−   TABLE { ⟨expression⟩ } =
+                 ⟨ ⟨(input) value⟩, ⟨(output) value⟩ ⟩*
```

The ⟨expression⟩ is evaluated, and that value is used to look up an entry in the table. The table itself consists of pairs of values. The first value in each pair is an input, and the second value is the corresponding output. Linear interpolation is done between entries. For values of ⟨expression⟩ outside the table's range, the device's output is a constant with value equal to the entry with the smallest (or largest) input.

Typical Statements

The TABLE can be used to represent the voltage–current characteristics of a diode,

```
EDIODE   5    6     TABLE { I(VSENSE) } =
+     (0.0,   0.5)    (10E-3, 0.870)   (20E-3, 0.98)   (30E-3, 1.058)
+     (40E-3, 1.115)  (50E-3, 1.173)   (60E-3, 1.212)  (70E-3, 1.250)
```

The TABLE can be used to represent a constant power load $P = 400$ W with a voltage-controlled current source,

```
GCONST   2  3   TABLE   {400/V(2,3)} = (−400, −400) (400, 400)
```

GCONST tries to dissipate 400 W of power regardless of the voltage across it. But for a very small voltage, the formula 400/V(2,3) can lead to unreasonable values of current. The TABLE limits the current to be between −400 and 400 A.

Notes

1. TABLE *must* be followed by a space.
2. The input to the table is ⟨expression⟩, which *must* fit in one line.
3. TABLE's input *must* be in order from lowest to highest.

4-5.3 LAPLACE

The LAPLACE extension to the controlled sources allows a transfer function to be described by a Laplace transform function. The general forms are:

```
E⟨name⟩  N+  N−   LAPLACE { ⟨expression⟩ } = { ⟨transform⟩ }
G⟨name⟩  N+  N−   LAPLACE { ⟨expression⟩ } = { ⟨transform⟩ }
```

The input to the transform is the value of ⟨expression⟩, which follows the same rules as in Section 4-5.1. The ⟨transform⟩ is an expression in the Laplace variable, *s*.

Typical Statements

The output voltage of a lossless integrator with a time constant of 1 ms and an input voltage V(5) can be described by

```
ERC    4    0    LAPLACE  {V(5)} = {1/(1 + 0.001*s)}
```

Frequency-dependent impedances can be simulated with a capacitor, which can be written as

```
GCAP   5    4    LAPLACE  {V(5,4)} = {s}
```

> *Notes*
>
> 1. LAPLACE *must* be followed by a space.
> 2. ⟨expression⟩ and ⟨transform⟩ *must* each fit on one line.
> 3. Voltages, currents, and TIME must not appear in a Laplace transform.
> 4. The LAPLACE device uses much more computer memory than does the built-in capacitor (C) device and should be avoided if possible.

4-5.4 FREQ

The FREQ extension to the controlled sources allows a transfer function to be described by a frequency response table. The general forms are:

```
E⟨name⟩   N+   N−    FREQ { ⟨expression⟩ } =
+    ⟨ ⟨(frequency) value⟩, ⟨(magnitude in dB) value⟩, ⟨(phase) value⟩ ⟩*
G⟨name⟩   N+   N−    FREQ { ⟨expression⟩ } =
+    ⟨ ⟨(frequency) value⟩, ⟨(magnitude in dB) value⟩, ⟨(phase) value⟩ ⟩*
```

The input to the table is the value of ⟨expression⟩, which follows the same rules as in Section 4-5.1. The table contains the magnitude [in decibels (dB)] and the phase (in degrees) of the response for each frequency. Linear interpolation is done between entries. Phase is interpolated linearly, and the magnitude is interpolated logarithmically with frequencies. For frequencies outside the table's range, the entry with the smallest (or largest) frequency is used.

Typical Statements

The output voltage of a low-pass filter with input voltage V(2) can be expressed by

```
ELOWPASS   2    0    FREQ {V(2)} = (0,    0,    0)
+                   (5 kHz,    0,   −57.6)   (6kHz,   40,   −69.2)
```

Notes

1. FREQ *should* be followed by a space.
2. ⟨expression⟩ *must* fit on one line.
3. TABLE's frequencies *must* be in order from lowest to highest.

SUMMARY

The PSpice variables can be summarized as follows:

PULSE   Pulse source

        PULSE (V1   V2   TD   TR   TF   PW   PER)

PWL   Piecewise linear source

        PWL (T1   V1   T2   V2 ... TN   VN)

SIN   Sinusoidal source

        SIN (VO   VA   FREQ   TD   ALP   THETA)

EXP   Exponential source

        EXP (V1   V2   TRD   TRC   TFD   TFC)

SFFM   Single-frequency frequency modulation

        SFFM (VO   VA   FC   MOD   FS)

POLY   Polynomial source

        POLY(n) ⟨(controlling) nodes⟩ ⟨(coefficients) values⟩

E   Voltage-controlled voltage source

        E⟨name⟩   N+   N−   NC+   NC−   ⟨(voltage gain) value⟩

F   Current-controlled current source

        F⟨name⟩   N+   N−   vN   ⟨(current gain) value⟩

G   Voltage-controlled current source

        G⟨name⟩   N+   N−   NC+   NC−   ⟨(transconductance) value⟩

H	Current-controlled voltage source

```
H(name)  N+  N-  VN  ((transresistance) value)
```

I	Independent current source

```
I(name)  N+  N-  [dc (value)]  [ac ((magnitude) value)
+     ((phase) value)]  [(transient specifications)]
```

V	Independent voltage source

```
V(name)  N+  N-  [dc (value)]  [ac ((magnitude) value)
+      ((phase) value)]  [(transient specifications)]
```

VALUE	Arithmetical function

```
E(name)  N+  N-  VALUE = {(expression)}
G(name)  N+  N-  VALUE = {(expression)}
```

TABLE	Look-up table

```
E(name) N+  N-   TABLE { (expression) } =
        + ( ((input) value), ((output) value) )*
G(name) N+  N-   TABLE { (expression) } =
        + ( ((input) value), ((output) value) )*
```

LAPLACE	Laplace's transfer function

```
E(name)  N+  N-  LAPLACE { (expression) } = { (transform) }
G(name)  N+  N-  LAPLACE { (expression) } = { (transform) }
```

FREQ	Frequency response transfer function

```
E(name)  N+   N-   FREQ { (expression) } =
        + (((frequency) value), ((magnitude) value),
        ((phase) value) )*
G(name)  N+   N-   FREQ { (expression) } =
        + (((frequency) value), ((magnitude) value),
        ((phase) value) )*
```

PROBLEMS

Write PSpice statements for the following questions. Assume that the first node is the positive terminal and the second node is the negative terminal.

4-1. The various voltage or current waveforms that are connected between nodes 4 and 5 are shown in Fig. P4-1.

4-2. A voltage source that is connected between nodes 10 and 0 has a dc voltage of 12 V for dc analysis, a peak voltage of 2 V

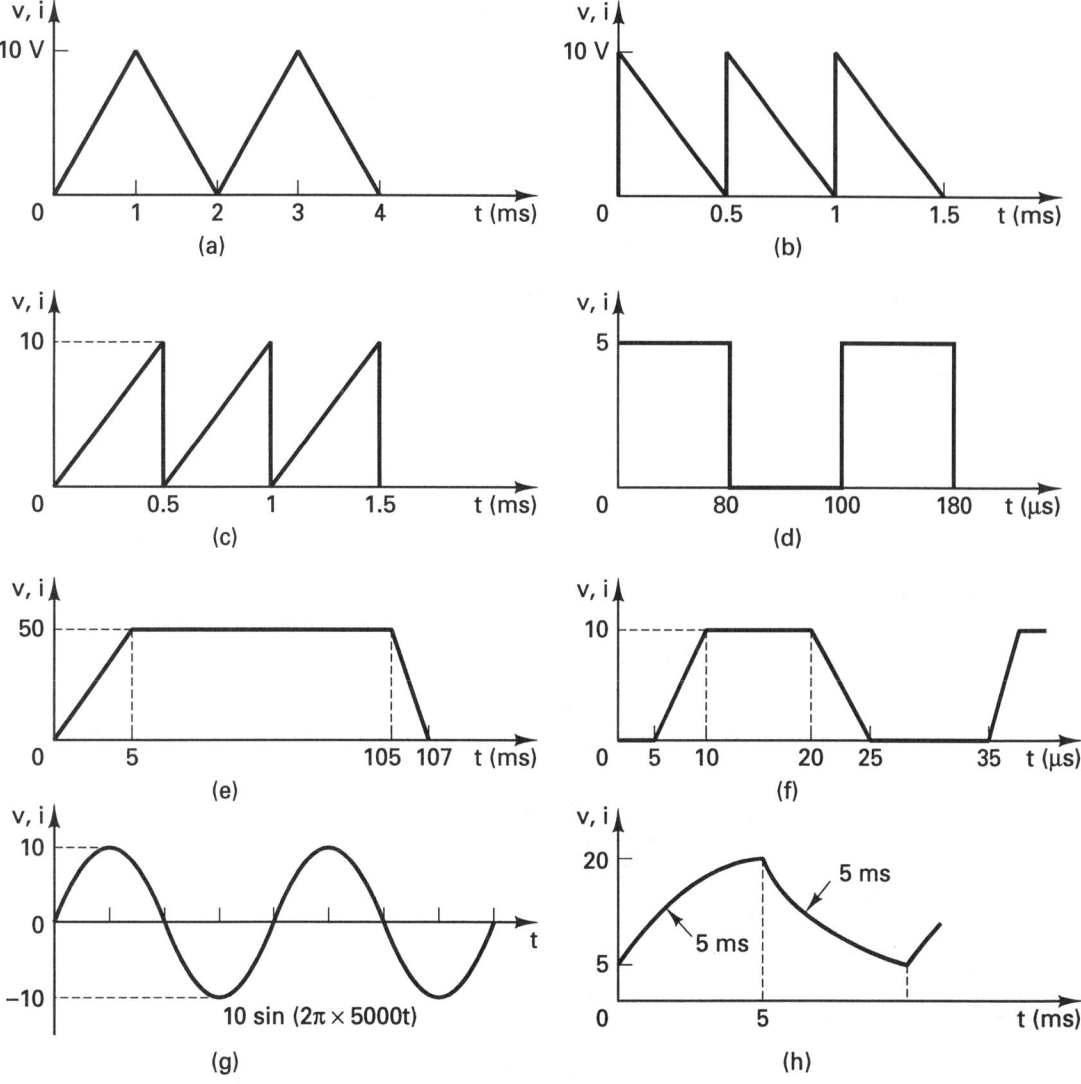

Figure P4-1

with 60° phase shift for ac analysis, and a sinusoidal peak voltage of 0.1 V at 1 MHz for transient analysis.

4-3. A current source that is connected between nodes 5 and 0 has a dc current of 0.1 A for dc analysis, a peak current of 1 A with 60° phase shift for ac analysis, and a sinusoidal current of 0.1 A at 1 kHz for transient analysis.

4-4. A voltage source that is connected between nodes 4 and 5 is given by

$$v = 2 \sin[(2\pi \times 50,000t) + 5 \sin(2\pi \times 1000t)]$$

4-5. A polynomial voltage source Y that is connected between nodes 1 and 2 is controlled by a voltage source V_1 connected between nodes 4 and 5. The source is given by

$$Y = 0.1V_1 + 0.2V_1^2 + 0.05V_1^2$$

4-6. A polynomial current source I that is connected between nodes 1 and 2 is controlled by a voltage source V_1 connected between nodes 4 and 5. The source is given by

$$Y = 0.1V_1 + 0.2V_1^2 + 0.05V_1^2$$

4-7. A voltage source V_0 that is connected between nodes 5 and 6 is controlled by a voltage source V_1 and has a voltage gain of 25. The controlling voltage is connected between nodes 10 and 12. The source is expressed as $V_0 = 25V_1$.

4-8. A current source I_0 that is connected between nodes 5 and 6 is controlled by a current source I_1 and has a current gain of 10. The voltage through which the controlling current flows is V_C. The current source is given by $I_0 = 10I_1$.

4-9. A current source I_0 that is connected between nodes 5 and 6 is controlled by a voltage source V_1 between nodes 8 and 9. The transconductance is 0.05 Ω^{-1}. The current source is given by $I_0 = 0.051V_i$.

4-10. A voltage source V_0 that is connected between nodes 5 and 6 is controlled by a current source I_1 and has a transresistance of 150 Ω. The voltage through which the controlling current flows is V_C. The voltage source is expressed as $V_0 = 150I_1$.

4-11. A nonlinear resistance R that is connected between nodes 4 and 6 is controlled by a voltage source V_1 and has a resistance of the form

$$R = V_1 + 0.2V_1^2$$

4-12. A nonlinear transconductance G_m that is connected between nodes 4 and 6 is controlled by a current source. The voltage through the controlling current flows is V_1. The transconductance has the form

$$G_m = V_1 + 0.2V_1^2$$

4-13. The v–i characteristic of a diode is described by

$$I_D = I_S e^{v_D/nV_T}$$

where $I_S = 2.2 \times 10^{-15}$ A, $n = 1$, and $V_T = 26.8$ mV. Use VALUE to simulate the diode voltage between nodes 3 and 4 as a function of diode current.

4-14. The v–i characteristic of a diode is described by

$$I_D = I_S e^{v_D/nV_T}$$

where $I_S = 2.2 \times 10^{-15}$ A, $n = 1$, and $V_T = 26.8$ mV. Use VALUE to simulate the diode current between nodes 3 and 4 as a function of diode voltage.

4-15. The v–i characteristic of a diode is given by

i_D (A)	10	20	30	40	50	100	150	500	900
v_D (V)	0.55	0.65	0.7	0.75	0.8	0.9	1.0	1.3	1.4

Use TABLE to simulate the diode voltage between nodes 3 and 4 as a function of diode current.

4-16. The current through a varistor depends on its voltage, which is given by

i (mA)	0	0.1	0.3	1	2.5	5	15	45	100	200
v (V)	0	50	100	150	200	250	300	350	400	450

Use TABLE to simulate the varistor current between nodes 3 and 4 as a function of varistor current.

Chapter 5
■■■■■■■■■

Passive elements

5-1 INTRODUCTION

PSpice recognizes passive elements by their symbols and their models. The elements can be resistor (R), inductor (L), capacitor (C), magnetic, or switches. The models are necessary to take into account the parameter variations (e.g., the value of a resistance depends on the operating temperature). The simulation of passive elements on PSpice requires that the following be specified:

Modeling of elements
Operating temperature
RLC elements
Magnetic elements and transformers
Lossless transmission lines
Switches

5-2 MODELING OR ELEMENTS

A model that specifies a set of parameters for an element is specified in PSpice by the .MODEL command. The same model can be used by one or more elements in the same circuit. The various . (dot) commands are explained in Chapter 6. The general form of the model statement is

```
.MODEL   MNAME   TYPE(P1=V1 P2=V2
+               P3=V3 ...... PN=VN  [(tolerance specification)]* )
```

MNAME is the name of the model and must start with a letter. Although not necessary, it is advisable to make the first letter the symbol of the element (e.g., R for resistor, L for inductor). The list of symbols for elements is shown in Table 2-1. P1, P2, . . . are the element parameters, and V1, V2, . . . are their values, respectively. TYPE is the type name of the elements and must have the correct type as shown in Table 5-1. An element must have the correct model type name. That is, a resistor must have the type name RES, not type IND or CAP. However, there can be more than one model of the same type in a circuit with different model names. (tolerance specification) is used with .MC analysis only, and it may be appended to each parameter with the format

```
[ DEV/⟨distribution name⟩ ⟨value in % from 0 to 9⟩ ]
[ LOT/⟨distribution name⟩ ⟨value in % from 0 to 9⟩ ]
```

where ⟨distribution name⟩ is one of the following:

UNIFORM   Generates uniformly distributed deviations over the range of ± ⟨value⟩

GAUSS   Generates deviations with Gaussian distribution over the range ±4, and ⟨value⟩ specifies the ±1 deviation

TABLE 5-1 TYPE NAME OF ELEMENTS

Type name	Element
RES	Resistor
CAP	Capacitor
D	Diode
IND	Inductor
NPN	NPN bipolar junction transistor
PNP	PNP bipolar junction transistor
NJF	n-Channel junction FET
PJF	p-Channel junction PET
NMOS	n-Channel MOSFET
PMOS	p-Channel MOSFET
GASFET	n-Channel GaAs MESFET
VSWITCH	Voltage-controlled switch
ISWITCH	Current-controlled switch
CORE	Nonlinear magnetic core (transformer)

Some Model Statements

```
.MODEL  RLOAD    RES  (R=1   TC1=0.02   TC2=0.005)
.MODEL  RLOAD    RES  (R=1   DEV/GAUSS 0.5%   LOT/UNIFORM 10%)
.MODEL  CPASS    CAP  (C=1   VC1=0.01   VC2=0.002   TC1=0.02   TC2=0.005)
.MODEL  LFILTER  IND  (L=1   IL1=0.1    IL2=0.002   TC1=0.02   TC2=0.005)
.MODEL  DNOM     D    (IS=1E−9)
.MODEL  DLOAD    D    (IS=1E−9   DEV 0.5%    LOT 10%)
.MODEL  QOUT     NPN  (BF=50   IS=1E−9)
```

5-3 OPERATING TEMPERATURE

The operating temperature of an analysis can be set to any value desired by the .TEMP command. The general form of the statement is

```
.TEMP   ⟨(one or more temperature) values⟩
```

The temperatures are in degrees Celsius. If more than one temperature is specified, the analysis is performed for each temperature. The model parameters are assumed to be measured at a nominal temperature, which is, by default, 27°C. The default nominal temperature of 27°C can be changed by the TNOM option in the .OPTIONS statements, which are discussed in Section 6-5.

Some Temperature Statements

```
.TEMP    50
.TEMP    25   50
.TEMP     0   25   50   100
```

5-4 *RLC* ELEMENTS

The voltage and current relationships of resistor, inductor, and capacitor are shown in Fig. 5-1.

5-4.1 Resistor

The symbol for a resistor is R. The name of a resistor must start with R, and it takes the general form

```
R⟨name⟩   N+   N−   RNAME   RVALUE
```

A resistor does not have a polarity, and the order of the nodes does not matter. However, by defining N+ as the positive node and N− as the negative node, the current is assumed to flow from node N+ through the resistor to node N−.

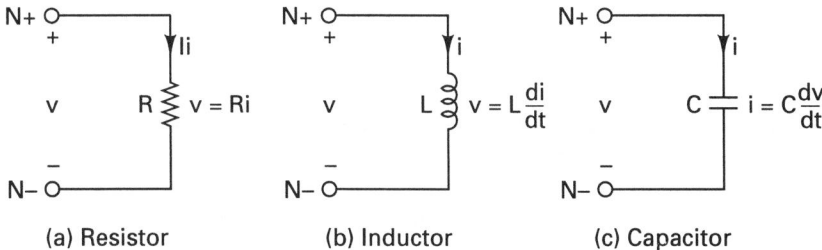

Figure 5-1   Voltage and current relationships.

RNAME is the model name that defines the parameters of the resistor. RVALUE is the nominal value of the resistance.

Note. Some versions of PSpice or SPICE do not recognize the polarity of resistors and do not allow referring currents through the resistor [e.g., I(R1)].

The model parameters are shown in Table 5-2. If RNAME is omitted, RVALUE is the resistance in ohms, and RVALUE can be positive or negative but *must* not be zero. If RNAME is included and TCE is *not* specified, the resistance as a function of temperature is calculated from

```
RES = RVALUE * R * [1 + TC1 * (T - T0) + TC2 * (T - T0)² ]
```

If RNAME is included and TCE is specified, the resistance as a function of temperature is calculated from

```
                   TCE*(T - T0)
RES = RVALUE * R * 1.01
```

where T and T0 are the operating temperautre and room temperature respectively, in degrees Celsius.

TABLE 5-2 MODEL PARAMETERS FOR RESISTORS

Name	Meaning	Unit	Default
R	Resistance multiplier		1
TC1	Linear temperature coefficient	$°C^{-1}$	0
TC2	Quadratic temperature coefficient	$°C^{-2}$	0
TCE	Exponential temperature coefficient	%/°C	0

Some Resistor Statements

```
R1          6    5      10K
RLOAD       12   11     ARES   2MEG
.MODEL   ARES   RES  (R=1   TC1=0.02   TC2=0.005)
RINPUT      15   14     RRES   5K
.MODEL   RRES   RES  (R=1   TCE=2.5)
```

5-4.2 Capacitor

The symbol for a capacitor is C. The name of a capacitor must start with C, and it takes the general form

```
C⟨name⟩   N+   N-   CNAME   CVALUE   IC=V0
```

N+ is the positive node and N− is the negative node. The voltage of node N+ is assumed positive with respect to node N− and the current flows from node N+

through the capacitor to node N−. CNAME is the model name, and CVALUE is the nominal vaue of the capacitor. IC defines the initial (time-zero) voltage of the capacitor, V0.

The model parameters are shown in Table 5-3. If CNAME is omitted, CVALUE is the capacitance in farads. The CVALUE can be positive or negative but *must* not be zero. If CNAME is included, the capacitance, which depends on the voltage and temperature, is calculated from

```
CAP = CVALUE * C * (1 + VC1 * V + VC2 * V²)
            * [1 + TC1 * (T − T0) + TC2 * (T − T0)²]
```

where T is the operating temperature and T0 is the room temperature in degrees Celsius.

TABLE 5-3 MODEL PARAMETERS FOR CAPACITORS

Name	Meaning	Unit	Default
C	Capacitance multiplier		1
VC1	Linear voltage coefficient	V^{-1}	0
VC2	Quadratic voltage coefficient	V^{-2}	0
TC1	Linear temperature coefficient	$°C^{-1}$	0
TC2	Quadratic temperature coefficient	$°C^{-2}$	0

Some Capacitor Statements

```
C1          6    5     10UF
CLOAD       12   11    5PF     IC=2.5V
CINPUT      15   14    ACAP    10PF
C2          20   19    ACAP    20NF    IC=1.5V
.MODEL   ACAP   CAP  (C=1   VC1=0.01   VC2=0.002   TC1=0.02   TC2=0.005)
```

Note. The initial conditions (if any) apply only if the UIC (use initial condition) option is specified on the .TRAN command that is described in Section 6-9.

5-4.3 Inductor

The symbol for an inductor is L. The name of an inductor must start with L, and it takes the general form

```
L⟨name⟩ N+  N− LNAME   LVALUE   IC=I0
```

N+ is the positive node and N− is the negative node. The voltage of N+ is assumed positive with respect to node N−, and the current flows from node N+ through the inductor to node N−. LNAME is the model name, and LVALUE is the nominal value of the inductor. IC defines the initial (time-zero) current of the inductor I0.

The model parameters of an inductor are shown in Table 5-4. If LNAME is omitted, LVALUE is the inductance in henries. LVALUE can be positive or negative but *must* not be zero. If LNAME is included, the inductance, which depends on the current and temperature, is calculated from

```
IND = LVALUE * L * (1 + IL1 * I + IL2 * I²)
            * [1 + TC1 * (T − T0) + TC2 * (T − T0)²]
```

where T is the operating temperautre and T0 is the room temperature in degrees Celsius.

Note. The initial conditions (if any) apply only if the UIC (use initial condition) option is specified on the .TRAN command that is described in Section 6-9.

TABLE 5-4 MODEL PARAMETERS FOR INDUCTORS

Name	Meaning	Unit	Default
L	Inductance multiplier		1
IL1	Linear current coefficient	A^{-1}	0
IL2	Quadratic current coefficient	A^{-2}	0
TC1	Linear temperature coefficient	$°C^{-1}$	0
TC2	Quadratic temperature coefficient	$°C^{-2}$	0

Some Inductor Statements

```
L1       6    5      10MH
LLOAD    12   11     5UH       IC=0.2MA
LLINE    15   14     LMOD      5MH
LCHOKE   20   19     LMOD      2UH       IC=0.5A
.MODEL   LMOD  IND  (L=1   IL1=0.1   IL2=0.002   TC1=0.02   TC2=0.005)
```

Example 5-1

The *RLC* circuit of Fig. 5-2(a) is supplied with the input voltage as shown in Fig. 5-2(b). Calculate and plot the transient response from 0 to 1 ms with a time increment of 5 μs. The output voltage is taken across resistor R_2. The results should be available for display and hard copy by Probe. The model parameters are, for the

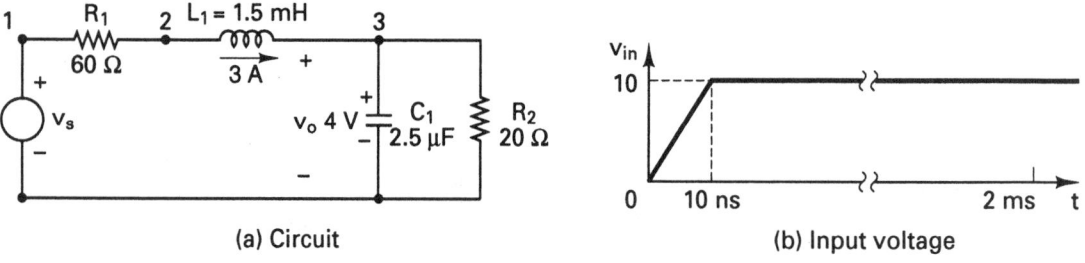

(a) Circuit

(b) Input voltage

Figure 5-2 *RLC* circuit for Example 5-1.

resistor, R=1, TC1=0.02, and TC2=0.005; for the capacitor, C=1, VC1=0.01, VC2=0.002, TC1=0.02, and TC2=0.005; and for the inductor, L=1, IL1=0.1, IL2=0.002, TC1=0.02, and TC2=0.005. The operating temperature is 50°C.

Solution The circuit file contains the following statements:

■ ■

Example 5-1 *RLC* circuit

SOURCE ■ * Input step voltage represented as a PWL waveform:
```
        VS  1   0    PWL (0  0  10NS  10V  2MS  10V)
```
CIRCUIT ■ ■ * R₁ has a value of 60 Ω with model RMOD:
```
        R1   1   2    RMOD    60
        *    Inductor of 1.5 mH with an initial current of 3 A and model name LMOD:
        L1   2   3   LMOD   1.5MH   IC=3A
        *    Capacitor of 2.5 µF with an initial voltage of 4 V and model name CMOD:
        C1   3   0    CMOD    2.5UF   IC=4V
        R2   3   0    RMOD    20
        *    Model statements for resistor, inductor, and capacitor:
        .MODEL   RMOD   RES (R=1   TC1=0.02   TC2=0.005)
        .MODEL   CMOD   CAP (C=1   VC1=0.01   VC2=0.002   TC1=0.02   TC2=0.005)
        .MODEL   LMOD   IND (L=1   IL1=0.1    IL2=0.002   TC1=0.02   TC2=0.005)
        *  The operating temperature in 50°C.
        .TEMP    50
```
ANALYSIS ■ ■ ■ * Transient analysis from 0 to 1 ms with a 5-µs time increment and
```
        *    using initial conditions (UIC):
        .TRAN   5US   1MS   UIC
        *   Plot the results of transient analysis with the voltage at nodes 3 and 1.
        .PLOT    TRAN   V(3)   V(1)
        .PROBE
.END
```
■ ■

The results of the simulation that are obtained by Probe are shown in Fig. 5-3. The inductor has an initial current of 3 A, which is taken into consideration by UIC in the .TRAN command.

5-5 MAGNETIC ELEMENTS AND TRANSFORMERS

The magnetic elements are mutual inductors (transformer). PSpice allows simulating two types of magnetic circuits:

Linear magnetic circuits
Nonlinear magnetic circuits

5-5.1 Linear Magnetic Circuits

The symbol for mutual coupling is K. The general form of coupled inductors is

```
K⟨name⟩   L⟨(first inductor) name⟩   L⟨(second inductor) name⟩
+         ⟨(coupling) value⟩
```

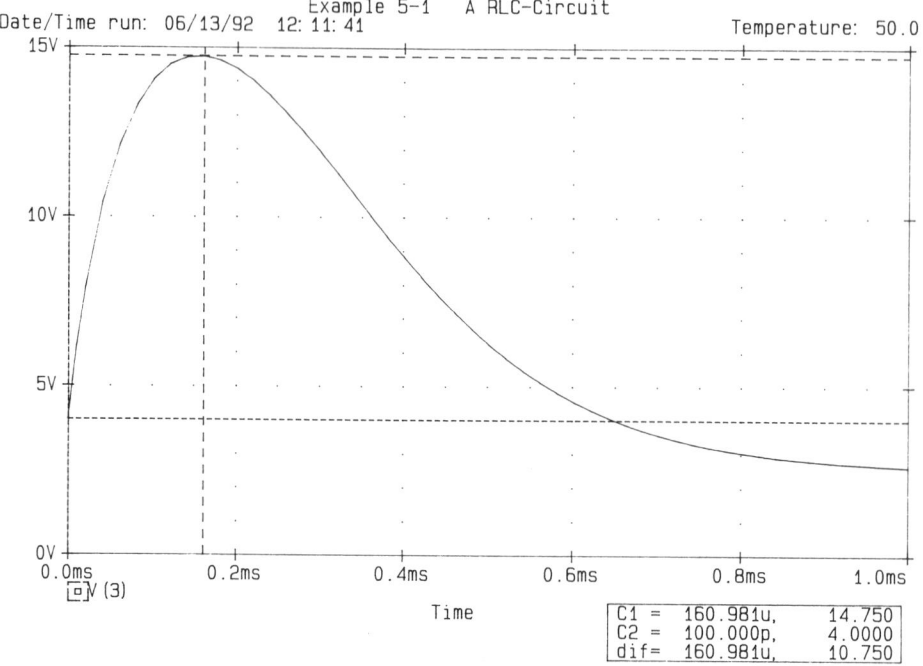

C1 =	160.981u,	14.750
C2 =	100.000p,	4.0000
dif=	160.981u,	10.750

Figure 5-3   Transient response for Example 5-1.

For linear coupled inductors, K⟨name⟩ couples two or more inductors. ⟨(coupling) value⟩ is the coefficient of coupling, k. The value of coefficient of coupling must be greater than 0 and less than or equal to 1: $0 < k \leq 1$.

The inductors can be coupled in either order positively or negatively as shown Fig. 5-4. In terms of the dot convention shown in Fig. 5-4(a), PSpice assumes a dot on the first node of each inductor. The mutual inductance is determined from

$$M = k\sqrt{L_1 L_2}$$

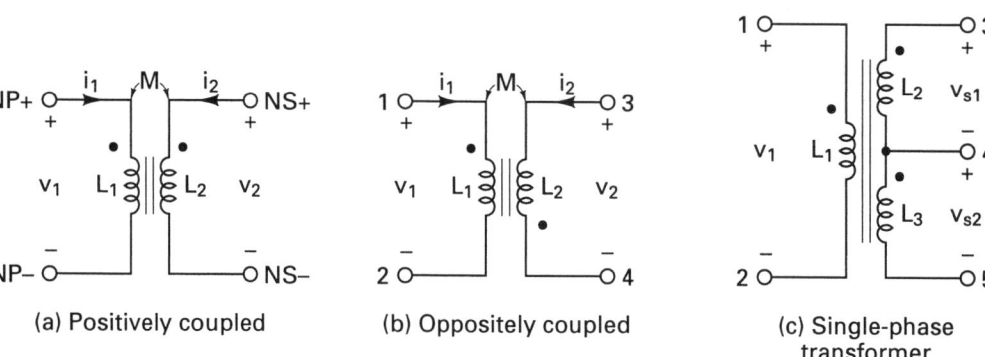

(a) Positively coupled

(b) Oppositely coupled

(c) Single-phase transformer

Figure 5-4   Coupled inductors.

In the time domain, the voltages of coupled inductors are expressed as

$$v_1 = L_1 \frac{di_1}{dt} + M \frac{di_2}{dt}$$

$$v_2 = M \frac{di_1}{dt} + L_2 \frac{di_2}{dt}$$

In the frequency domain, the voltages are expressed as

$$V_1 = j\omega L_1 I_1 + j\omega M I_2$$
$$V_2 = j\omega M I_1 + j\omega L_2 I_2$$

where ω is the frequency in rad/s.

Some Coupled-Inductor Statements

```
KTR     LA   LB   0.9
KIND    L1   L2   0.98
```

The coupled inductors of Fig. 5-4(a) can be written as a single-phase transformer (with $k = 0.9999$):

```
*   PRIMARY
L1          1    2    0.5MH
*   SECONDARY
L2          3    4    0.5MH
*   MAGNETIC COUPLING
KXFRMER  L1   L2   0.9999
```

If the dot in the second coil is changed as shown in Fig. 5-4(b), the coupled inductors are written as

```
L1          1    2    0.5MH
L2          4    3    0.5MH
KXFRMER  L1   L2   0.9999
```

A transformer with a single primary coil and center-tapped secondary as shown in Fig. 5-4(c) can be written as

```
*   PRIMARY
L1      1    2    0.5MH
*   SECONDARY
L2      3    4    0.5MH
L3      4    5    0.5MH
*   MAGNETIC COUPLING
K12   L1   L2   0.9999
K13   L1   L3   0.9999
K23   L2   L3   0.9999
```

The three statements above (K12, K13, K23) can be written in PSpice as

```
KALL   L1   L2   L3   0.9999
```

Notes

1. The name Kxx need not be related to the names of the inductors it is coupling. However, it is a good practice, because it is convenient to identify the inductors involved in the coupling.
2. The polarity (or dot) is determined by the order of the nodes in the L . . . statements and not by the order of the inductors in the K . . . statement [e.g., (K12 L1 L2 0.9999) has the same result as (K12 L2 L1 0.9999)].

Example 5-2

A circuit with two coupled inductors is shown in Fig. 5-5. The input voltage is 120 V peak. Calculate the magnitude and phase of the output current for frequencies from 60 to 120 Hz with a linear increment. The total number of points in the sweep is 2. The coefficient of coupling for the transformer is 0.999.

Solution It is important to note that the primary and the secondary have a common node. Without this, PSpice will give an error message, because there is no dc path from the nodes of the secondary to the ground.

The circuit file contains the following statements:

■ ■

Example 5-2   Coupled linear inductors

```
SOURCE   ■ *    Input voltage is 120 V peak and 0° phase for ac analysis:
             VIN   1    0    AC    120V
CIRCUIT  ■ ■ R1    5    2    0.5
             *    A dummy voltage source of VY=0 is added to measure the load current:
             VY    1    5    DC    0V
             *    The dot convention is followed in inductors L₁ and L₂:
             L1    2    0    0.5MH
             L2    0    4    0.5MH
             *    Magnetic coupling coefficient is 0.999. The order of L₁ and L₂ is
             *    not significant.
             K12   L1   L2   0.999
             R2    4    6    0.5
             RL    6    7    150
             *    A dummy voltage source of VX=0 is added to measure the load current:
             VX    7    0    DC    0V
ANALYSIS ■ ■ ■ *    Ac analysis where the frequency is varied linearly from
             *    60 Hz to 120 Hz with two points:
             .AC   LIN   2   60HZ   120HZ
             *    Print the magnitude and phase of output current. Some versions of
             *    Pspice and SPICE do not permit reference to currents through
             *    resistors [e.g., IM(RL), IP(RL)].
             .PRINT   AC   IM(VX)   IP(VX)   IM(RL)   IP(RL)   ; Prints in the output file
             .END
```
■ ■

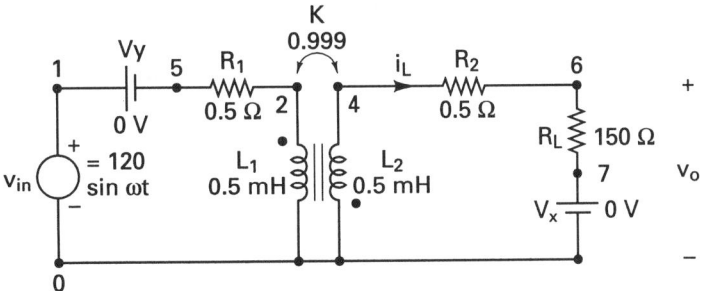

Figure 5-5 Circuit with two coupled inductors.

The transformer is considered to be linear and its inductances remain constant. The results of the simulation, which are stored in output file EX5-2.OUT, are

```
FREQ          IM(VX)       IP(VX)       IM(RL)       IP(RL)
  6.000E+01    2.809E-01   -1.107E+02    2.809E-01   -1.107E+02
  1.200E+02    4.790E-01   -1.271E+02    4.790E-01   -1.271E+02
```

5-5.2 Nonlinear Magnetic Circuits

For a nonlinear inductor, the general form is

```
K⟨name⟩   L⟨(inductor) name⟩   ⟨(coupling) value⟩
+         ⟨(model) name⟩   [(size) value]
```

For an iron-core transformer, k is very high, greater than 0.999. The model type name for a nonlinear magnetic inductor is CORE, and the model parameters are shown in Table 5-5. The [(size) value] scales the magnetic cross section and defaults to 1. It represents the number of lamination layers, so that only one model statement can be used for a particular lamination type of core.

If the ⟨(model) name⟩ is specified, the mutual coupling inductor becomes a nonlinear magnetic core and the inductor specifies the "number of turns" instead of the inductance. The list of the coupled inductors may be just one inductor.

TABLE 5-5 MODEL PARAMETERS FOR NONLINEAR MAGNETIC

Name	Meaning	Unit	Default
AREA	Mean magnetic cross section	cm²	0.1
PATH	Mean magnetic path length	cm	1.0
GAP	Effective air-gap length	cm	0
PACK	Pack (stacking)		1.0
MS	Magnetic saturation	A/m	1E+6
ALPHA	Mean field parameter		1E-3
A	Shape parameter		1E+3
C	Domain wall-flexing constant		0.2
K	Domain wall-pinning constant		500

The magnetic core's B–H characteristics are analyzed using the Jiles–Atherton model [2].

If the inductors of Fig. 5-4(a) use the nonlinear core, the statements would be as follows:

```
*   Inductor L₁ of 100 turns:
L1         1     2     100
*   Inductor L₂ of 10 turns:
L2         3     4     10
*   Nonlinear coupled inductors with model CMOD:
K12    L1    L2    0.9999    CMOD
*   Model for the nonlinear inductors:
.MODEL   CMOD   CORE  (AREA=2.0  PATH=62.8  GAP=0.1  PACK=0.98)
```

The model parameters can be adjusted to specify a B–H characteristic. The nonlinear magnetic model uses MKS (metric) units. However, the results for Probe are converted to gauss and oersted and may be displayed using B(Kxx) and H(Kxx). The B–H curve can be drawn by a transient run with a slowly rising current through a test inductor and then by displaying B(Kxx) against H(Kxx).

Characterizing core materials may be done by trial by using PSpice and Probe. The procedures for setting parameters to obtain a particular characteristic are as follows:

1. Set the domain wall-pinning constant, K = 0. The curve should be centered in the B–H loop, like a spine. The slope of the curve at $H = 0$ should be approximately equal to that when it crosses the H-axis at $B = 0$.
2. Set the magnetic saturation, MS $= B_{max}/0.01257$.
3. ALPHA sets the slope of the curve. Start with the mean field parameter ALPHA $= 0$, and then vary its value to get the desired slope of the curve. It may be necessary to change MS slightly to get the desired saturation value.
4. Change K to a nonzero value to create the desired hysteresis. K affects the opening of the hysteresis loop.
5. Set C to obtain the initial permeability. Probe displays the permeability, which is $\Delta B/\Delta H$. Since Probe calculates differences, not derivatives, the curves will not be smooth. The initial value of $\Delta B/\Delta H$ is the initial permeability.

Note. A nonlinear magnetic model is not available in SPICE2.

Example 5-3

The coupled inductors of Fig. 5-4(a) are nonlinear. The parameters of the inductors are $L_1 = 200$ turns, $L_2 = 100$ turns, and $k = 0.9999$. Plot the B–H characteristic of the core from the results of transient analysis if the input current is varied very slowly from 0 to -15 A, -15 to $+15$ A, and $+15$ to -15 A. The load resistance of $R_L = 1$ kΩ is connected to the secondary of the transformer. The model parameters of the core are AREA=2.0, PATH=62.73, GAP=0.1, MS=1.6E+6, ALPHA=1E−3, A=1E+3, C=0.5, and K=1500.

Solution The circuit file for the coupled inductors in Fig. 5-4(a) would be:

■■

Example 5-3 Typical *B–H* characteristic

SOURCE   ■ * PWL waveform for transient analysis:
```
         IN   1   0    PWL (0   0   1  −15   2   15   3   −15)
```
CIRCUIT ■ ■ * Inductors represent the number of turns:
```
         L1   1   0    200
         L2   2   0    100
         R2   2   0    1000              ; Load resistance
         *   Coupled inductors with k = 0.999 and model CMOD:
         K12  L1  L2   0.9999   CMOD
         *   Model parameters for CMOD:
         .MODEL CMOD  CORE (AREA=2.0 PATH=62.73 GAP=0.1 MS=1.6E+6 ALPHA=1E−3
         +    A=1E+3  C=0.5 K=1500)
```
ANALYSIS ■ ■ ■ * Transient analysis from 0 to 3 s in steps of 0.05 s:
```
         .TRAN   0.05S   3S
         .PROBE
.END
```
■■

The *B–H* characteristic obtained by Probe is shown in Fig. 5-6. Note that by using the Probe menu, the *x*-axis has been changed to H(K12).

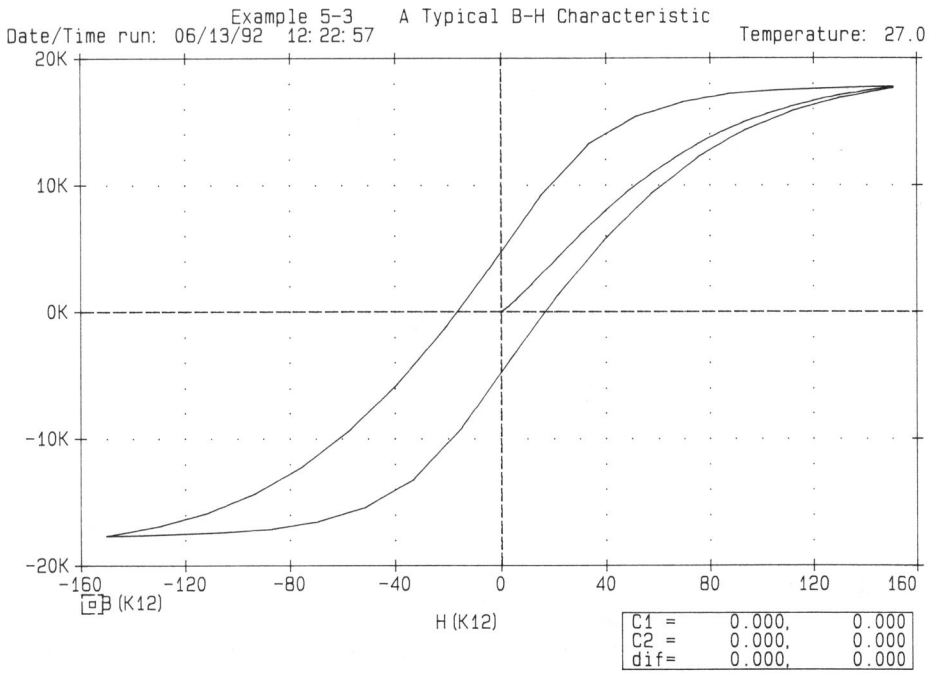

Figure 5-6 Typical *B–H* characteristic.

Example 5-4

The coupled inductor in Example 5-2 is replaced by a nonlinear core with the B–H characteristic of Fig. 5-6. This is shown in Fig. 5-7. The parameters of the inductors are $L_1 = 200$ turns, $L_2 = 100$ turns, and $k = 0.9999$. The input voltage is $v_s = 170 \sin(2\pi \times 60t)$. Plot the instantaneous values of the secondary voltage and current from 0 to 35 ms with a 100-μs increment. The results should be available for display and hard copy by Probe. The model parameters of the core are AREA=2.0, PATH=62.73, GAP=0.1, MS=1.6E+6, ALPHA=1E−3, A=1E+3, C=0.5, and K=1500.

Solution The primary and the secondary have a common node. Without this, PSpice will give an error message, because there is no dc path from the nodes of the secondary to the ground. The circuit file contains the following statements:

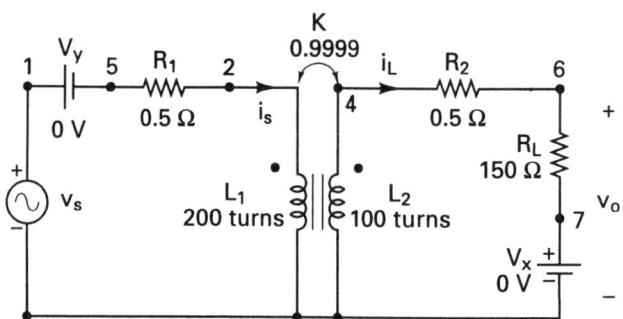

Figure 5-7 Circuit with nonlinear coupled inductors.

■ ■

Example 5-4 Nonlinear coupled inductors

SOURCE   ■ *   Input sinusoidal voltage of 170 V peak and 0° phase:
```
            VS    1    0    SIN (0    170V    60HZ)
```
CIRCUIT ■ ■ *   A dummy voltage source of VY = 0 is added to measure the load current:
```
            VY    1    5    DC    0V
            R1    5    2    0.5
            *   Inductors represent the number of turns:
            L1    2    0    200
            L2    4    0    100
            *   Coupled inductors with k = 0.9999 and model CMOD:
            K12   L1   L2    0.9999    CMOD
            *   Model parameters for CMOD:
            .MODEL CMOD   CORE (AREA=2.0 PATH=62.73 GAP=0.1 MS=1.6E+6 ALPHA=1E−3
            +       A=1E+3  C=0.5 K=1500)
            R2    4    6    0.5
            RL    6    7    150
            *   A dummy voltage source of VX = 0 is added to measure the load current:
            VX    7    0    DC    0V
```
ANALYSIS ■ ■ ■ *   Transient analysis from 0 to 35 ms with a 100-μs increment:
```
            .TRAN 100US    35MS
            *   Print the magnitude and phase of output current. Some versions of
            *   PSpice and SPICE do not permit reference to currents through
            *   resistors [e.g., IM(RL), IP(RL)].
```

```
.PRINT   TRAN   V(4)   I(VX)           ; Prints in the output file
.PROBE                                 ; Graphics post-processor
.END
```
■■

The results of the simulation that are obtained by *Probe* are shown in Fig. 5-8. Due to nonlinear *B–H* characteristics, the transformer current, which becomes nonlinear and contains harmonics, deviates from the sinusoidal form.

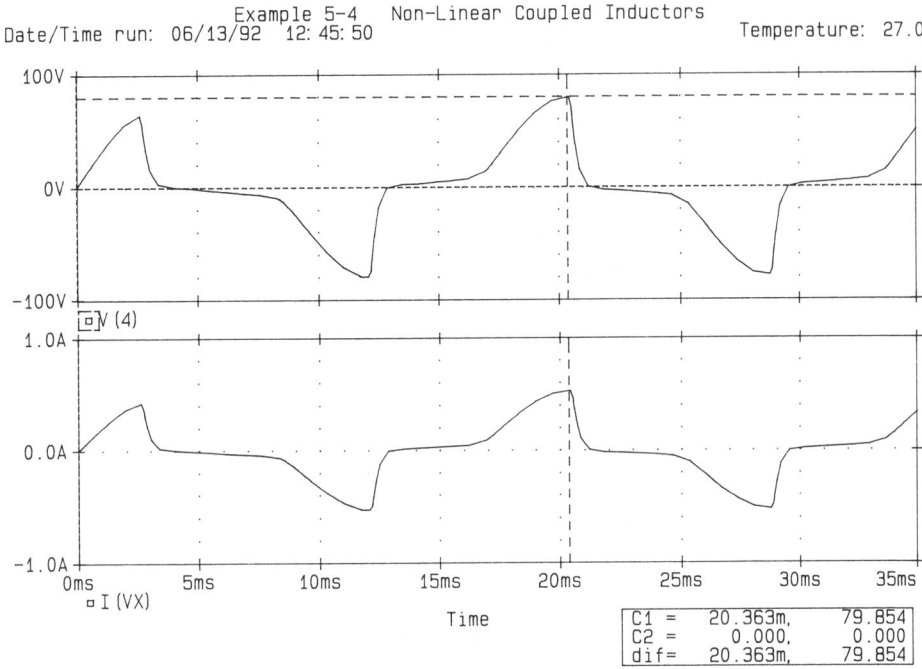

Figure 5-8 Transient response for Example 5-4.

5-6 LOSSLESS TRANSMISSION LINES

The symbol for a lossless transmission line is T. A transmission line has two ports: input and output. The general form of a transmission line is

```
T〈name〉  NA+  NA−  NB+  NB−  Z0=〈value〉  [TD=〈value〉]
+         [F=〈value〉  NL=〈value〉]
```

T〈name〉 is the name of the transmission line. NA+ and NA− are the nodes at the input port. NB+ and NB− are the nodes at the output port. NA+ and NB+ are defined as the positive nodes. NA− and NB− are defined as the negative nodes. The positive current flows from NA+ to NA− and from NB+ to NB−. Z0 is the characteristic impedance.

The length of the line can be expressed in either of two forms: (1) the transmission delay TD may be specified, or (2) the frequency F may be specified together with NL, which is the normalized electrical length of the transmission line with respect to wavelength in the line at frequency F. If the frequency F is specified but not NL, the default value of NL is 0.25; that is, F has the quarter-wave frequency. It should be noted that one of the options for expressing the length of the line must be specified. That is, TD or at least F must be specified. The block diagram of transmission line is shown in Fig. 5-9(a).

Some Transmission Statements

```
T1    1   2    3    4    Z0=50   TD=10NS
T2    4   5    6    7    Z0=50   F=2MHZ
TTRM  9   10   11   12   Z0=50   F=2MHZ   NL=0.4
```

The coaxial line shown in Fig. 5-9(b) can be represented by two propagating lines, where the first line (T1) models the inner conductor with respect to the shield, and the second line (T2) models the shield with respect to the outside. This is shown below,

```
T1    1   2   3   4    Z0=50    TD=1.5NS
T2    2   0   4   0    Z0=150   TD=1NS
```

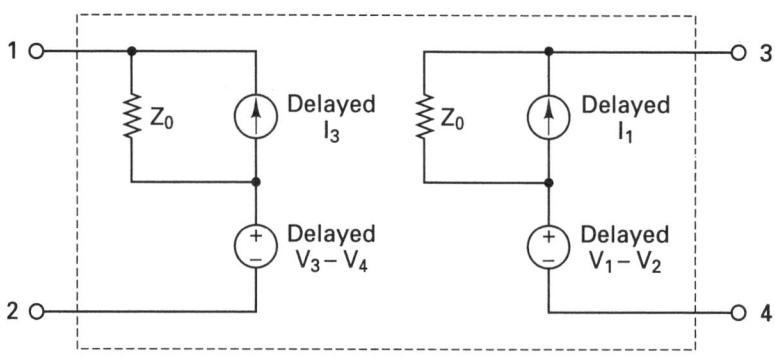

(a) Transmission line

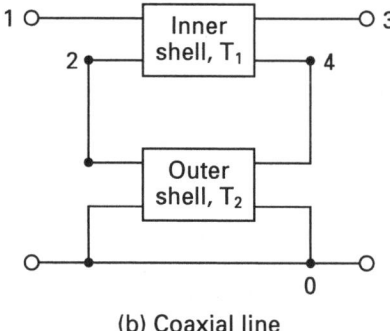

(b) Coaxial line

Figure 5-9   Transmission line.

Note. During the transient (.TRAN) analysis, the internal time step of PSpice is limited to be no more than one-half of the smallest transmission delay. Thus short transmission lines will cause long run times.

5-7 SWITCHES

PSpice allows simulating a special type of switch, as shown in Fig. 5-10, whose resistance varies continuously depending on the voltage or current. When the switch S_1 is on, the resistance is R_{ON}; and when it is off, the resistance becomes R_{OFF}. Two types of switches are permitted in PSpice:

Voltage-controlled switch
Current-controlled switch

Note. The voltage- and current-controlled switches are not available in SPICE2. However, they are available in SPICE3.

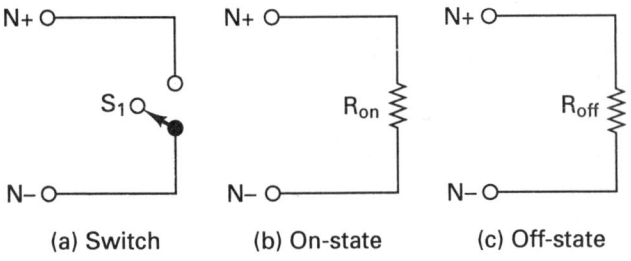

| (a) Switch | (b) On-state | (c) Off-state | **Figure 5-10** Switch with a variable resistance. |

5-7.1 Voltage-Controlled Switch

The symbol for a voltage-controlled switch is S. The name of this switch must start with S, and it takes the general form

S⟨name⟩   N+   N−   NC+   NC−   SNAME

N+ and N− are the two nodes of the switch. The current is assumed to flow from N+ through the switch to node N−. NC+ and NC− are the positive and negative nodes of the controlling voltage source, as shown in Fig. 5-11. SNAME is the

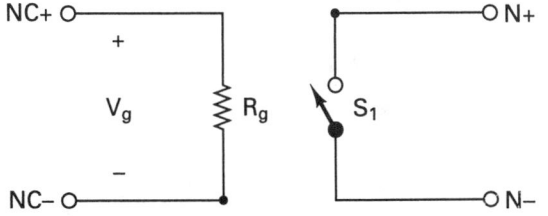

Figure 5-11 Voltage-controlled switch.

TABLE 5-6 MODEL PARAMETERS FOR
VOLTAGE-CONTROLLED SWITCH

Name	Meaning	Unit	Default
VON	Control voltage for on-state	V	1.0
VOFF	Control voltage for off-state	V	0
RON	On resistance	Ω	1.0
ROFF	Off resistance	Ω	10^6

model name. The resistance of the switch varies depending on the voltage across
the switch. The type name for a voltage-controlled switch is VSWITCH and the
model parameters are shown in Table 5-6.

Voltage-Controlled Switch Statement

```
S1        6    5    4    0      SMOD
.MODEL    SMOD    VSWITCH  (RON=0.5 ROFF=10E+6 VON=0.7 VOFF=0.0)
```

Notes

1. R_{ON} and R_{OFF} must be greater than zero and less than 1/GMIN. The value of
   GMIN can be defined as an option as described in .OPTIONS command in
   Section 6-5. The default value of conductance, GMIN, is $1E-12 \ \Omega^{-1}$.
2. The ratio of R_{OFF} to R_{ON} should be less than $1E+12$.
3. The difficulty due to the high gain of an ideal switch can be minimized by
   choosing the value of R_{OFF} as high as permissible and that of R_{ON} as low as
   possible compared to other circuit elements, within the limits of allowable
   accuracy.

Example 5-5

A circuit with a voltage-controlled switch is shown in Fig. 5-12. The input voltage is
$v_s = 200 \ \sin(2000\pi t)$. Plot the instantaneous voltage at node 3 and the current
through the load resistor R_L for a time duration of 0 to 1 ms with an increment of
5 μs. The model parameters of the switch are RON=5M, ROFF=10E+9,
VON=25M, and VOFF=0.0. The results should be available for display by Probe.
Solution The voltage source VX = 0 V is inserted to monitor the output current.
The listing of the circuit file is as follows:

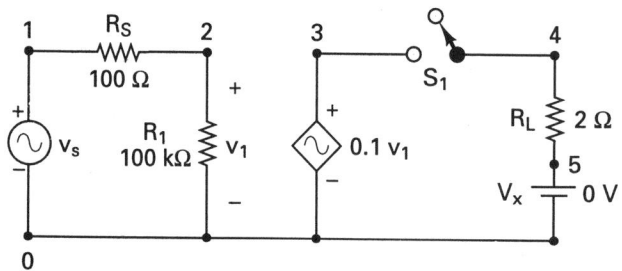

Figure 5-12 Circuit with a voltage-controlled switch.

■ ■

Example 5-5   Voltage-controlled switch

SOURCE   ■ * Sinusoidal input voltage of 200 V peak with 0° phase delay:
```
        VS  1    0     SIN (0  200V  1KHZ)
```
CIRCUIT   ■ ■ RS 1   2   100OHM
```
        R1  2    0     100KOHM
        *  Voltage-controlled voltage source with a voltage gain of 0.1:
        E1  3    0     2   0   0.1
        RL  4    5     2OHM
        *  Dummy voltage source of VX = 0 to measure the load current:
        VX  5    0     DC  0V
        *  Voltage-controlled switch controlled by voltage across nodes 3 and 0:
        S1  3    4     3   0   SMOD
        *  Switch model descriptions:
        .MODEL  SMOD  VSWITCH (RON=5M ROFF=10E+9 VON=25M  VOFF=0.0)
```
ANALYSIS   ■ ■ ■ * Transient analysis from 0 to 1 ms with a 5-μs increment:
```
        .TRAN    5US    1MS
        *   Plot the current through VS and the input voltage.
        .PLOT    TRAN   V(3) I(VX)                    ; On the output file
        .PROBE                                        ; Graphics post-processor
     .END
```
■ ■

The results of the simulation that are obtained by Probe are shown in
Fig. 5-13, which is the output of a diode rectifier. Switch S_1 behaves as a diode.

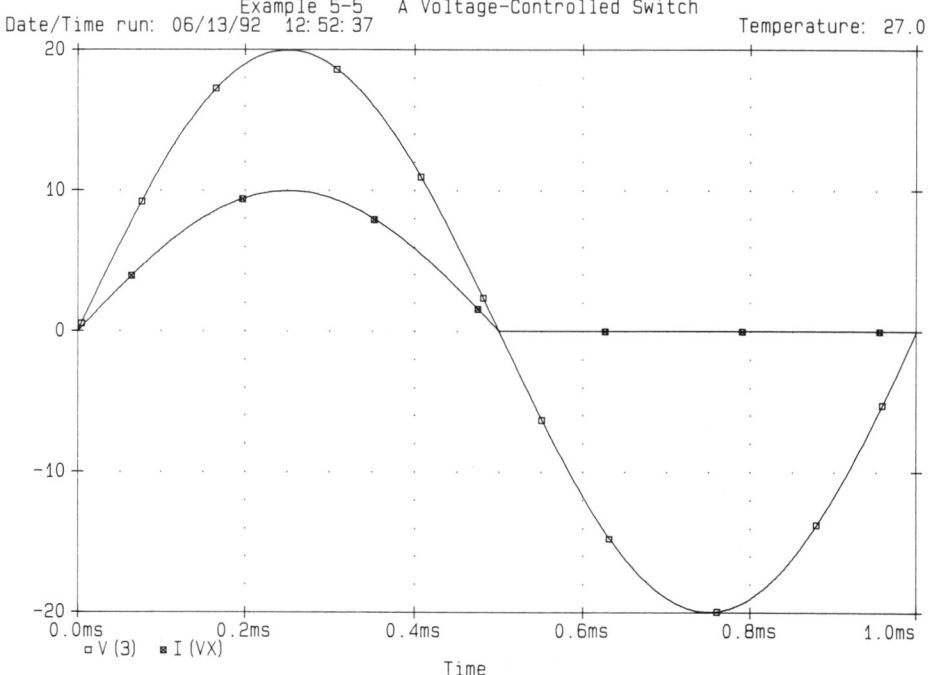

Figure 5-13   Transient response for Example 5-5.

5-7.2 Current-Controlled Switch

The symbol for a current-controlled switch is W. The name of the switch must start with W, and it takes the general form

```
W⟨name⟩   N+   N−   VN   WNAME
```

N+ and N− are the two nodes of the switch. VN is a voltage source through which the controlling current flows as shown in Fig. 5-14. WNAME is the model name. The resistance of the switch depends on the current through the switch. The type name for a current-controlled switch is ISWITCH and the model parameters are shown in Table 5-7.

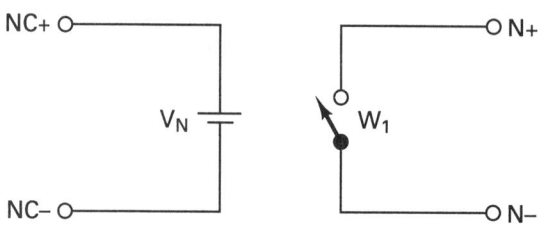

Figure 5-14 Current-controlled switch.

TABLE 5-7 MODEL PARAMETERS FOR CURRENT-CONTROLLED SWITCH

Name	Meaning	Unit	Default
ION	Control current for on-state	A	$1E-3$
IOFF	Control current for off-state	A	0
RON	On resistance	Ω	1.0
ROFF	Off resistance	Ω	10^6

Current-Controlled Switch Statement

```
W1       6  5  VN  RELAY
.MODEL   RELAY ISWITCH (RON=0.5 ROFF=10E+6 ION=0.07 IOFF=0.0)
```

Note. The current through voltage source VN controls the switch. The voltage source VN must be an independent source, and it can have a zero or a finite value. The limitations of the parameters are similar to those for the voltage-controlled switch.

Example 5-6

A circuit with a current-controlled switch is shown in Fig. 5-15. Plot the capacitor voltage and the inductor current for a time duration of 0 to 160 μs with an increment of 1 μs. The model parameters of the switch are RON=1E+6, ROFF=0.001, ION=1MA, and IOFF=0. The results should be available for display by Probe.
Solution The voltage source VX = 0 V is inserted to monitor the controlling current. The listing of the circuit file is as follows:

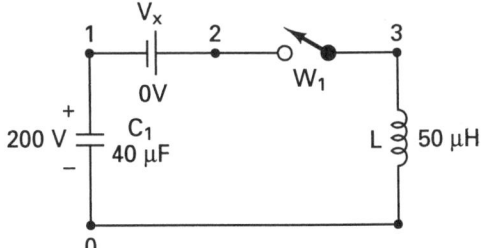

Figure 5-15 Circuit with a current-controlled switch.

■ ■

Example 5-6   Current-controlled switch

SOURCE ■ ■ * C1 of 40 μF with an initial voltage of 200 V:

C1   1   0   40UF   IC=200

*   Dummy voltage source of VX = 0:

VX   2   1   DC   0V

*   Current-controlled switch with model name SMOD:

W1   2   3   VX   SMOD

*   Model parameters:

.MODEL   SMOD   ISWITCH (RON=1E+6 ROFF=0.001 ION=1MA IOFF=0)

L1   3   0   50UF

CIRCUIT ■ ■ ■ * Transient analysis with UIC (use initial condition) option:

.TRAN   1US   160US   UIC

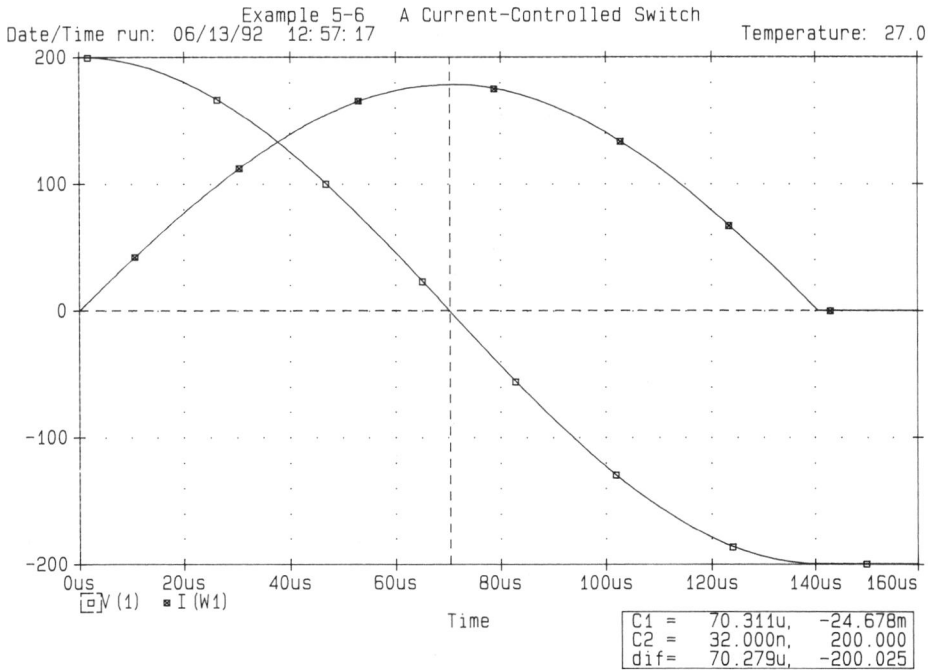

Figure 5-16 Transient response for Example 5-6.

```
          * Plot the voltage at node 1 and the current through VX.
          .PLOT  TRAN   V(1)    I(VX)                    ; On the output file
          .PROBE                                         ; Graphics post-processor
     .END
■■■■■■■■■■■■■■■■■■■■■■■■■■■■■■■■■■■■■■■■■■■■■■■■■■■■■
```

The results of the simulation that are obtained by Probe are shown in Fig. 5-16. Switch S_1 acts as diode and allows only positive current flow. The initial voltage on the capacitor is the driving source.

SUMMARY

The symbols for the passive elements are:

C   Capacitor

```
C⟨name⟩   N+   N−   CNAME   CVALUE   IC=V0
```

L   Inductor

```
L⟨name⟩   N+   N−   LNAME   LVALUE   IC=I0
```

K   Linear mutual inductors (transformer)

```
K⟨name⟩   L⟨(first inductor) name⟩
L⟨(second inductor)name⟩ ⟨value⟩]
```

K   Nonlinear inductor

```
K⟨name⟩   L⟨(inductor) name⟩  ⟨(coupling) value⟩
+         ⟨(model) name⟩  [(size) value]
```

R   Resistor

```
R⟨name⟩   N+   N−   RNAME   RVALUE
```

S   Voltage-controlled switch

```
S⟨name⟩   N+   N−   NC+   NC−   SNAME
```

T   Lossless transmission lines

```
T⟨name⟩   NA+   NA−   NB+   NB−   Z0=⟨value⟩   [TD=⟨value⟩]
+         [F=⟨value⟩    NL=⟨value⟩]
```

W   Current-controlled switch

```
W⟨name⟩   N+   N−   VN   WNAME
```

SUGGESTED READING

1. *PSpice Manual.* Irvine, Calif.: MicroSim Corporation, 1992.

2. D. C. Jiles and D. L. Atherton, "Theory of ferromagnetic hysteresis," *Journal of Magnetism and Magnetic Material*, Vol. 61, No. 48, 1986, pp. 48–60.

3. J. F. Lindsay and M. H. Rashid, *Electromechanics and Electrical Machinery.* Englewood Cliffs, N.J.: Prentice Hall, 1986.

PROBLEMS

Write the PSpice statements for the following problems. If applicable, the output should also be available for display and hard copy by Probe.

5-1. A resistor R_1, which is connected between nodes 3 and 4, has a nominal value of $R = 10$ kΩ. The operating temperature is 55°C, and it has the form

$$R_1 = R[1 + 0.2(T - T_0) + 0.002(T - T_0)^2]$$

5-2. A resistor R_1, which is connected between nodes 3 and 4, has a nominal value of $R = 10$ kΩ. The operating temperature is 55°C and it has the form

$$R_1 = R \times 1.01^{4.5(T-T_0)}$$

5-3. A capacitor C_1, which is connected between nodes 5 and 6, has a value of 10 pF and an initial voltage of -20 V.

5-4. A capacitor C_1, which is connected between nodes 5 and 6, has a nominal value of $C = 10$ pF. The operating temperature is $T = 55$°C. The capacitance, which is a function of its voltage and the operating temperature, is given by

$$C_1 = C(1 + 0.01V + 0.002V^2) \times [1 + 0.03(T - T_0) + 0.05(T - T_0)^2]$$

5-5. An inductor L_1, which is connected between nodes 5 and 6, has a value of 0.5 mH and carries an initial current of 0.04 mA.

5-6. An inductor L_1, which is connected between nodes 3 and 4, has a nominal value of $L = 1.5$ mH. The operating tempera-

ture is $T = 55$°C. The inductance is a function of its current and the operating temperature, and it is given by

$$L_1 = L(1 + 0.01I + 0.002I^2) \times [1 + 0.03(T - T_0) + 0.05(T - T_0)^2]$$

5-7. The two inductors, which are oppositely coupled as shown in Fig. 5-4(b), are $L_1 = 1.2$ mH and $L_2 = 0.5$ mH. The coefficients of coupling are $K_{12} = K_{21} = 0.999$.

5-8. Plot the transient response of the circuit in Fig. P5-8 from 0 to 5 ms with a time increment of 25 μs. The output voltage is taken across the capacitor. Use Probe for graphical output.

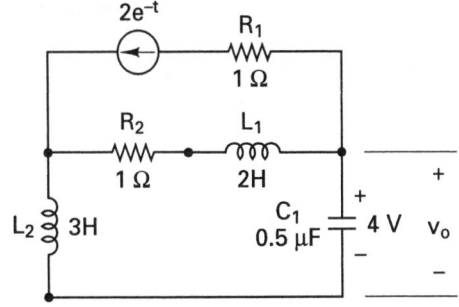

Figure P5-8

5-9. Repeat Problem 5-8 for the circuit of Fig. P5-9.

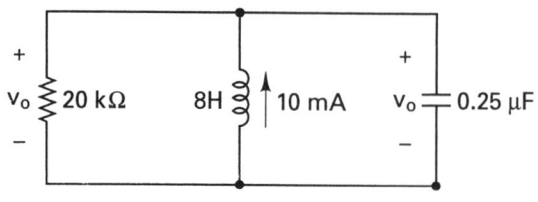

Figure P5-9

5-10. Plot the frequency response of the circuit in Fig. P5-10 from 10 Hz to 100 kHz with a decade increment and 10 points per decade. The output voltage is taken across the capacitor. Print and plot the magnitude and phase angle of the output voltage. Assume a source voltage of 1 V peak.

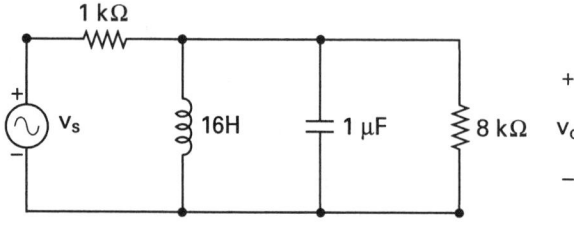

Figure P5-10

5-11. As shown in Fig. P5-11, a single-phase transformer has a center-tapped primary, where $L_p = 1.5$ mH, $L_s = 1.3$ mH, and $K_{ps} = K_{sp} = 0.999$. The primary voltage is $v_p = 170 \sin(377t)$. Plot the instantaneous secondary voltage and load current from 0 to 35 ms with a 0.1-ms increment. The

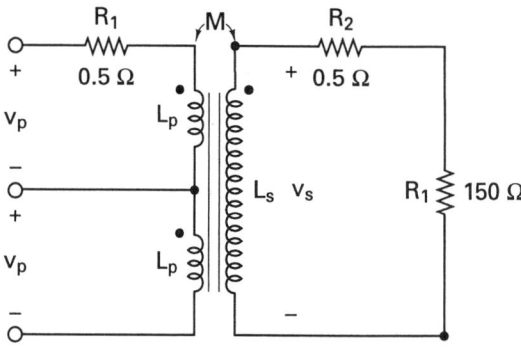

Figure P5-11

output should also be available for display and hard copy by Probe.

5-12. Repeat Problem 5-11 if the transformer is nonlinear with $L_p = 200$ turns and $L_s = 100$ turns. The model parameters of the core are AREA=2.0, PATH=62.73, GAP=0.1, MS=1.6E+6, ALPHA= 1E−3, A=1E+3, C=0.5, and K=1500.

5-13. A three-phase transformer, which is shown in Fig. P5-13, has $L_1 = L_2 = L_3 = 1.2$ mH and $L_4 = L_5 = L_6 = 0.5$ mH. The

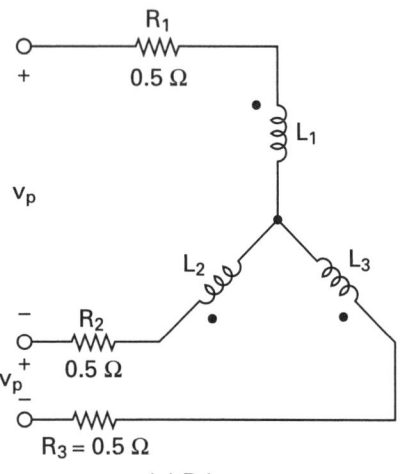

(a) Primary

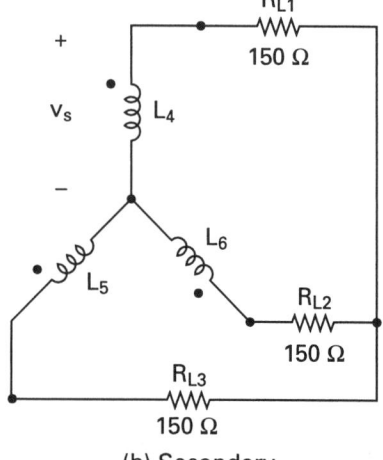

(b) Secondary

Figure P5-13

Passive Elements   Chap. 5

coupling coefficients between the primary and secondary of each phase are $K_{14} = K_{41} = K_{25} = K_{52} = K_{36} = K_{63} = 0.9999$. There is no cross-coupling with other phases. The primary phase voltage is $v_p = 170 \sin(377t)$. Plot the instantaneous secondary phase voltage and load phase current from 0 to 35 ms with a 0.1-ms increment. Assume balanced three-phase input voltages.

5-14. Repeat Problem 5-13 if the transformer is nonlinear with $L_1 = L_2 = L_3 = 200$ turns and $L_4 = L_5 = L_6 = 100$ turns. The model parameters of the core are AREA=2.0, PATH=62.73, GAP=0.1, MS=1.6E+6, ALPHA=1E−3, A=1E+3, C=0.5, and K=1500.

5-15. A switch that is connected between nodes 5 and 4 is controlled by a voltage source between nodes 3 and 0. The switch will

conduct if the controlling voltage is 0.5 V. The on-state resistance is 0.5 Ω, and the off-state resistance is 2E+6 Ω.

5-16. A switch that is connected between nodes 5 and 4 is controlled by a current. The voltage source V_1 through which the controlling current flows is connected between nodes 2 and 0. The switch will conduct if the controlling current is 0.55 mA. The on-state resistance is 0.5 Ω, and the off-state resistance is 2E+6 Ω.

5-17. For the circuit in Fig. P5-17, plot the transient response of the load and source currents for five cycles of the switching period with a time increment of 10 μs. The model parameters of the voltage-controlled switches are RON=0.025, ROFF=1E+8, VON=0.05, and VOFF=0. The output should also be available for display and hard copy by Probe.

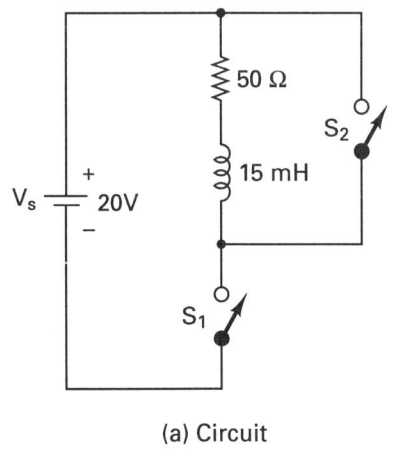

(a) Circuit

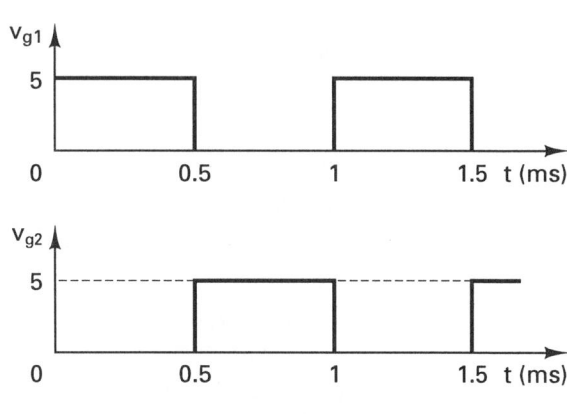

(b) Controlling switch voltages

Figure P5-17

Chapter 6

∎∎∎∎∎∎∎∎∎

Dot Commands

6-1 INTRODUCTION

PSpice has commands for performing various analyses, getting different types of output and modeling elements. These commands begin with a dot and are known as *dot commands*. These commands can be used to specify:

> Models
> Types of output
> Operating temperature and end of circuit
> Options
> Dc analysis
> Ac analysis
> Noise analysis
> Transient analysis
> Fourier analysis
> Monte Carlo analysis

Note. If you are not sure about a command and its effect, run a circuit file with the command and then look at the results. If there is a syntax error, PSpice will display a message identifying the problem.

6-2 MODELS

PSpice allows one (1) to model an element based on its parameters, (2) to model a small circuit that is repeated a number of times in the main circuit, (3) to use a model that is defined in another file, (4) to use a user-defined function, (5) to use parameters instead of number values, and (6) to use parameter variations. The commands are as follows:

.MODEL	Model
.SUBCKT	Subcircuit
.ENDS	End of subcircuit
.FUNC	Function
.GLOBAL	Global
.LIB	Library file
.INC	Include file
.PARAM	Parameter
.STEP	Parametric analysis

6-2.1 .MODEL (Model)

The .MODEL command was discussed in Section 5-2.

6-2.2 .SUBCKT (Subcircuit)

A subcircuit permits one to define a block of circuitry and then to use that block in several places. The general form for subcircuit definition (or description) is

```
.SUBCKT   SUBNAME   [⟨(two or more) nodes⟩]
```

The symbol for a subcircuit call is X. The general form of a call statement is

```
X⟨name⟩   [⟨(two or more) nodes⟩]   SUBNAME
```

SUBNAME is the name of the subcircuit definition, and ⟨(two or more) nodes⟩ are the nodes of the subcircuit. X⟨name⟩ causes the referenced subcircuit to be inserted into the circuit with given nodes replacing the argument nodes in the definition. The subcircuit name SUBNAME may be considered as equivalent to a subroutine name in FORTRAN programming, where X⟨name⟩ is the call statement and ⟨(two or more) nodes⟩ are the variables or arguments of the subroutine.

Subcircuits may be nested. That is, subcircuit A may call other subcircuits. But the nesting cannot be circular, which means that if subcircuit A contains a call to subcircuit B, subcircuit B must not contain a call to subcircuit A. There must be the same number of nodes in the subcircuit calling statement as in its definition. The subcircuit definition should contain only element statements (statements without a dot) and may contain .MODEL statements.

6-2.3 .ENDS (End of Subcircuit)

A subcircuit must end with an .ENDS statement. The end of a subcircuit definition has the general form

```
.ENDS   SUBNAME
```

SUBNAME is the name of the subcircuit, and it indicates which subcircuit description is to be terminated. If the .ENDS statement is missing, all subcircuit descriptions are terminated.

End of Subcircuit Statements

```
. ENDS     OPAMP
. ENDS
```

Note. The name of the subcircuit can be omitted. However, it is advisable to identify the name of the subcircuit to be terminated, especially if there is more than one subcircuit.

Example 6-1

Write the subcircuit call and subcircuit description for the op-amp circuit of Fig. 6-1.
Solution The list of statements for subcircuit call and description is as follows:

```
*   The call statement X1 to be connected to input nodes 1 and 4 and the
output nodes 7 and 9: The subcircuit name is OPAMP. Nodes 1, 4, 7, and 9
are referred to the main circuit file and do not interact with the nodes
of the subcircuit.
X1       1     4     7     9       OPAMP
*        vi−   vi+   vo+   vo−     model name
*   The subcircuit definition: Nodes 1, 2, 3, and 4 are referred to the
subcircuit and do not interact with the nodes of the main circuit.
. SUBCKT    OPAMP     1    2    3    4
*           model name   vi−  vi+  vo+  vo−
RIN     1    2     2MEG
ROUT    5    3     75
E1      5    4     2    1    0. 2MEG  ; Voltage-controlled voltage source
. ENDS     OPAMP                      ; End of subcircuit definition OPAMP
```

Note. There is no interaction between the nodes in the main circuit and the subcircuit. Node numbers in the subcircuit are independent of those in the main circuit. However, the subcircuit should not have node 0, because node 0, which is considered global by PSpice, is the ground.

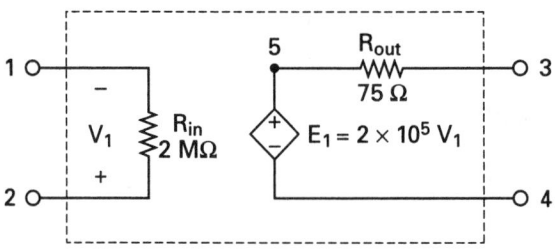

Figure 6-1 Op-amp subcircuit.

6-2.4 .FUNC (Function)

A function statement can be used to define "functions" that may be used in expressions discussed in Section 4-5.1. The functions are defined by users and are flexible to meet one's need. They are also useful for overcoming the restriction of expressions to a single line and where there are several similar subexpressions in a single circuit file. The general form of a function statement is

```
.FUNC    FNAME (arg)      ⟨function⟩
```

FNAME (arg) is the name of the function with argument arg. FNAME must not be the same as built-in functions described in Section 4-5.1: for example, "sin." Up to 10 arguments may be used in a definition. The number of arguments in the use of a function must agree with the number in the definition. A function may be defined with no arguments, but the parentheses are still required. The ⟨function⟩ may refer to other functions defined previously. The .FUNC statement must precede the first use of FNAME. The users can create a file of frequently used .FUNC definitions and access them with an .INC statement (Section 6-2.7) near the beginning of the circuit file.

Some Function Statements

```
.FUNC    E(x)            exp(x)
.FUNC    Sinh(x)         (E(x)+E(−x))/2
.FUNC    MIN(C,D)        (C+D−ABS(C−D))/2
.FUNC    MAX(C,D)        (C+D+ABS(C+D))/2
.FUNC    IND(I(Vsense))  (A0+A1*I(V(Sense))+A2*I(V(Sense))+I(V(Sense)))
```

Notes

1. The definition of the ⟨function⟩ *must* fit on one line.
2. In-line comments *must* not be used after the ⟨function⟩ definition.
3. The last statement illustrates a current dependent nonlinear inductor.

6-2.5 .GLOBAL (Global)

PSpice has the capability of defining global nodes. These nodes are accessible by all subcircuits without being passed in as arguments. Global nodes may be handy for such applications as power supplies, power converters, and clock lines. The general statement is

```
.GLOBAL   N
```

where N is the node number. For example,

```
.GLOBAL    4
```

makes node 4 global to the circuit file and subcircuit(s).

6-2.6 .LIB (Library File)

A library file may be referenced into the circuit file by using the statement

```
.LIB   FNAME
```

FNAME is the name of the library file to be called. A library file may contain comments, .MODEL statements, subcircuit definition, .LIB statements, and .END statements. No other statements are permitted. If FNAME is omitted, PSpice looks for the default file, EVAL.LIB, that comes with PSpice programs. The library file FNAME may call for another library file. When a .LIB command calls for a file, it does not bring the entire text of the library file into the circuit file. It simply reads in those models or subcircuits that are called by the main circuit file. As a result, only those models or subcircuit descriptions that are needed by the main circuit file take up the main memory (RAM) space.

Some Library File Statements

```
.LIB
.LIB   NOM.LIB   (library file NOM.LIB is on the default drive)
.LIB   B:\LIB\NOM.LIB (library file NOM.LIB is on directory file LIB
            in drive B)
.LIB   C:\LIB\NOM.LIB (library file NOM.LIB is in directory file LIB
            on drive C)
```

6-2.7 .INC (Include File)

The contents of another file may be included in the circuit file using the statement

```
.INC   NFILE
```

NFILE is the name of the file to be included and can be any character string that is a legal file name for computer systems. It may include a volume, directory, and version number.

Included files may contain any statements except a title line. However, a comment line may be used instead of a title line. If an .END statement is present, it only marks the end of an included file. An .INC statement may be used only up to four levels of "including." The include statement simply brings everything of the included file into the circuit file and takes up space in main memory (RAM).

Some Include File Statements

```
.INC   OPAMP.CIR
.INC   a:INVERTER.CIR
.INC   c:\LIB\NOR.CIR
```

6-2.8 .PARAM (Parameter)

In many applications it is convenient to use a parameter instead of a numerical value, such that the parameter can be combined into arithmetic expressions. The parameter definition is one of the forms

```
.PARAM    ⟨ PNAME = ⟨value⟩ or { ⟨expression⟩ } ⟩*
```

The keyword .PARAM is followed by a list of names with values or expressions. The ⟨value⟩ must be a constant and does not need "{" and "}". The ⟨expression⟩ must contain only parameters defined previously. PNAME is the parameter name, and it cannot be a predefined parameter such as TIME (time) or TEMP (temperature) or VT (thermal voltage) or GMIN (shunt conductance for semiconductor *p-n* junctions).

Some PARAM Statements

```
.PARAM   VSUPPLY = 12V
.PARAM   VCC = 15 V, VEE = −15V
.PARAM   BANDWIDTH = {50kHz/5}
.PARAM   PI = 3.14159, TWO_PI ={2*3.14159}
```

Once a parameter is defined, it can be used in place of numerical value; for example,

```
VCC    1    0    {SUPPLY}
VEE    0    5    {SUPPLY}
```

will change the value of SUPPLY in both statements. For example,

```
.FUNC    IND(I(Vsense))    (A0+A1*I(V(Sense))+A2*I(V(Sense))+I(V(Sense)))
.PARAM   INDUCTOR = IND(I(Vsense))
L1    1    3    {INDUCTOR}
```

will change the value of INDUCTOR.

Parameters defined in .PARAM statement are global; they can be used anywhere in the circuit, including inside subcircuits. The parameters can be made local to subcircuits by having parameters as arguments to subcircuits. For example,

```
.SUBCKT   FILTER   1   2   PARAMS: CENTER=100kHz, WIDTH=10kHZ
```

is a subcircuit definition for a bandpass filter with nodes 1 and 2 and with parameters CENTER (center frequency) and WIDTH (bandwidth). PARAMs separate the nodes list from the parameter list. When calling this subcircuit FILTER, the parameters can be given new values; for example,

```
X1    4    6    FILTER    PARAMS: CENTER=200kHz
```

will override the default value of 100 kHz with a CENTER value of 200 kHz.

A defined parameter can be used in the following cases:

1. All model parameters
2. Device parameters, such as AREA, L, NRD, Z0, and IC values on capacitors and inductors, and TC1 and TC2 for resistors
3. All parameters of independent voltage and current sources (V and I devices)
4. Values on .NODESET statement (Section 6-6.2) and .IC statement (Section 6-9.1)

A defined parameter *cannot* be used for:

1. Transmission line parameters NL and F
2. In-line temperature coefficients for resistors
3. E, F, G, and H device polynomial coefficient values
4. Replacing node numbers
5. Values on analysis statements (.TRAN, .AC, .DC, etc.)

6-2.9 .STEP (Parametric Analysis)

The effects of parameter variations can be evaluated by the .STEP command, whose general statement takes the general forms

```
.STEP   LIN     SWNAME      SSTART      SEND    SINC
.STEP   OCT     SWNAME      SSTART      SEND    NP
.STEP   DEC     SWNAME      SSTART      SEND    NP
.STEP   SWNAME  LIST ⟨value⟩*
```

SWNAME is the sweep variable name. SSTART, SEND, and SINC are the start value, the end value, and the increment value of the sweep variables, respectively. NP is the number of steps. LIN, OCT, or DEC specifies the type of sweep as follows:

LIN   *Linear sweep:* SWNAME is swept linearly from SSTART to SEND. SINC is the step size.

OCT   *Sweep by octave:* SWNAME is swept logarithmically by octave, and NP becomes the number of steps per octave. The next variable is generated by multiplying the present value by a constant larger than unity. OCT is used if the variable range is wide.

DEC   *Sweep by decade:* SWNAME is swept logarithmically by decade, and NP becomes the number of steps per decade. The next variable is generated by multiplying the present value by a constant larger than unity. DEC is used if the variable range is widest.

LIST   *List of values:* There are no start and end values. The values of the sweep variables are listed after the keyword LIST.

The SWNAME can be one of the following types:

Source: name of an independent voltage, or current, source. During the sweep, the source's voltage or current is set to the sweep value.

Model parameter: model name type and model name followed by a model parameter name in parentheses. The parameter in the model is set to the sweep value. The model parameters L and W for MOS device and any temperature parameters such as TC1 and TC2 for the resistor *cannot* be swept.

Temperature: keyword TEMP followed by the keyword LIST. The temperature is set to the sweep value. For each value of sweep, the model parameters of all circuit components are updated to that temperature.

Global parameter: keyword PARAM followed by parameter name. The parameter is set to sweep. During the sweep, the global parameter's value is set to the sweep value and all expressions are evaluated.

Some Step Statements

```
.STEP     VCE    0V     10V     −5V
```

sweeps the voltage VCE linearly.

```
.STEP     LIN    IS     −10mA    5mA     0.1mA
```

sweeps the current IS linearly.

```
.STEP     RES    RMOD(R)    0.9     1.1     0.001
```

sweeps linearly the model parameter R of the resistor model RMOD.

```
.STEP     DEC    NPN    QM(IS)    1E−18    1E−14    10
```

sweeps with a decade increment the parameter IS of the NPN transistor.

```
.STEP     TEMP   LIST   0    50    80    100    150
```

sweeps the temperature TEMP as listed.

```
.STEP     PARAM  Centerfreq  8.5kHz   10.5kHz   50Hz
```

sweeps linearly the parameter PARAM Centerfreq.

Notes

1. The sweep start value SSTART may be greater than or less than the sweep end value SEND.

2. The sweep increment SINC must be greater than zero.

3. The number of points NP must be greater than zero.

4. If the .STEP command is included in a circuit file, all analyses specified (.DC, .AC, .TRAN, etc.) are done for each step.

6-3 TYPES OF OUTPUT

The commands that are available to obtain output from the results of simulations are

```
.PRINT          Print
.PLOT           Plot
.PROBE          Probe
Probe output
.WIDTH          Width
```

6-3.1 .PRINT (Print)

The results from dc, ac, transient (TRAN), and noise (NOISE) analyses can be obtained in the form of tables. The print statement takes one of the forms

```
.PRINT  DC     ⟨output variables⟩
.PRINT  AC     ⟨output variables⟩
.PRINT  TRAN   ⟨output variables⟩
.PRINT  NOISE  ⟨output variables⟩
```

The maximum number of output variables is eight in any .PRINT statement. However, more than one .PRINT statement can be used to print all the output variables desired. The values of the output variables are printed as a table, with each column corresponding to one output variable. The number of digits for output values can be changed by the NUMDGT option on the .OPTIONS statement in Section 6-4. The results of the .PRINT statement are stored in the output file.

Some Print Statements

```
.PRINT  DC V(2), V(3,5), V(R1), VCE(Q2), I(VIN), I(R1), IC(Q2)
.PRINT  AC VM(2), VP(2), VM(3,5), V(R1), VG(5), VDB(5), IR(5), II(5)
.PRINT  NOISE INOISE ONOISE DB(INOISE) DB(ONOISE)
.PRINT  TRAN V(5) V(4,7) (0,10V) IB(Q1) (0,50MA) IC(Q1) (−50MA, 50MA)
```

Note. Having two .PRINT statements for the same variable will not produce two tables. PSpice will ignore the first statement and produce output for the second statement.

6-3.2 .PLOT (Plot)

The results from dc, ac, transient (TRAN), and noise (NOISE) analyses can be obtained in the form of line printer plots. The plots are drawn by using characters and the results can be obtained in any type of printer. The plot statement takes one of the forms

```
.PLOT  DC    ⟨output variables⟩
        + (⟨(lower limit) value⟩, ⟨(upper limit) value⟩)
.PLOT  AC    ⟨output variables⟩
        + [⟨(lower limit) value⟩, ⟨(upper limit) value⟩]
.PLOT TRAN  ⟨output variables⟩
        + [⟨(lower limit) value⟩, ⟨(upper limit) value⟩]
.PLOT NOISE ⟨output variables⟩
        + [⟨(lower limit) value⟩, ⟨(upper limit) value⟩]
```

The maximum number of output variables is eight in any .PLOT statement. More than one .PLOT statement can be used to plot all the output variables desired.

The range and increment of the x-axis is fixed by the type of analysis command (e.g., .DC or .AC or .TRAN or .NOISE). The range of the y-axis is set by adding (⟨(lower limit) value⟩, ⟨(upper limit) value⟩) at the end of a .PLOT statement. The y-axis range, (⟨(lower limit) value⟩, ⟨(upper limit) value⟩), can be placed in the middle of a set of output variables. The output variables will follow the range specified, which comes immediately to the left.

If the y-axis range is omitted, PSpice assigns a default range determined by the range of the output variable. If the ranges of output variables vary widely, PSpice assigns the ranges corresponding to the different output variables.

Some Plot Statements

```
.PLOT DC V(2), V(3,5), V(R1), VCE(Q2), I(VIN), I(R1), IC(Q2)
.PLOT AC VM(2), VP(2), VM(3,5), V(R1), VG(5), VDB(5), IR(5), II(5)
.PLOT NOISE INOISE ONOISE DB(INOISE) DB(ONOISE)
.PLOT TRAN V(5) V(4,7) (0,10V) IB(Q1) (0, 50MA) IC(Q1) (−50MA, 50MA)
```

Note. In the first three statements, the y-axis is by default. In the last statement, the range for voltages V(5) and V(4,7) is 0 to 10 V, that for current IB(Q1) is 0 to 50 mA, and that for the current IC(Q1) is −50 to 50 mA.

6-3.3 .PROBE (Probe)

Probe is a graphics post-processor, and it is available as an option for the professional version of PSpice. However, Probe comes with the student's version of PSpice. The results from the dc, ac, and transient (TRAN) analysis cannot be used directly by Probe. First, the results have to be processed by the .PROBE command, which writes the processed data on a file, PROBE.DAT, for use by

---

Probe. The command takes one of the forms

```
.PROBE
.PROBE  ⟨one or more output variables⟩
```

In the first form, where no output variable is specified, the .PROBE command writes all the node voltages and all the element currents into the PROBE.DAT file. The element currents are written in the forms that are permitted as output variables and are discussed in Section 3-3.2. In the second form, where the output variables are specified, PSpice writes only the specified output variables to the PROBE.DAT file. This form is suitable for users without a fixed disk and for limiting the size of the PROBE.DAT file.

Probe Statements

```
.PROBE
.PROBE  V(5), V(4,3), V(C1), VM(2), I(R2), IB(Q1), VBE(Q1)
```

6-3.4 Probe Output

It is very easy to use Probe. Once the results of the simulations are processed by the .PROBE command, the results are available for graphics output. Probe comes with a first menu as shown in Fig. 6-2 from which to choose the type of analysis. After the first choice, the second level is the choice for the plots and coordinates

```
                          ┌────────┐
                          ‖ Probe  ‖
                          └────────┘
              Graphics Post-Processor for PSpice
                   Version 1.13 - October 1987
          © Copyright 1985, 1986, 1987 by MicroSim Corporation
                            --------
                        Classroom Version
           Copying of this program is welcomed and encouraged



            Circuit:  EXAMPLE1 - An Illustration of all Commands
          Date/Time run:  10/31/88 16:15:19        Temperature: 35.0




        0)Exit Program 1)DC Sweep 2)AC Sweep 3)Transient Analysis :  1
```

Figure 6-2   Select analysis display for Probe.

of output variables as shown in Fig. 6-3. After the choices, the output is displayed as shown in Fig. 6-4.

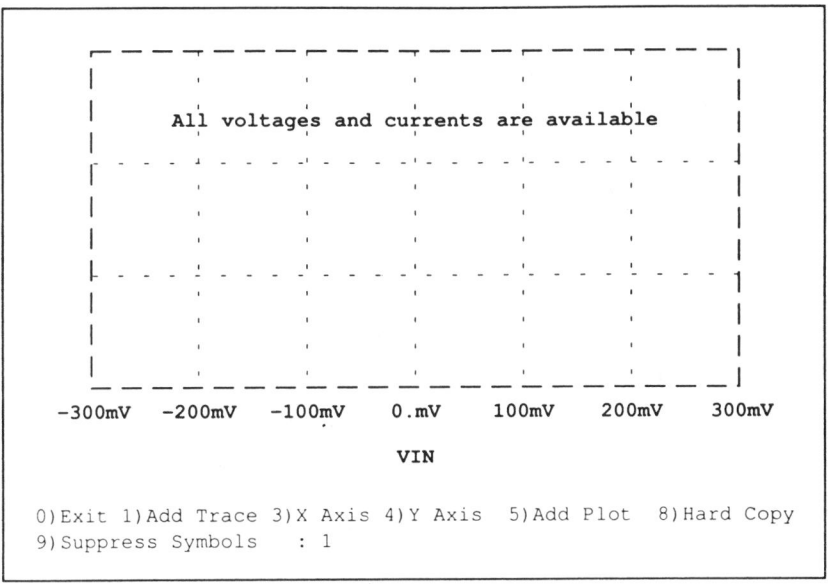

Figure 6-3 Select plot/graphics output.

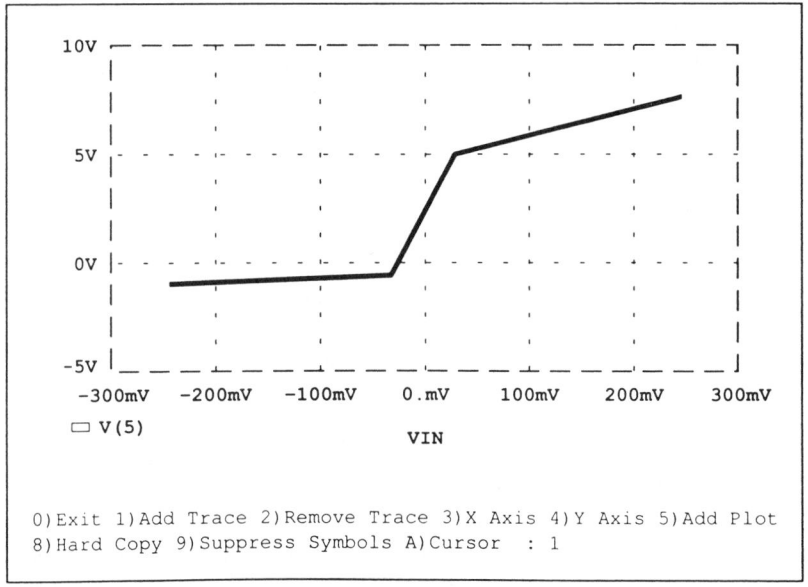

Figure 6-4 Output display.

With one exception, Probe disregards upper/lower case: "V(4)" and "v(4)" are equivalent. The exception is the "m" scale suffix for numbers: "m" means "milli" (1E−3), whereas "M" means "mega" (1E+6). The suffixes "MEG" and "MIL" are not available. The units that are recognized by Probe are:

V   Volts
A   Amperes
W   Watts
d   Degrees (of phase)
s   Seconds
H   Hertz

Probe also recognizes that W = V × A, V = W/A, and A = W/V. Therefore, the addition of a trace, such as

```
VCE(Q1)*IC(Q1)
```

gives the power dissipation of transistor Q_1 and will be labeled with a "W." Arithmetic expressions of output variables are also allowed and the available operators are "+", "−", "*", and "/", along with parentheses. The available functions are as follows:

FUNCTION	MEANING
ABS(x)	$\|x\|$ (absolute value)
B(Kxy)	Flux density of coupled inductor K_{xy}
H(Kxy)	Magnetization of coupled inductor K_{xy}
SGN(x)	+1 (if $x > 0$), 0 (if $x = 0$), −1 (if $x < 0$)
EXP(x)	e^x
DB(x)	$20 \log(\|x\|)$ (log of base 10)
LOG(x)	$\ln(x)$ (log of base e)
LOG10(x)	$\log(x)$ (log of base 10)
PWR(x,y)	$\|x\|^y$
SQRT(x)	$x^{1/2}$
SIN(x)	$\sin(x)$ (x in radians)
COS(x)	$\cos(x)$ (x in radians)
TAN(x)	$\tan(x)$ (x in radians)
ARCTAN(x)	$\tan^{-1}(x)$ (result in radians)
d(y)	Derivative of y with respect to the x-axis variable
s(y)	Integral of y over the x-axis variable

AVG(x)	Running average of x
RMS(x)	Running rms average of x

For derivatives and integrals of simple variables (not expressions), the shorthand notations that are available are

    dV(4) is equivalent to d(V(4))

    sIC(Q3) is equivalent to s(IC(Q3))

and the plot of

    dIC(Q2)/dIB(Q2)

will give the small-signal beta value of Q_2.

    Two or more traces can be added with only one Add Trace command, where all the expressions are separated by " " or ",". For instance,

```
V(2)   V(4),  IC(M1),  RMS(I(VIN))
```

will add four traces. This gives the same result as using Add Trace four times with only one trace at a time, but is faster since the plot is not redrawn between adding each trace.

    The PROBE.DAT file can contain more than one of any kind (e.g., two transient analyses with three temperatures). If PSpice was run for a transient analysis at three temperatures, the expression

```
V(1)
```

will result in Probe drawing three curves instead of the usual one curve. Entering the expression

```
V(1) @ n
```

will result in drawing the curve of V(1) for the nth transient analysis. Entering the expression

```
V(1)@1-V(1)@2
```

will display the difference between the waveforms from the first and second temperatures, whereas the expression

```
V(1)-V(2)@2
```

will display three curves, one for each V(1).

Notes

1. The .PROBE command requires a math co-processor for the professional version of PSpice, but it is not required for the student version.
2. Probe is not available on SPICE. However, the newest version of SPICE (SPICE3) has a post-processor similar to Probe called *Nutmeg*.
3. It is required to specify the type of display and the type of hard-copy devices on the PROBE.DEV file as follows:
   Display = ⟨display name⟩
   Hard copy = ⟨port name⟩, ⟨device name⟩
   The details of names for display, port, and device (printer) can be found in the README.DOC file that comes with the PSpice programs or in the PSpice manual.
4. The display and hard-copy devices can be set from the display/printer setup menu.

6-3.5 .WIDTH (Width)

The width of the output in columns can be set by the .WIDTH statement, which has the general form

```
.WIDTH  OUT=⟨value⟩
```

The ⟨value⟩ is in columns and must be either 80 or 132. The default value is 80.

6-4 OPERATING TEMPERATURE AND END OF CIRCUIT

The temperature command is discussed in Section 4-2. The last statement for the end of a circuit is

```
.END
```

All data and commands must come before the .END command. The .END command instructs PSpice to perform all the circuit analysis specified. After processing the results of the analysis specified, PSpice resets itself to perform further computations, if required.

Note. An input file may have more than one circuit, where each circuit has its .END command. PSpice will perform all the analysis specified and processes the results of each circuit one by one. PSpice resets everything at the beginning of each circuit. Instead of running PSpice separately for each circuit, this is a convenient way to perform the analysis of many circuits with one run statement.

PSpice allows various options to control and to limit parameters for the various analysis. The general form is

```
.OPTIONS [(options) name)] [((options) name)=⟨value⟩]
```

The options can be listed in any order. There are two types of options: (1) those without values, and (2) those with values. The options without values are used as flags of various kinds and only the option name is mentioned. Table 6-1 shows the options without values. The options with values are used to specify certain optional parameters. The option names and their values are specified. Table 6-2 shows the options with values. The commonly used options are NOPAGE, NOECHO, NOMOD, TNOM, CPTIME, NUMDGT, GMIN, and LIMPTS.

Options Statements

```
.OPTIONS    NOPAGE  NOECHO  NOMOD  DEFL=20U  DEFW=15U  DEFAD=50P  DEFAS=50P
.OPTIONS    ACCT   LIST  RELTOL=.005
```

Job statistics summary. If the option ACCT is specified in the .OPTIONS statement, PSpice will print various statistics about the run at the end. This option is not required for most circuit simulations. This list follows the format of the output.

ITEM	MEANING
NUNODS	Number of distinct circuit nodes before sub-circuit expansion.
NCNODS	Number of distinct circuit nodes after sub-circuit expansion. If there are no subcircuits, NCNODS = NUNODS.

TABLE 6-1 OPTIONS WITHOUT VALUES

Option	Effects
NOPAGE	Suppresses paging and printing of a banner for each major section of output
NOECHO	Suppresses listing of the input file
NODE	Causes output of net list (node table)
MONOD	Suppresses listing of model parameters
LIST	Causes summary of all circuit elements (devices) to be output
OPTS	Causes values for all options to be output
ACCT	Summary and accounting information is output at the end of all the analysis
WIDTH	Same as .WIDTH OUT= statement

TABLE 6-2 OPTIONS WITH VALUES

Option	Effects	Unit	Default
DEFL	MOSFET channel length (L)	m	100u
DEFW	MOSFET channel width (W)	m	100u
DEFAD	MOSFET drain diffusion area (AD)	m^{-2}	0
DEFAS	MOSFET source diffusion area (AS)	m^{-2}	0
TNOM	Default temperature (also the temperature at which model parameters are assumed to have been measured)	°C	27
NUMDGT	Number of digits output in print tables		4
CPTIME	Central processing unit (CPU) time allowed for a run	s	1E6
LIMPTS	Maximum points allowed for any print table or plot		201
ITL1	Dc and bias-point "blind" iteration limit		40
ITL2	Dc and bias-point "educated guess" iteration limit		20
ITL4	Iteration limit at any point in transient analysis		10
ITL5	Total iteration limit for all points in transient analysis (ITL5 = 0 means ITL5 = infinite)		5000
RELTOL	Relative accuracy of voltages and currents		0.001
TRTOL	Transient analysis accuracy adjustment		7.0
ABSTOL	Best accuracy of currents	A	1pA
CHGTOL	Best accuracy of charges	C	0.01pC
VNTOL	Best accuracy of voltages	V	1uV
PIVREL	Relative magnitude required for pivot in matrix solution		$1E-13$
GMIN	Minimum conductance used for any branch	Ω^{-1}	$1E-12$

ITEM	MEANING
NUMNOD	Total number of distinct nodes in circuit. This is NCNODS plus the internal nodes generated by parasitic resistances. If no device has parasitic resistances, NUMNOD = NCNODS.
NUMEL	Total number of devices (elements) in circuit after subcircuit expansion. This includes all statements that do not begin with "." or "X".
DIODES	Number of diodes after subcircuit expansion.
BJTS	Number of bipolar transistors after subcircuit expansion.
JFETS	Number of junction FETs after subcircuit expansion.
MFETS	Number of MOSFETs after subcircuit expansion.
GASFETS	Number of GaAs MESFETs after subcircuit expansion.
NUMTEM	Number of different temperatures.
ICVFLG	Number of steps of dc sweep.
JTRFLG	Number of print steps of transient analysis.

ITEM	MEANING
JACFLG	Number of steps of ac analysis.
INOISE	1 or 0: noise analysis was/was not done.
NOGO	1 or 0: run did/did not have an error.
NSTOP	The circuit matrix is conceptually (not physically) of dimension NSTOP $\times$ NSTOP.
NTTAR	Actual number of entries in circuit matrix at beginning of run.
NTTBR	Actual number of entries in circuit matrix at end of run.
NTTOV	Number of terms in circuit matrix that come from more than one device.
IFILL	Difference between NTTAR and NTTBR.
IOPS	Number of floating-point operations needed to do one solution of circuit matrix.
PERSPA	Percent sparsity of circuit matrix.
NUMTTP	Number of internal time steps in transient analysis.
NUMRTP	Number of times in transient analysis that a time step was too large and had to be cut back.
NUMNIT	Total number of iterations for transient analysis.
MEMUSE/MAXMEM	Amount of circuit memory used/available in bytes. There are two memory pools. Exceeding either one will abort the run.
COPYKNT	Number of bytes that were copied in the course of doing memory management for this run.
READIN	Time spent reading and error checking the input file.
SETUP	Time spent setting up the circuit matrix pointer structure.
DCSWEEP	Time spent and iteration count for calculating dc sweep.
BIASPNT	Time spent and iteration count for calculating bias point and bias point for transient analysis.
MATSOL	Time spent solving circuit matrix (this time is also included in the time for each analysis). The iteration count is the number of times the rows or columns were swapped in the course of solving it.
ACAN	Time spent and iteration count for ac analysis.

ITEM	MEANING
TRANAN	Time spent and iteration count for transient analysis.
OUTPUT	Time spent preparing .PRINT tables and .PLOT plots.
LOAD	Time spent evaluating device equations (this time is also included in the time for each analysis).
OVERHEAD	Other time spent during run.
TOTAL JOB TIME	Total run time, excluding the time to load the program files PSPICE1.EXE and PSPICE2.EXE into memory.

6-6 DC ANALYSIS

In dc analysis, all the independent and dependent sources are of dc types. The inductors and capacitors in a circuit are considered as short circuits and open circuits, respectively. This is due to the fact that at zero frequency, the impedance represented by an inductor is zero and that by a capacitor is infinite. The commands that are available for dc analyses are

.OP	Dc operating point
.NODESET	Node set
.SENS	Small-signal sensitivity
.TF	Small-signal transfer function
.DC	Dc sweep

6-6.1 .OP (Operating Point)

Electronic and electrical circuits contain nonlinear devices (e.g., diodes, transistors), whose parameters depend on the *operating point*. The operating point is also known as a *bias point* or *quiescent point*. The operating point is always calculated by PSpice for calculating the small-signal parameters of nonlinear devices during the dc sweep and transfer function analysis. The command takes the form

.OP

The .OP command controls the output of the bias point but not the method of bias analysis and the results of bias point. If the .OP command is omitted, PSpice prints only a list of the node voltages. If the .OP command is present, PSpice prints the currents and power dissipations of all the voltages. The small-signal parameters of all nonlinear controlled sources and all semiconductor devices are also printed.

6-6.2 .NODESET (Nodeset)

In calculating the operating bias point, some or all of the nodes of the circuit may be assigned initial guesses to help dc convergence by the statement

```
.NODESET  V(1)=V1 V(2)=V2 ... V(N)=VN
```

$V(1)$, $V(2)$, . . . are the node voltages, and V1, V2, . . . are their respective values of the initial guesses. Once the operating point is found, the .NODESET command has no effect during the dc sweep or transient analysis. This command may be necessary for convergence: for example, on flip-flop circuits to break the tie-in condition. In general, this command should not be necessary. One should not confuse it with the .IC command, which sets the initial conditions of the circuits during the operating point calculations for transient analysis. The .IC command is discussed in Section 6-9.1.

Statement for Nodeset

```
.NODESET  V(4)=1.5V  V(6)=0  V(25)=1.5V
```

6-6.3 .SENS (Sensitivity Analysis)

The sensitivity of output voltages or currents with respect to every circuit and device parameter can be calculated by the .SENS statement, which has the general form

```
.SENS  ⟨(one or more output) variables⟩
```

The .SENS statement calculates the bias point and the linearized parameters around the bias point. In this analysis the inductors are assumed to be short circuits and the capacitors to be open circuits. If the output variable is a current, that current *must* be through a voltage source. The sensitivity of each output variable with respect to all the device values and model parameters are calculated. The .SENS statement prints the results automatically. Therefore, it should be noted that a .SENS statement may generate a huge amount of data if many output variables are specified.

Statement for Sensitivity Analysis

```
.SENS  V(5)  V(2,3)  I(V2)  I(V5)
```

Example 6-2

An op-amp circuit is shown in Fig. 6-5. The op-amp is represented by the subcircuit of Fig. 6-1. Calculate and print the sensitivity of output voltage $V(4)$ with respect to each circuit element. The operating temperature is 40°C.

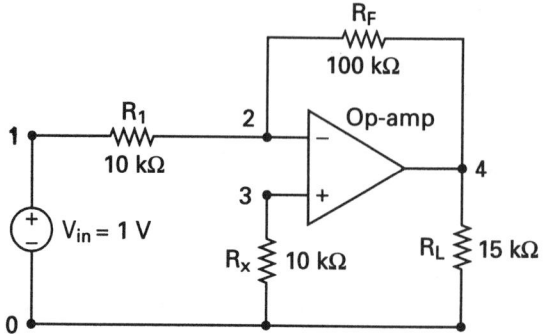

Figure 6-5 Op-amp circuit for Example 6-2.

Solution The list of the statements for the circuit file is as follows:

■■■

Example 6-2   Dc sensitivity analysis

SOURCE ■ * Dc input voltage of 1 V:
```
          VIN   1    0    DC    1V
```
CIRCUIT ■■ R1   1   2   10K
```
          RF    2    4    100K
          RL    4    0    15K
          RX    3    0    10K
          *    Subcircuit call OPAMP:
          X1    2    3    4    0    OPAMP
          *    Subcircuit definition:
          .SUBCKT     OPAMP     1    2    3    4
          *          model name   vi-   vi+   vo+   vo-
          RIN   1    2    2MEG
          ROUT  5    3    75
          E1    5    4    2    1    2MEG  ; Voltage-controlled voltage source
          .ENDS    OPAMP                 ; End of subcircuit definition OPAMP
```
ANALYSIS ■■■ * Operating temperature is 40° Celsius:
```
          .TEMP   40
          *   Options:
          .OPTIONS    NOPAGE    NOECHO
          *  It calculates and prints the sensitivity analysis of output
          *  voltage V(4) with respect to all elements in the circuit.
          .SENS    V(4)
     .END
```
■■■

The results of the sensitivity analysis are shown next. The node voltages are also printed automatically.

```
 ****     SMALL SIGNAL BIAS SOLUTION      TEMPERATURE =   40.000 DEG C
  NODE    VOLTAGE      NODE    VOLTAGE     NODE    VOLTAGE     NODE    VOLTAGE
 (   1)    1.0000    (    2) 5.054E-06  (    3) 25.14E-09  (    4)   -9.9999
 ( X1.5)  -10.0570
```

```
VOLTAGE SOURCE CURRENTS
NAME            CURRENT
VIN            -1.000E-04
    TOTAL POWER DISSIPATION    1.00E-04   WATTS
 ****      DC SENSITIVITY ANALYSIS            TEMPERATURE =    40.000 DEG C
DC SENSITIVITIES OF OUTPUT V(4)
        ELEMENT          ELEMENT          ELEMENT         NORMALIZED
        NAME             VALUE            SENSITIVITY     SENSITIVITY
                                          (VOLTS/UNIT)   (VOLTS/PERCENT)
        R1               1.000E+04         1.000E-03        1.000E-01
        RF               1.000E+05        -1.000E-04       -1.000E-01
        RL               1.500E+04        -1.851E-11       -2.776E-09
        RX               1.000E+04         2.766E-11        2.766E-09
        X1.RIN           2.000E+06        -2.640E-13       -5.280E-09
        X1.ROUT          7.500E+01         4.257E-09        3.193E-09
        VIN              1.000E+00        -1.000E+01       -1.000E-01
```

6-6.4 .TF (Small-Signal Transfer Function)

The small-signal transfer function capability of PSpice can be used to find the small-signal dc gain, the input resistance, and the output resistance of a circuit. If V(1) and V(4) are the input and output variables, respectively, PSpice will calculate the small-signal dc gain between nodes 1 and 4, defined by

$$A_v = \frac{\Delta V_o}{\Delta V_i} = \frac{V(4)}{V(1)}$$

as well as the input resistance between nodes 1 and 0 and the small-signal dc output resistance between nodes 4 and 0.

PSpice calculates the small-signal dc transfer function by linearizing the circuit around the operating point. The statement for the transfer function has one of the forms

```
. TF   VOUT   VIN
. TF   IOUT   IIN
```

where VIN is the input voltage. VOUT (or IOUT) is the output voltage (or output current). If the output is a current, that current *must* be through a voltage source. The output variable VOUT (or IOUT) has the same format and meaning as in a .PRINT statement. If there are inductors and capacitors in a circuit, the inductors are treated as short circuits and capacitors as open circuits.

The .TF command calculates the parameters of the Thévenin's (or Norton's) equivalent circuit of the circuit file. It prints the output automatically and does not require .PRINT or .PLOT or .PROBE statements.

Statements for Transfer Function Analysis

```
. TF   V(10)    VIN
. TF   I(VN)    IIN
```

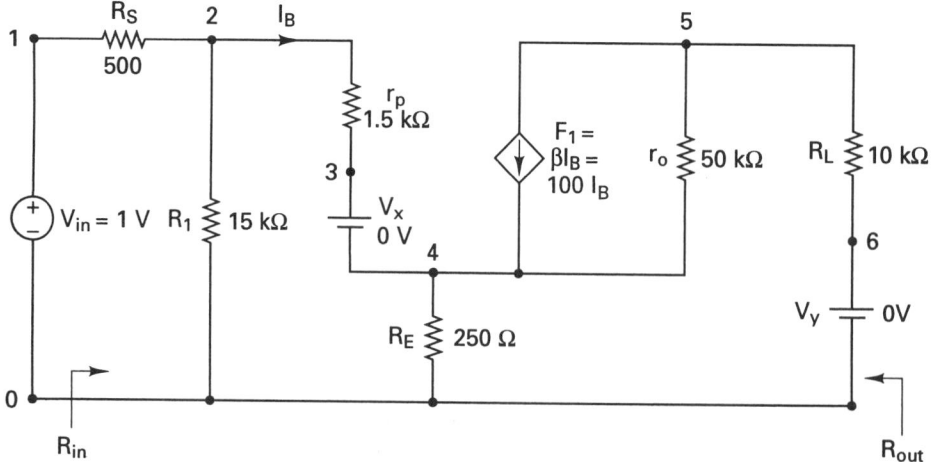

Figure 6-6 Amplifier circuit for Example 6-3.

Example 6-3

An amplifier circuit is shown in Fig. 6-6. Calculate and print (a) the voltage gain, $A_v = V(5)/v_{in}$, (b) the input resistance, R_{in}, and (c) the output resistance, R_o.
Solution The list of the circuit file is as follows:

■■

Example 6-3   Transfer function analysis

SOURCE ■ *   Dc input voltage of 1 V:
```
          VIN   1    0    DC     1V
```
CIRCUIT ■■ RS   1   2   500
```
          R1    2    0    15K
          RP    2    3    1.5K
          RE    4    0    250
          *  Current-controlled current source:
          F1    5    4    VX    100
          R0    5    4    50K
          RL    5    6    10K
          *  Dummy voltage source to measure the controlling current:
          VX    3    4    DC    0V
          *  Dummy voltage source to measure the output current:
          VY    6    0    DC    0V
```
ANALYSIS ■■■ *   The .TF command calculates and prints the dc gain and the input
```
          *    and output resistances.  The input voltage is VIN and
          *    the output voltage is V(5).
          .TF    V(5)    VIN
      .END
```
■■

The results of the .TF command are shown next.

```
VOLTAGE SOURCE CURRENTS
NAME          CURRENT
VIN           -1.053E-04
```

```
VX            4.211E-05
VY           −3.495E-03
TOTAL POWER DISSIPATION   1.05E-04  WATTS
****      SMALL-SIGNAL CHARACTERISTICS
     V(5)/VIN = −3.495E+01
     INPUT RESISTANCE AT VIN =  9.499E+03
     OUTPUT RESISTANCE AT V(5) =  9.839E+03
```

6-6.5 .DC (DC Sweep)

Dc sweep is also known as the *dc transfer characteristic*. The input variable is varied over a range of values. For each value of input variables, the dc operating point and the small-signal dc gain are computed by calling the small-signal transfer function capability of PSpice. The dc sweep (or dc transfer characteristic) is obtained by repeating the calculations of small-signal transfer function for a set of values. The statement for performing dc sweep takes one of the following general forms:

```
.DC   LIN    SWNAME    SSTART    SEND    SINC
+            [(nested sweep specification)]
.DC   OCT    SWNAME    SSTART    SEND    NP
+            [(nested sweep specification)]
.DC   DEC    SWNAME    SSTART    SEND    NP
+            [(nested sweep specification)]
.DC   SWNAME    LIST (value)*
+            [(nested sweep specification)]
```

SWNAME is the sweep variable name. SSTART, SEND, and SINC are the start value, the end value, and the increment value of the sweep variables, respectively. NP is the number of steps. LIN, OCT, or DEC specifies the type of sweep, as follows:

LIN — *Linear sweep:* SWNAME is swept linearly from SSTART to SEND. SINC is the step size.

OCT — *Sweep by octave:* SWNAME is swept logarithmically by octave, and NP becomes the number of steps per octave. The next variable is generated by multiplying the present value by a constant larger than unity. OCT is used if the variable range is wide.

DEC — *Sweep by decade:* SWNAME is swept logarithmically by decade, and NP becomes the number of steps per decade. The next variable is generated by multiplying the present value by a constant larger than unity. DEC is used if the variable range is widest.

LIST — *List of values:* There are no start and end values. The values of the sweep variables are listed after the keyword LIST.

The SWNAME can be one of the following types:

Source: name of an independent voltage, or current, source. During the sweep, the source's voltage or current is set to the sweep value.

Model parameter: model name type and model name followed by a model parameter name in parentheses. The parameter in the model is set to the sweep value. The model parameters L and W for a MOS device and any temperature parameters such as TC1 and TC2 for the resistor *cannot* be swept.

Temperature: keyword TEMP followed by the keyword LIST. The temperature is set to the sweep value. For each value of sweep, the model parameters of all circuit components are updated to that temperature.

Global parameter: keyword PARAM followed by parameter name. The parameter is set to sweep. During the sweep, the global parameter's value is set to the sweep value and all expressions are evaluated.

The dc sweep can be nested, similar to a DO loop within a DO loop in FORTRAN programming. The first sweep will be the inner loop and the second sweep is the outer loop. The first sweep will be done for each value of the second sweep. ⟨nested sweep specification⟩ follows the same rules as those for the main ⟨sweep variable⟩.

Statements for DC Sweep

```
.DC    VIN    −5V    10V    0.25V
```

sweeps the voltage VIN linearly.

```
.DC    LIN    IIN    50MA    −50MA    1MA
```

sweeps the current IIN linearly.

```
.DC    VA    0    15V    0.5V    IA    0    1MA    0.05MA
```

sweeps the current IA linearly within the linear sweep of VA.

```
.DC    RES    RMOD(R)    0.9    1.1    0.001
```

sweeps the model parameter R of the resistor model RMOD linearly.

```
.DC    DEC    NPN    QM(IS)    1E−18    1E−14    10
```

sweeps with a decade increment the parameter IS of the NPN transistor.

```
.DC    TEMP    LIST    0    50    80    100    150
```

sweeps the temperature TEMP as listed values.

```
.DC    PARAM   Vsupply   -15V    15V    0.5V
```

sweeps the parameter PARAM Vsupply linearly.

PSpice does not print or plot any output by itself for dc sweep. The results of dc sweep are obtained by .PRINT, .PLOT, or .PROBE statements. Probe allows nested sweeps to be displayed as a family of curves.

Notes

1. If the source has a dc value, its value is set by the sweep overriding the dc value.
2. In the third statement, the current source IA is the inner loop and the voltage source VA is the outer loop. PSpice will vary the value of current source IA from 0 to 1 mA with an increment of 0.05 mA for each value of voltage source VA, and generate an entire print table or plot for each value of voltage sweep.
3. The sweep-start value SSTART may be greater than or less than the sweep-end value SEND.
4. The sweep increment SINC must be greater than zero.
5. The number of points NP must be greater than zero.
6. After the dc sweep is finished, the sweep variable is set back to the value it had before the sweep started.

Example 6-4

For the amplifier circuit of Fig. 6-6, calculate and plot the dc transfer characteristic, V_o versus V_{in}. The input voltage is varied from 0 to 100 mV with an increment of 2 mV. The load resistance is varied from 10 kΩ to 30 kΩ with a 10-kΩ increment. **Solution**   The list of the circuit file is as follows:

■ ■

Example 6-4   Dc sweep

```
SOURCE    ■ VIN    1     0     DC     100MV
CIRCUIT   ■ ■ *    Parameter definition for VAL:
             .PARAM   VAL = 10K
             *    Step variation for VAL:
             .STEP    PARAM    VAL    10K    30K    10K
             *    Vary the load resistance R_L:
          RL    5     6     {VAL}
          RS    1     2     500
          R1    2     0     15K
          RP    2     3     1.5K
          RE    4     0     250
          *  Current-controlled current source:
          F1    5     4     VX     100
          R0    5     4     50K
          *RL   5     6     10K
```

```
       *  Dummy voltage source to measure the controlling current:
       VX    3    4    DC    0V
       *  Dummy voltage source to measure the output current:
       VY    6    0    DC    0V
```
ANALYSIS ■ ■ ■ ```* Dc sweep from 0 to 100 mV with an increment of 2 mV:
       .DC   VIN   0   100MV   2MV
       *   PSpice plots the results of dc sweep.
       .PLOT   DC   V(5)
       .PROBE   ; Graphics post-processor
.END
```

The transfer characteristic is shown in Fig. 6-7.

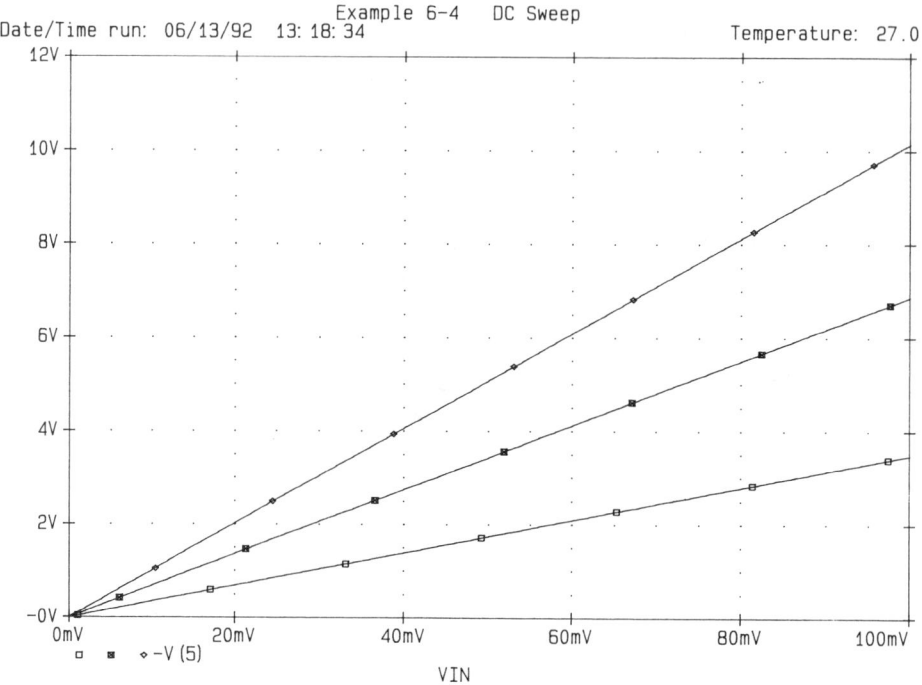

**Figure 6-7** Dc transfer characteristic for Example 6-4.

## 6-7 AC ANALYSIS

The ac analysis calculates the frequency response of a circuit over a range of frequencies. If the circuit contains nonlinear devices or elements, it is necessary to obtain the small-signal parameters of the elements before calculating the frequency response. Prior to the frequency response (or ac analysis), PSpice determines the small-signal parameters of the elements. The method for calculation of
```

bias point for ac analysis is identical to that for dc analysis. The details of the bias points can be printed by an .OP command.

The command for performing frequency response takes one of the general forms

```
.AC   LIN   NP   FSTART   FSTOP
.AC   OCT   NP   FSTART   FSTOP
.AC   DEC   NP   FSTART   FSTOP
```

NP is the number of points in a frequency sweep. FSTART is the starting frequency, and FSTOP is the ending frequency. Only one of LIN, OCT, or DEC must be specified in the statement. LIN, OCT, or DEC specifies the type of sweep as follows:

LIN   *Linear sweep:* The frequency is swept linearly from the starting frequency to the ending frequency, and NP becomes the total number of points in the sweep. The next frequency is generated by adding a constant to the present value. LIN is used if the frequency range is narrow.

OCT   *Sweep by octave:* The frequency is swept logarithmically by octave, and NP becomes the number of points per octave. The next frequency is generated by multiplying the present value by a constant larger than unity. OCT is used if the frequency range is wide.

DEC   *Sweep by decade:* The frequency is swept logarithmically by decade, and NP becomes the number of points per decade. DEC is used if the frequency range is the widest.

PSpice does not print or plot any output by itself for ac analyses. The results of ac sweep are obtained by .PRINT, .PLOT, or .PROBE statements.

Some Statements for AC Analysis

```
.AC   LIN   20    100HZ   300HZ
.AC   LIN   1     60HZ    120HZ
.AC   OCT   10    100HZ   10KHZ
.AC   DEC   100   1KHZ    1MEGHZ
```

Notes

1. FSTART must be less than FSTOP and must *not* be zero.
2. NP = 1 is permissible and the second statement calculates the frequency response at 60 Hz only.
3. Before performing the frequency response analysis, PSpice calculates, automatically, the biasing point to determine the linearized circuit parameters around the bias point.
4. All independent voltage and current sources that have ac values are inputs to

the circuit. At least one source must have ac value; otherwise, the analysis would not be meaningful.

5. If a group delay output is required by a "G" suffix, as noted in Section 3-3, the frequency steps should be small, so that the output changes smoothly.

Example 6-5

An *RLC* circuit is shown in Fig. 6-8. Plot the frequency response of the current through the circuit and the magnitude of the input impedance. The frequency of the source is varied from 100 Hz to 100 kHz with a decade increment and 10 points per decade. The values of the inductor *L* are 5, 15, and 25 mH.

Solution The list of the circuit file is as follows:

■ ■

Example 6-5   Input impedance characteristics

```
SOURCE    ■ VIN    1    0    AC    1V
CIRCUIT   ■ ■ *    Parameter definition for VAL:
            .PARAM    VAL = 1MH
            *    Step variation for VAL:
            .STEP    PARAM    VAL    5MH    25MH    10MH
            R    2    3    50
            *    Vary the inductor:
            L    3    4    {VAL}
            C    4    0    1UF
            *    Dummy voltage source to measure the controlling current:
            VX    1    2    DC    0V
ANALYSIS  ■ ■ ■ *    Ac   sweep from 100 Hz to 100 kHz with 10 points per decade
            .AC    DEC    10    100HZ    100KHZ
            *    PSpice plots the results of a dc sweep.
            .PLOT    AC    V(1)              ; Plot on the output file
            .PROBE                           ; Graphics post-processor
          .END
```

■ ■

The frequency response of the current through the circuit and the magnitude of the input impedance are shown in Fig. 6-9. As the inductance is increased, the resonant frequency is decreased.

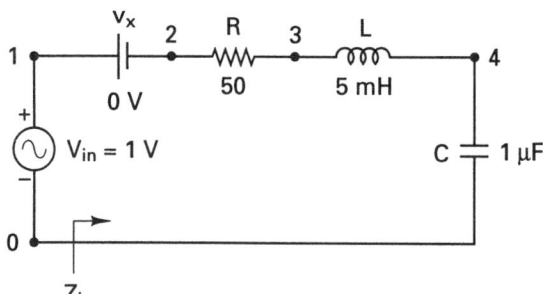

Figure 6-8   *RLC* circuit for Example 6-5.

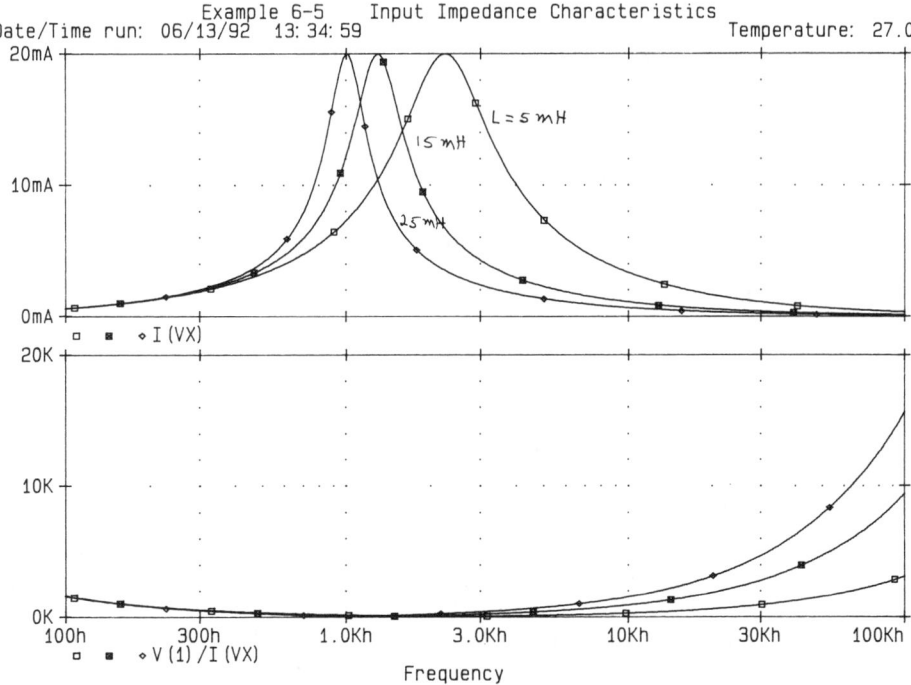

Figure 6-9   Frequency response for Example 6-5.

6-8 NOISE ANALYSIS

The resistors and semiconductor devices generate noise. The level of the noise depends on the frequency. The various types of noise are generated by resistors and semiconductor devices [3]. Noise analysis is done in conjunction with ac analysis and requires an .AC command. For each frequency of the ac analysis, the noise level of each generator in a circuit (e.g., resistors, transistor) is calculated and their contributions to the output nodes are computed by summing the rms noise values. The gain from the input source to the output voltage is calculated. From this gain the equivalent input noise level at the specified source is calculated by PSpice.

The statement for performing noise analysis is of the form

```
.NOISE  V(N+, N-)   SOURCE  M
```

where V(N+, N−) is the output voltage across nodes N+ and N−. The output could be at a node N, such as V(N). SOURCE is the name of an independent voltage or current source at which the equivalent input noise will be generated. It should be noted that SOURCE is not a noise generator—rather, it is a place at

---

which to compute the equivalent noise input. For a voltage source, the equivalent input is in $V/\sqrt{Hz}$, and for a current source, it is in $A/\sqrt{Hz}$.

M is the print interval that permits one to print a table for the individual contributions of all generators to the output nodes for every mth frequency. The output noise and equivalent noise of individual contributions are printed by a .PRINT or .PLOT command. If the value of M is not specified, PSpice does not print a table of individual contributions. But PSpice prints automatically a table of total contributions rather than individual contributions.

Statements for Noise Analysis

```
.NOISE   V(4,5)   VIN
.NOISE   V(6)     IIN
.NOISE   V(10)    V1    10
```

Note. The .PROBE command cannot be used for noise analysis.

Example 6-6

For the circuit of Fig. 6-5, calculate and print the equivalent input and output noise. The frequency of the source is varied from 1 Hz to 100 kHz. The frequency should be increased by a decade with 1 point per decade.

Solution The input source is of ac type. The list of the circuit file is as follows:

■ ■

Example 6-6   Noise analysis

SOURCE
```
■ *   Ac input voltage of 1 V:
  VIN   1   0   AC   1V
```

CIRCUIT
```
■ ■ R1    1   2   10K
    RF    2   4   100K
    RL    4   0   15K
    RX    3   0   10K
    *   Subcircuit call OPAMP:
    X1    2   3   4   0     OPAMP
    *   Subcircuit definition:
    .SUBCKT     OPAMP     1    2    3    4
    *         model name   vi-  vi+  vo+  vo-
    RIN   1   2     2MEG
    ROUT  5   3     75
    E1    5   4   2   1      2MEG  ; Voltage-controlled voltage source
    .ENDS   OPAMP                  ; End of subcircuit definition OPAMP
```

ANALYSIS
```
■ ■ ■ *  Ac  sweep from 1 Hz to 100 kHz with a decade increment and 1 point
    *   per decade:
    .AC    DEC    1    1HZ    100kHZ
    *  Noise analysis without printing details of individual contributions:
    .NOISE   V(4)    VIN
    *  PSpice prints the details of equivalent input and output noise.
    .PRINT   NOISE   ONOISE   INOISE
  .END
```

■ ■

The results of the noise analysis are shown next.

```
****      AC ANALYSIS              TEMPERATURE =  27.000 DEG C
  FREQ          ONOISE        INOISE
  1.000E+00   1.966E-07    1.966E-08
  1.000E+01   1.966E-07    1.966E-08
  1.000E+02   1.966E-07    1.966E-08
  1.000E+03   1.966E-07    1.966E-08
  1.000E+04   1.966E-07    1.966E-08
  1.000E+05   1.966E-07    1.966E-08
```

Note. We could combine the .AC, .NOISE, and .SEN V(4) commands in the circuit file of Example 6-2 by modifying

```
VIN   1   0   AC   1V   DC   1V
```

6-9 TRANSIENT ANALYSIS

A transient response determines the output in the time domain in response to an input signal in the time domain. The method for the calculation of transient analysis bias point differs from that of dc analysis bias point. The dc bias point is also known as the *regular bias point*. In the regular (dc) bias point, the initial values of the circuit nodes do not contribute to the operating point or to the linearized parameters. The capacitors and inductors are considered open- and short-circuited, respectively, whereas in the transient bias point, the initial values of the circuit nodes are taken into account in calculating the bias point and the small-signal parameters of the nonlinear elements. The capacitors and inductors, which may have initial values, therefore remain as parts of the circuit. Determination of the transient analysis requires statements involving

.IC   Initial transient conditions
.TRAN   Transient analysis

6-9.1 .IC (Initial Transient Conditions)

The various nodes can be assigned to initial voltages during transient analysis, and the general form for assigning initial values is

```
.IC   V(1)=V1   V(2)=V2 ...  V(N)=VN
```

where V1, V2, V3, . . . are the initial voltages for nodes V(1), V(2), V(3), . . . , respectively. These initial values are used by PSpice to calculate the transient analysis bias point and the linearized parameters of nonlinear devices for transient analysis. After the transient analysis bias point has been calculated, the transient analysis starts and the nodes are released. It should be noted that these initial conditions do not affect the regular bias-point calculation during dc analysis or dc sweep. For the .IC statement to be effective, UIC (use initial conditions) *should not* be specified in the .TRAN command.

Statement for Initial Transient Conditions

    .IC   V(1)=2.5   V(5)=1.7V   V(7)=0.5

6-9.2 .TRAN (Transient Analysis)

Transient analysis can be performed by the .TRAN command, which has one of the general forms

    .TRAN   TSTEP   TSTOP   [TSTART TMAX]   [UIC]
    .TRAN[/OP]   TSTEP   TSTOP   [TSTART TMAX]   [UIC]

TSTEP is the printing increment, TSTOP is the final time (or stop time), and TMAX is the maximum step size of internal time step. TMAX allows the user to control the internal time step. TMAX could be smaller or larger than the printing time, TSTEP. The default value of TMAX is TSTOP/50.

Transient analysis always starts at time = 0. However, it is possible to suppress the printing of the output for a time of TSTART. TSTART is the initial time at which the transient response is printed. In fact, PSpice analyzes the circuit from $t = 0$ to TSTART, but it does not print or store the output variables. Although PSpice computes the results with an internal time step, the results are generated by interpolation for a printing step of TSTEP. Figure 6-10 shows the relationships of TSTART, TSTOP, and TSTEP.

In transient analysis, only the node voltages of the transient analysis bias point are printed. However, the .TRAN command can control the output for the transient response bias point. An .OP command with a .TRAN command (i.e., .TRAN/OP) will print the small-signal parameters during transient analysis.

If UIC is not specified as optional at the end of a .TRAN statement, PSpice calculates the transient analysis bias point before the beginning of transient analysis. PSpice uses the initial values specified with the .IC command. If UIC (use initial conditions) is specified as an option at the end of a .TRAN statement. PSpice does not calculate the transient analysis bias point before the beginning of transient analysis. However, PSpice uses the initial values specified with the "IC=" initial conditions for capacitors and inductors, which are discussed in Chapter 5. Therefore, if UIC is specified, the initial values of the capacitors and inductors must be supplied.

The .TRAN statements require .PRINT or .PLOT or .PROBE statements to get the results of the transient analysis.

Statements for Transient Analysis

    .TRAN   5US   1MS
    .TRAN   5US   1MS 200US 0.1NS
    .TRAN   5US   1MS 200US 0.1NS   UIC
    .TRAN/OP   5US   1MS 200US 0.1NS   UIC

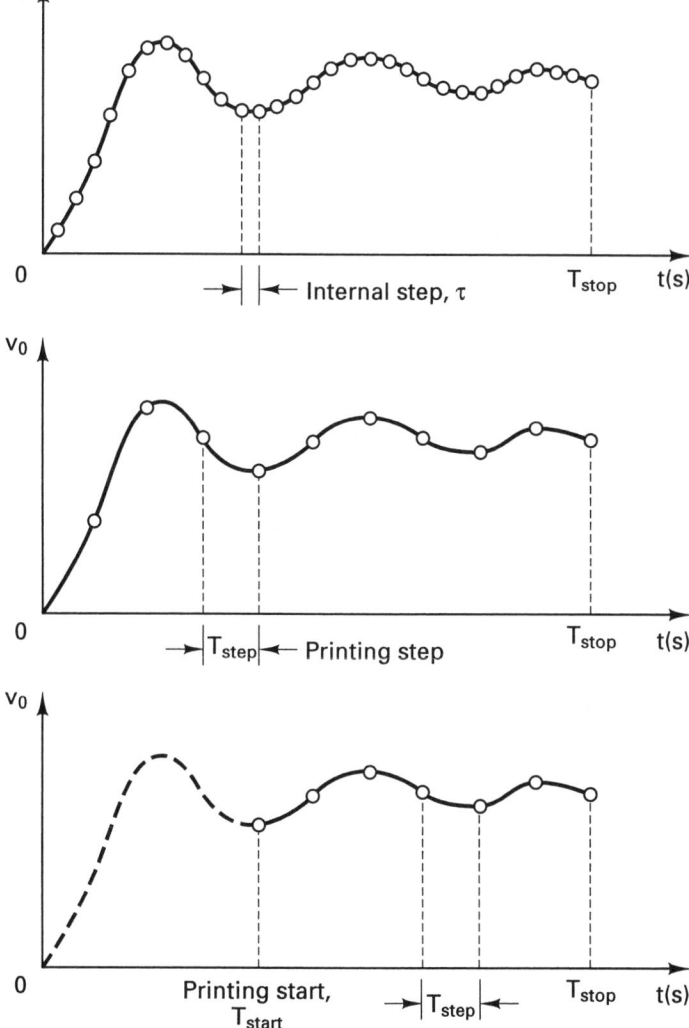

Figure 6-10 Response of transient analysis.

Example 6-7

Repeat Example 5-1, if the voltage across the capacitor is set by an .IC command instead of an IC condition and UIC is not specified.

Solution The list of the circuit file with .IC statement and without UIC is as follows:

■■■

Example 6-7   Transient response of *RLC* circuit

SOURCE ■ * Input step voltage represented as an PWL waveform:
```
       VS  1   0   PWL (0  0  10NS  10V  2MS  10V)
```

CIRCUIT ■ ■ * <u>R</u>₁ has a value of 60 Ω with model RMOD:

R1   1   2   RMOD   60

* Inductor of 1.5 mH with an initial current of 3 A and model name LMOD:

L1   2   3   LMOD   1.5MH   IC=3A

* Capacitor of 2.5 μF with an initial voltage of 4 V and model name CMOD:

C1   3   0   CMOD   2.5UF   IC=4V

R2   3   0   RMOD   20

* The initial voltage at node 3 is 4 V:

.IC   V(3)=4V

* Model statements for resistor, inductor, and capacitor:

.MODEL   RMOD   RES   (R=1   TC1=0.02   TC2=0.005)

.MODEL   CMOD   CAP   (C=1   VC1=0.01   VC2=0.002   TC1=0.02   TC2=0.005)

.MODEL   LMOD   IND   (L=1   IL1=0.1   IL2=0.002   TC1=0.02   TC2=0.005)

* The operating temperature is 50°C:

.TEMP   50

ANALYSIS ■ ■ ■ * Transient analysis from 0 to 1 ms with a 5-μs time increment and

* without using initial conditions (UIC): that is, IC-4V has no effect.

.TRAN   5US   1MS

* Plot the results of transient analysis with voltage at nodes 3 and 1.

.PLOT   TRAN   V(3)   V(1)

.PROBE

.END

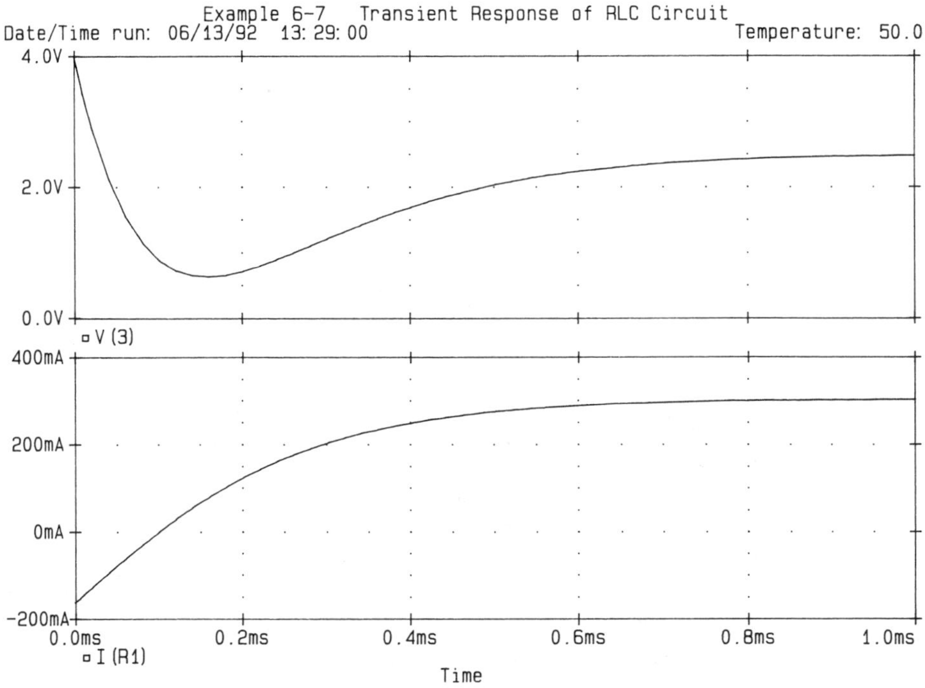

■ ■

Figure 6-11   Transient response for Example 6-7.

The results of the simulation for the circuit of Fig. 5-3 are shown in Fig. 6-11. It can be noted that the response is completely different, due to assigning an initial node voltage of 4 V on the capacitor.

6-10 FOURIER ANALYSIS

The output variables from the transient analysis are in discrete forms. These sampled data can be used to calculate the coefficients of Fourier series. A periodic waveform can be expressed in a Fourier series as

$$v(\theta) = C_0 + \sum_{n=1}^{\infty} C_n \sin(n\theta + \phi_n)$$

where $\theta = 2\pi ft$
$\quad f$ = frequency, in hertz
$\quad C_0$ = dc component
$\quad C_n$ = nth harmonic component

PSpice uses the results of the transient analysis to perform the Fourier analysis up to ninth harmonics, or 10 coefficients. The statement takes one of the general forms

```
.FOUR   FREQ   N   V1   V2   V3 ... VN
.FOUR   FREQ   N   I1   I2   I3 ... IN
```

FREQ is the fundamental frequency; V1, V2, . . . (or I1, I2, . . .) are the output voltages (or currents) for which the Fourier analysis is desired. N is the number of harmonics to be calculated. A .FOUR statement must have a .TRAN statement. The output voltages (or currents) must have the same forms as in the .TRAN statement for transient analysis.

Fourier analysis is performed over the interval (TSTOP-PERIOD) to TSTOP, where TSTOP is the final (or stop) time for the transient analysis and PERIOD is one period of the fundamental frequency. Therefore, the duration of the transient analysis must be at least one period long, PERIOD. At the end of the analysis, PSpice determines the dc component and the amplitudes of up to ninth harmonics (or 10 co-efficients) by default, unless N is specified.

PSpice does print a table automatically for the results of Fourier analysis and does not require .PRINT, .PLOT, or .PROBE statements.

Statement for Fourier Analysis

```
.FOUR   100KHZ   V(2,3),   V(3),   I(R1),   I(VIN)
```

Example 6-8

For Example 5-5, calculate the coefficients of the Fourier series if the fundamental frequency is 1 kHz.

Solution The list of the circuit file is as follows:

■ ■

Example 6-8   Fourier analysis

SOURCE   ■ *   Sinusoidal input voltage of 200 V peak with 0° phase delay:
```
         VS   1   0    SIN (0  200V  1KHZ)
```
CIRCUIT ■ ■ RS   1   2   100OHM
```
         R1   2   0    100KOHM
         *    Voltage-controlled voltage source with a voltage gain of 0.1:
         E1   3   0    2   0   0.1
         RL   4   5    20HM
         *    Dummy voltage source of VX = 0 to measure the load current:
         VX   5   0    DC   0V
         *    Voltage-controlled switch controlled by voltage across nodes 3 and 0:
         S1   3   4    3   0    SMOD
         *    Switch model descriptions:
         .MODEL   SMOD   VSWITCH (RON=5M ROFF=10E+9 VON=25M  VOFF=0.0)
```
ANALYSIS ■ ■ ■ *   Transient analysis from 0 to 1 ms with a 5-μs increment
```
         .TRAN   5US    1MS
         *   Plot the current through VX and the input voltage.
         .PLOT    TRAN   V(3) I(VX)                    ; On the output file
         *   Fourier analysis of load current at a fundamental frequency of 1 kHz:
         .FOUR  1KHZ  I(VX)
         .PROBE                                        ; Graphics post-processor
       .END
```
■ ■

It should be noted that we could add the statement .FOUR 1KHZ I(VX), in the circuit file of Example 5-5. The results of Fourier analysis are shown next.

FOURIER COMPONENTS OF TRANSIENT RESPONSE I(VX)

DC COMPONENT =   3.171401E+00

HARMONIC NO	FREQUENCY (HZ)	FOURIER COMPONENT	NORMALIZED COMPONENT	PHASE (DEG)	NORMALIZED PHASE (DEG)
1	1.000E+03	4.982E+00	1.000E+00	2.615E−05	0.000E+00
2	2.000E+03	2.115E+00	4.245E−01	−9.000E+01	−9.000E+01
3	3.000E+03	9.002E−08	1.807E−08	6.216E+01	6.216E+01
4	4.000E+03	4.234E−01	8.499E−02	−9.000E+01	−9.000E+01
5	5.000E+03	9.913E−08	1.990E−08	8.994E+01	8.994E+01
6	6.000E+03	1.818E−01	3.648E−02	−9.000E+01	−9.000E+01
7	7.000E+03	7.705E−08	1.547E−08	5.890E+01	5.890E+01
8	8.000E+03	1.012E−01	2.032E−02	−9.000E+01	−9.000E+01
9	9.000E+03	9.166E−08	1.840E−08	7.348E+01	7.348E+01

TOTAL HARMONIC DISTORTION =   4.349506E+01 PERCENT

6-11 MONTE CARLO ANALYSIS

PSpice allows one to perform the Monte Carlo (statistical) analysis of a circuit. The general form of the Monte Carlo analysis is

```
.MC  ⟨(# runs) value⟩    ⟨(analysis)⟩   ⟨(output variable)⟩   ⟨(function)⟩
+       [(option)]*
```

This command performs multiple runs of the analysis selected (dc, ac, transient). The first run is done with the nominal values of all components. Subsequent runs are done with variations on model parameters as specified by DEV and LOT tolerances on each .MODEL parameter (Section 6-2).

⟨(# runs) value⟩ is the total number of runs to do. For printed results, the upper limit is 2000. For the output to be viewed with Probe, the limit is 100. ⟨(analysis)⟩ must be specified from one of dc, ac, or transient. This analysis will be repeated in subsequent passes of the analysis. All analyses that the circuit contains are performed during the normal pass. Only the analysis selected is performed during subsequent passes. ⟨(output variable)⟩ is identical in format to that of a .PRINT output variable (in Chapter 3).

⟨(function)⟩ specifies the operation to be performed on the values of the ⟨(output variable)⟩. This value is the basis for comparisons between the nominal and subsequent runs. ⟨(function)⟩ must be one of:

YMAX	Finds the greatest difference in each waveform from the nominal run.
MAX	Finds the maximum value of each waveform.
MIN	Finds the minimum value of each waveform.
RISE_EDGE (⟨value⟩)	Finds the first occurrence of the waveform crossing above the threshold ⟨value⟩. The waveform must have one or more points at or below ⟨value⟩ followed by one above. The output will be the value where the waveform increases above ⟨value⟩.
FALL_EDGE (⟨value⟩)	Finds the first occurrence of the waveform crossing below the threshold ⟨value⟩. The waveform must have one or more points at or below ⟨value⟩ followed by one below. The output will be the value where the waveform decreases below ⟨value⟩.

[(option)]* may be zero for one of the following:

LIST	At the beginning of each run, prints out the model parameter values actually used for each component during that run.
OUTPUT	(output type) requests output from subsequent runs after the nominal (first) run. The output for any run follows the .PRINT, .PLOT, and .PROBE statements in the circuit file. If OUTPUT is omitted, only the nominal run produces out-

put. (output type) is one of the following:

ALL	Forces all output to be generated, including the nominal run.
FIRST ⟨value⟩	Generates output only during first *n* runs.
EVERY ⟨value⟩	Generates output every *n*th run.
RUN ⟨value⟩*	Does analysis and generates output only for the runs listed. Up to 25 values may be specified in the list.

RANGE   (⟨(low) value, ⟨(high) value⟩) restricts the range over which ⟨(function)⟩ will be evaluated. An "*" can be used in place of a ⟨value⟩ to indicate "for all values." Examples are:

YMAX RANGE (*, 0.5)	YMAX is evaluated for values of the sweep variable (time, frequency, etc.) of 0.5 or less.
MAX RANGE (−1, *)	The maximum value of the output variable is found for values of the sweep variable of −1 or more.

Some Statements for Monte Carlo Analysis

```
. MC  10  TRAN  V (2)    YMAX
. MC  40  DC    IC (Q3)  YMAX   LIST
. MC  20  AC    VP (3,4) YMAX   LIST   OUTPUT   ALL
```

SUMMARY

The PSpice dot commands can be summarized as follows:

.AC	Ac analysis
.DC	Dc analysis
.END	End of circuit
.ENDS	End of subcircuit
.FUNC	Function
.FOUR	Fourier analysis
.GLOBAL	Global
.IC	Initial transient conditions
.INC	Include file

.LIB	Library file
.MC	Monte Carlo analysis
.MODEL	Model
.NODESET	Nodeset
.NOISE	Noise analysis
.OP	Operating point
.OPTIONS	Options
.PARAM	Parameter
.PLOT	Plot
.PRINT	Print
.PROBE	Probe
.SENS	Sensitivity analysis
.STEP	Step
.SUBCKT	Subcircuit definition
.TEMP	Temperature
.TF	Transfer function
.TRAN	Transient analysis
.WIDTH	Width

SUGGESTED READING

1. *PSpice Manual.* Irvine, Calif.: MicroSim Corporation, 1992.

2. P. Antognetti and G. Massobri, *Semiconductor Device Modeling with SPICE*. New York: McGraw-Hill, 1988.

3. M. H. Rashid, *SPICE for Circuits and Electronics Using PSpice*. Englewood Cliffs, N.J.: Prentice Hall, 1990.

PROBLEMS

6-1. For the circuit in Fig. P6-1, calculate and print the sensitivity of output voltage V_o with respect to each circuit element. The operating temperature is 50°C.

6-2. For the circuit in Fig. P6-1, calculate and print **(a)** the voltage gain, $A_v = V_o/V_{in}$; **(b)** the input resistance, R_{in}; and **(c)** the output resistance, R_o.

6-3. For the circuit in Fig. P6-1, calculate and plot the dc transfer characteristic, V_o ver-

sus V_{in}. The input voltage is varied from 0 to 10 V with an increment of 0.5 V.

6-4. For the circuit in Fig. P6-1, calculate and print the equivalent input and output noise if the frequency of the source is varied from 10 Hz to 1 MHz. The frequency should be increased by a decade with 2 points per decade.

6-5. For the circuit in Fig. P6-5, the frequency response is to be calculated and printed

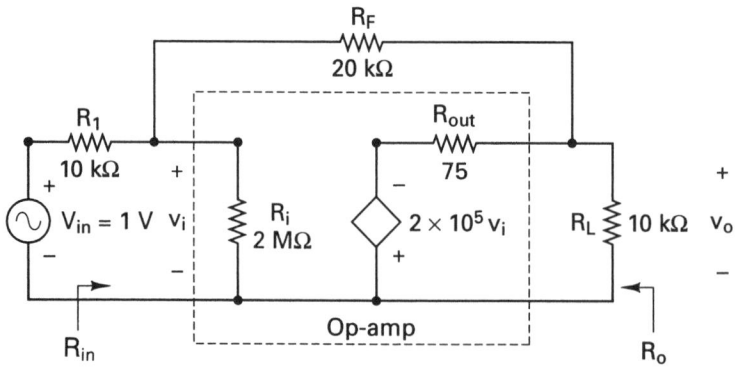

Figure P6-1

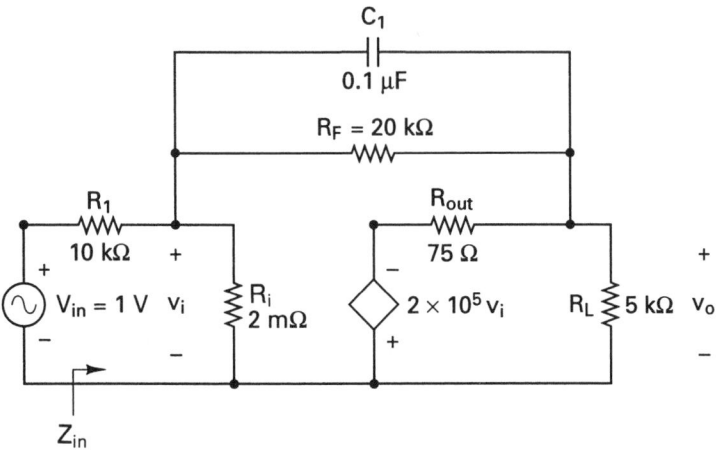

Figure P6-5

over the frequency range from 1 Hz to 100 kHz with a decade increment and 10 points per decade. The peak magnitude and phase angle of the output voltage is to be plotted on the output file. The results should also be available for display and hard copy by a .PROBE command.

6-6. Repeat Problem 6-5 for the circuit in Fig. P6-6.

6-7. For the circuit in Fig. P6-5, calculate and plot the transient response of the output voltage from 0 to 2 ms with a time increment of 5 μs. The input voltage is shown in Fig. P6-7. The results should be available for display and hard copy by Probe.

6-8. Repeat Problem 6-7 for the input voltage as shown in Fig. P6-8.

6-9. For the circuit of Fig. P6-6, calculate and

plot the transient response of the output voltage from 0 to 2 ms with a time increment of 5 μs. The input voltage is shown in Fig. P6-7. The results should be available for display and hard copy by Probe.

6-10. Repeat Problem 6-9 for the input voltage as shown in Fig. P6-8.

6-11. For Problem 6-7, calculate the coefficients of the Fourier series if the fundamental frequency is 500 Hz.

6-12. For Problem 6-8, calculate the coefficients of the Fourier series if the fundamental frequency is 500 Hz.

6-13. For Problem 6-9, calculate the coefficients of the Fourier series if the fundamental frequency is 500 Hz.

6-14. For Problem 6-10, calculate the coeffi-

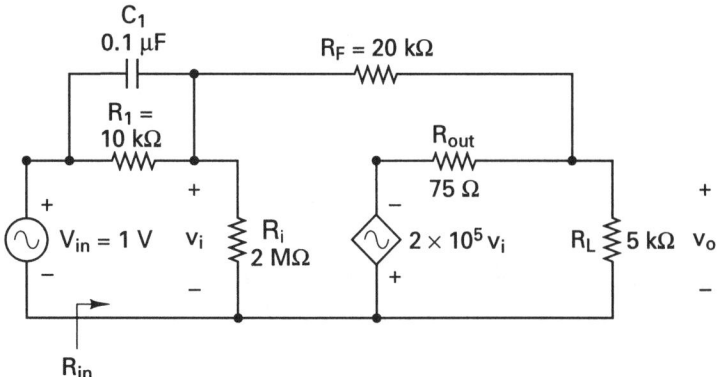

Figure P6-6

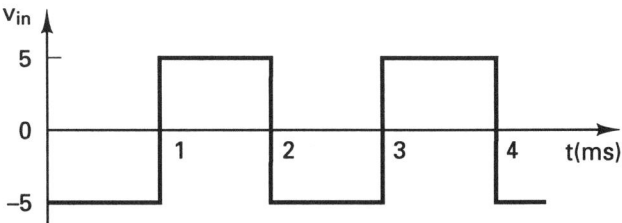

Figure P6-7

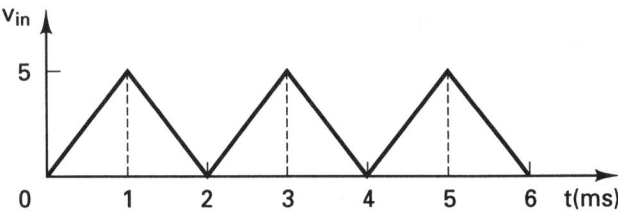

Figure P6-8

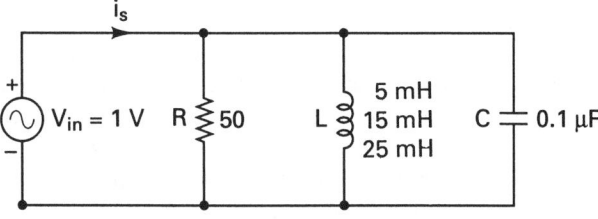

Figure P6-15

cients of the Fourier series if the fundamental frequency is 500 Hz.

6-15. For the *RLC* circuit of Fig. P6-15, plot the frequency response of the current i_s through the circuit and the magnitude of the input impedance. The frequency of the source is varied from 100 Hz to 100 kHz with a decade increment and 10 points per decade. The values of the inductor L are 5, 15, and 25 mH.

Chapter 7

■■■■■■■■

Diode Rectifiers

7-1 INTRODUCTION

A semiconductor diode may be modeled in SPICE by a diode statement in conjunction with a model statement. The diode statement specifies the diode name, the nodes to which the diode is connected, and its model name. The model incorporates an extensive range of diode characteristics: such as dc and small-signal behavior, temperature dependency, and noise generation. The model parameters take into account the temperature effects, the various capacitance, and the physical properties of semiconductors.

7-2 DIODE MODEL

The SPICE model for a reverse-biased diode is shown in Fig. 7-1 [1–3]. The small-signal and static models that are generated by SPICE are shown in Figs. 7-2 and 7-3, respectively. In the static model, the diode current, which depends on its voltage, is represented by a current source. The small-signal parameters are generated by SPICE from the operating point. SPICE generates a complex model for diodes. The model equations that are used by SPICE are described in Refs. 1 and 2. In many cases, especially the level at which this book is aimed, such complex models are not necessary. Many model parameters can be ignored by the users, and SPICE assigns default values to the parameters.

The model statement of a diode has the general form

```
.MODEL    DNAME    D  (P1=V1 P2=V2 P3=V3 ... PN=VN)
```

DNAME is the model name, and it can begin with any character; but its word size is normally limited to eight. D is the type symbol for diodes. P1, P2, . . . and V1,

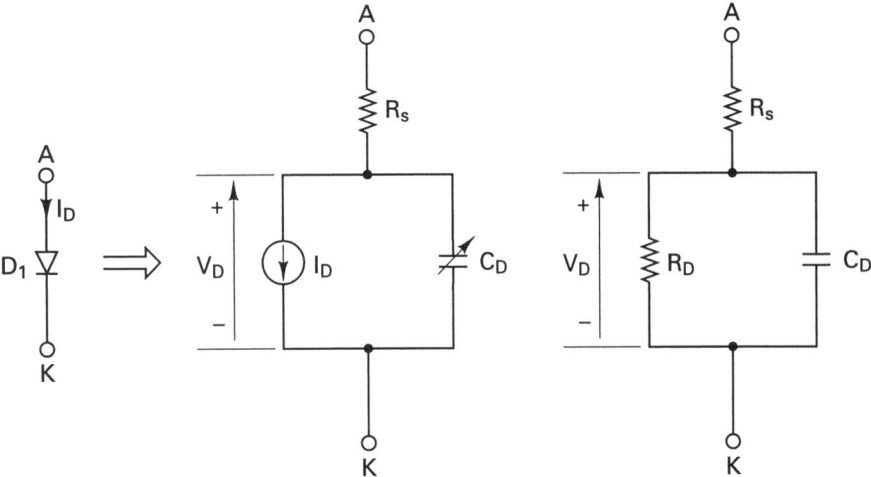

Figure 7-1 SPICE diode model
with reverse-biased condition.

Figure 7-2 SPICE small-signal
diode model.

V2, . . . are the model parameters and their values, respectively. The model
parameters are listed in Table 7-1. An area factor is used to determine the number
of equivalent parallel diodes of the model specified. The model parameters that
are affected by the area factor are marked by an asterisk (*) in the descriptions of
the model parameters.

The diode is modeled as an ohmic resistance (value = RS/area) in series with
an intrinsic diode. The resistance is attached between node NA and an internal
anode node. [(area value] scales IS, RS, CJO, and IBV, and defaults to 1. IBV
and BV are both specified as positive values.

The dc characteristic of a diode is determined by the reverse saturation
current IS, the emission coefficient N, and the ohmic resistance RS. Reverse
breakdown is modeled by an exponential increase in the reverse diode current,

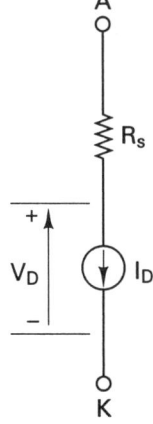

Figure 7-3 Static diode model
with reverse-biased condition.

TABLE 7-1 PARAMETERS OF DIODE MODEL

Name	Area	Model parameter	Unit	Default	Typical
IS	*	Saturation current	A	1E−14	1E−14
RS	*	Parasitic resistance	Ω	0	10
N		Emission coefficient	1	1	
TT		Transit time	s	0	0.1NS
CJO	*	Zero-bias *p-n* capacitance	F	0	2PF
VJ		Junction potential	V	1	0.6
M		Junction grading coefficient		0.5	0.5
EG		Activation energy	eV	1.11	11.1
XTI		IS temperature exponent		3	3
KF		Flicker noise coefficient		0	
AF		Flicker noise exponent		1	
FC		Forward-bias depletion capacitance coefficient		0.5	
BV		Reverse breakdown voltage	V	∞	50
IBV	*	Reverse breakdown current	A	1E−10	

and is determined by the reverse breakdown voltage BV, and the current at breakdown voltage IBV. The charge storage effects are modeled by the transit time TT and a nonlinear depletion layer capacitance, which depends on the zero-bias junction capacitance CJO, the junction potential VJ, and grading coefficient M.

The temperature of the reverse saturation current is defined by the gap activation energy (or gap energy) EG and the saturation temperature exponent XTI. The most important parameters for power electronics applications are IS, BV, IBV, TT, and CJO.

7-3 DIODE STATEMENT

The name of a diode must start with D, and it takes the general form

```
D⟨name⟩  NA  NK  DNAME  [⟨area⟩ value]
```

where NA and NK are the node and cathode nodes, respectively. The current flows from anode node NA, through the diode, to cathode node NK. DNAME is the model name.

Some Statements for Diode

```
D15  33   35  SWITCH  1.5
.MODEL  SWITCH  D (IS=100E-15 CJO=2PF TT=12NS BV=100 IBV=10E-3)
DCLAMP  0   8   D1N914
.MODEL  D1N914  D (IS=100E-15 CJO=2PF TT=12NS BV=100 IBV=10E-3)
```

The typical v–i characteristic of a diode is shown in Fig. 7-4. The characteristic can be expressed by an equation known as the *Schockley diode equation*, given by

$$I_D = I_s(e^{V_D/nV_T} - 1) \tag{7-1}$$

where I_D = current through the diode, A

V_D = diode voltage with anode positive with respect to cathode, V

I_s = leakage (or reverse saturation) current, typically in the range 10^{-6} to 10^{-20} A

n = empirical constant known as the *emission coefficient* (or ideality factor), whose value varies from 1 to 2

The emission coefficient, n, depends on the material and the physical construction of diodes. For germanium diodes, n is considered to be 1. For silicon diodes, the predicted value of n is 2, but for most silicon diodes the value of n falls in the range 1.1 to 1.8.

V_T in Eq. (7-1) is a constant called the *thermal voltage*, and it is given by

$$V_T = \frac{kT}{q} \tag{7-2}$$

where q = electron charge: 1.6022×10^{-19} C

T = absolute temperature, kelvin (K = 273 + °C)

k = Boltzmann's constant: 1.3806×10^{-23} J/K

At a junction temperature of 25°C, Eq. (7-2) gives

$$V_T = \frac{kT}{q} = \frac{1.3806 \times 10^{-23} \times (273 + 25)}{1.6022 \times 10^{-19}} \approx 25.8 \text{ mV}$$

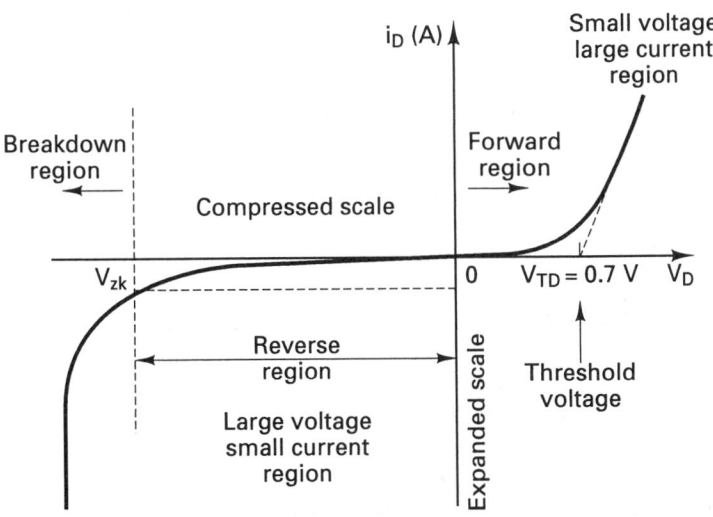

Figure 7-4 Diode characteristics.

At a specified temperature, the leakage current I_s remains constant for a given diode. For power diodes, the typical value of I_s is 10^{-15} A.

7-5 DIODE PARAMETERS

The data sheet of a diode IR-type R18C is shown in Fig. 7-5. Although PSpice allows specifying many parameters, we use only the parameters that significantly affect power converter output. From the data sheet, we get

Reverse breakdown voltage, BV = 1200
Reverse breakdown current, IBV = 13 mA
Instantaneous forward voltage, $v_F = 1$ V at $i_F = 150$ A
Reverse recovery charge, $Q_{RR} = 194$ μC at $I_{FM} = 200$ A

From the data for the $v-i$ forward characteristic of a diode, it is possible to determine the value of n, V_T, and I_s of the diode [4]. Assuming that $n = 1$ and $V_T = 25.8$ mV, we can apply Eq. (7-1) to find the saturation current I_s:

$$I_D = I_s(e^{V_D/nV_T} - 1)$$

$$150 = I_s(e^{1/(25.8 \times 10^{-3})} - 1)$$

which gives $I_s = 2.2\text{E}-15$ A.

Let us call the diode model name, DMOD. The values of TT and CJO are not available from the data sheet. Some versions of SPICE (e.g., PSpice) support device library files. The software PARTS of PSpice can generate SPICE models from the data sheet parameters of transistors and diodes. The transit time τ_T can be calculated approximated from

$$\tau_T = \frac{Q_{RR}}{I_{FM}} = \frac{194 \ \mu\text{C}}{200} \approx 1 \ \mu\text{s}$$

We shall assume the typical value CJO = 2 pF. Thus the PSpice model statement is

```
.MODEL  DMOD  D(IS=2.22E-15  BV=1200  IBV=13E-3  CJO=2PF  TT=1US)
```

7-6 DIODE RECTIFIERS

A rectifier converts an ac voltage to a dc voltage and uses diodes as the switching devices. The output voltage of an ideal rectifier should be pure dc and contain no harmonics or ripples. Similarly, the input current should be pure sine wave and contain no harmonics. That is, the total harmonic distortion (THD) of the input current and output voltage should be zero, and the input power factor should be unity.

R18C, R18S, R18CR & R18SR SERIES
1800 – 1200 VOLTS RANGE
185 AMP AVG STUD MOUNTED
DIFFUSED JUNCTION RECTIFIER DIODES

VOLTAGE RATINGS

VOLTAGE CODE [1]	V_{RRM}, V_R – (V) Max. rep. peak reverse and direct voltage		V_{RSM} – (V) Max. non-rep. peak reverse voltage
	$T_J = 0°$ to $200°C$	$T_J = -40°$ to $0°C$	$T_J = 25°$ to $200°C$
18A	1800	1710	1900
16A	1600	1520	1700
14B	1400	1330	1500
12B	1200	1200	1300

MAXIMUM ALLOWABLE RATINGS

PARAMETER		SERIES [2]	VALUE	UNITS	NOTES
T_J	Junction temperature	ALL	-40 to 200	°C	
T_{stg}	Storage temperature	ALL	-40 to 200	°C	
$I_{F(AV)}$	Average current	R18C/S	185	A	180° half sine wave, $T_C = 140°C$
$I_{F(AV)}$	Max. av. current (3)	ALL	200	A	180° half sine wave, $T_C = 133°C$, 18C/S $T_C = 145°C$, R18CR/SR
$I_{F(RMS)}$	Max. RMS current (3)	ALL	314	A	
I_{FSM}	Max. peak non-rep. surge current	ALL	3820	A	50Hz half cycle sine wave Initial $T_J = 200°C$, rated V_{RRM} applied after surge.
			4000		60Hz half cycle sine wave
			4550		50Hz half cycle sine wave Initial $T_J = 200°C$, no voltage applied after surge.
			4750		60Hz half cycle sine wave
I^2t	Max. I^2t capability	ALL	73	kA^2s	t = 10ms Initial $T_J = 200°C$, rated V_{RRM} applied after surge.
			67		t = 8.3ms
			104		t = 10ms Initial $T_J = 200°C$, no voltage applied after surge.
			95		t = 8.3ms
$I^2\sqrt{t}$	Max. $I^2\sqrt{t}$ capability	ALL	1040	$kA^2\sqrt{s}$	Initial $T_J = 200°C$, no voltage applied after surge. I^2t for time $t_x = I^2\sqrt{t} \cdot \sqrt{t_x}$, $0.1 \le t_x \le 10ms$.
T	Mounting torque	R18C/CR			
	Min.		14.1 [125]	N·m	Non-lubricated threads
	Max.		17.0 [150]	[lbf-in]	
	Min.		12.2 [108]		Lubricated threads
	Max.		15.0 [132]		
		R18S/SR			
	Min.		11.3 [100]	N·m	Non-lubricated threads
	Max.		14.1 [125]	[lbf-in]	
	Min.		9.5 [85]		Lubricated threads
	Max.		12.5 [110]		

(1) To complete the part number, refer to the Ordering Information table

(2) R18C & R18S series have cathode-to-case polarity. R18CR & R18SR series have anode-to-case polarity.

(3) For devices assembled in Europe, max. I$_{F(AV)}$ is 175A and max. I$_{F(RMS)}$ is 275A.

Figure 7-5 Data sheets for IR diodes type R18. (Courtesy of International Rectifier.)

CHARACTERISTICS

PARAMETER		SERIES	MIN.	TYP.	MAX.	UNITS	TEST CONDITIONS
V_{FM}	Peak forward voltage	ALL	—	1.30	1.42	V	Initial T_J = 25°C, 50-60Hz half sine, I_{peak} = 828A.
$V_{F(TO)1}$	Low-level threshold	ALL	—	—	0.703	V	T_j = 200°C
$V_{F(TO)2}$	High-level threshold		—	—	0.738		Av. power = $V_{F(TO)}$ · $I_{F(AV)}$ + r_F · $[I_{F(RMS)}]^2$
r_{F1}	Low-level resistance	ALL	—	—	1.100	mΩ	Use low level values for
r_{F2}	High-level resistance		—	—	1.080		$I_{FM} \leq \pi I_{F(AV)}$
t_a	Reverse current rise	ALL	—	18.0	—	μs	T_J = 175°C, I_{FM} = 200A, di_R/dt = 1.0A/μs.
t_b	Reverse current fall	ALL	—	3.5	—	μs	
$I_{RM(REC)}$	Reverse current	ALL	—	18	—	A	
Q_{RR}	Recovered charge	ALL	—	194	—	μC	
I_{RM}	Peak reverse current	ALL	—	13	20	mA	T_J = 175°C. Max. rated V_{RRM}.
R_{thJC}	Thermal resistance, junction-to-case	R18C/S	—	—	0.250	°C/W	DC operation
		R18CR/SR	—	—	0.200		
		R18C/S	—	—	0.271	°C/W	180° sine wave
		R18CR/SR	—	—	0.221		
		R18C/S	—	—	0.275	°C/W	120° rectangular wave
		R18CR/SR	—	—	0.225		
R_{thCS}	Thermal resistance, case-to-sink	ALL	—	—	0.10	°C/W	Mtg. surface smooth, flat and greased.
wt	Weight	ALL	—	100[3.5]	—	g[oz.]	
	Case Style	R18C/CR	DO-205AC (DO-30)			JEDEC	
		R18S/SR	DO-205AA (DO-8)				

Figure 7-5 *(Continued)*

Example 7-1

A single-phase half-wave rectifier is shown in Fig. 7-6. The input voltage is sinusoidal with a peak of 169.7 V, 60 Hz. The load inductance L is 6.5 mH, and the load resistance R is 0.5 Ω. Use PSpice (a) to plot the instantaneous output voltage v_o and the load current i_o, (b) to calculate the Fourier coefficient of the output voltage, and (c) to find the input power factor.

Solution f = 60 Hz and V_m = 169.7 V. The list of the circuit file is as follows:

■ ■

Example 7-1   Single-phase half-wave rectifier with *RL* load

```
SOURCE    ■ VS    1   0   SIN (0    169.7V    60HZ)
CIRCUIT   ■ ■ R   2   3   0.5
            L     3   4   6.5MH
```

R18C, R18S, R18CR & R18SR SERIES
1800 – 1200 VOLTS RANGE

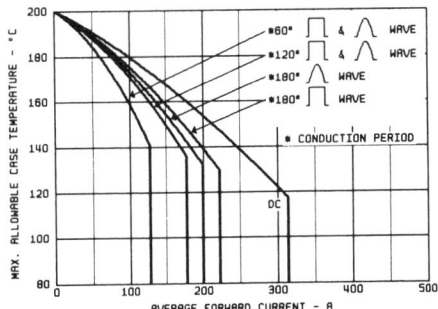

Fig. 1 — Case Temperature Ratings —
R18C and R18S Series

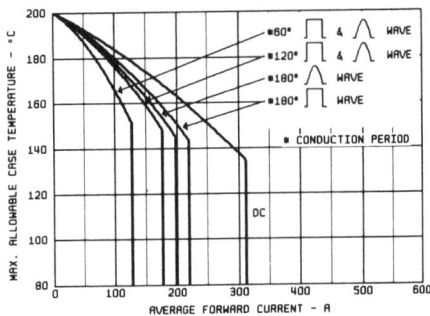

Fig. 1a — Case Temperature Ratings —
R18CR and R18SR Series

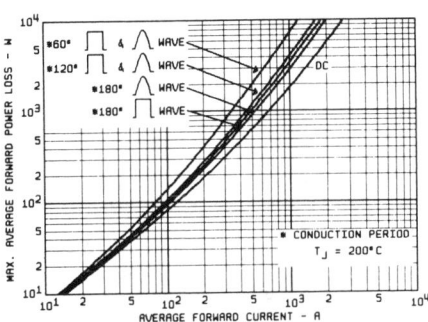

Fig. 2 — Power Loss Characteristics —
R18C, R18S, R18CR and R18SR Series

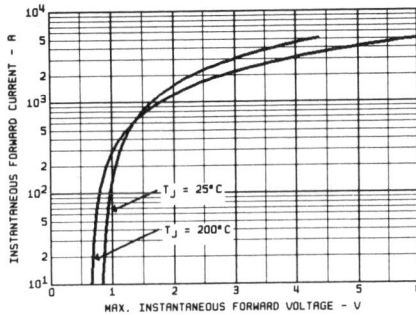

Fig. 3 — Forward Characteristics —
R18C, R18S, R18CR and R18SR Series

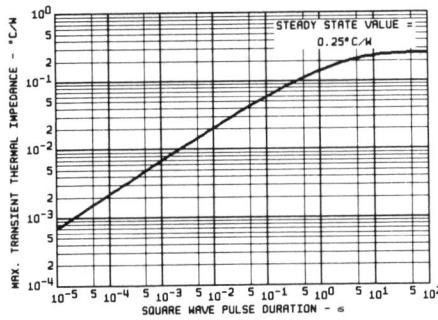

Fig. 4 — Transient Thermal Impedance,
Junction-to-Case — R18C and R18S Series

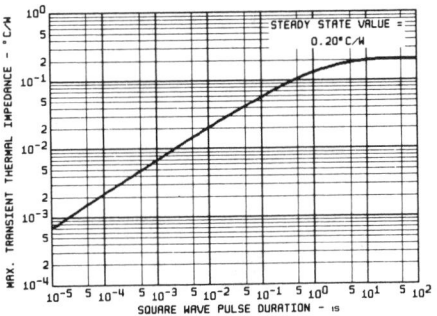

Fig. 4a — Transient Thermal Impedance,
Junction-to-Case — R18CR and R18SR Series

Figure 7-5 *(Continued)*

R18C, R18S, R18CR & R18SR SERIES
1800 – 1200 VOLTS RANGE

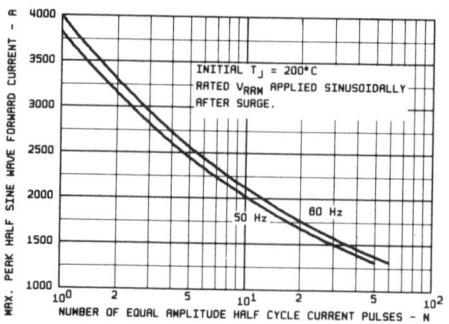

Fig. 5 — Non-Repetitive Surge Current Ratings —
R18C, R18S, R18CR and R18SR Series

ORDERING INFORMATION

TYPE	PACKAGE (1)		POLARITY		VOLTAGE	
	CODE	DESCRIPTION	CODE	DESCRIPTION	CODE	V_{RRM}
R18	C	1/2" stud, ceramic housing (Fig. 1)	———	cathode-to-case	18A	1800V
			R	anode-to-case	18A	1600V
	S	3/8" stud, ceramic housing (Fig. 2)			14B	1400V
					12B	1200V

(1) Other packages are also available:
  – stud base with flag terminal.
  – stud base with threaded top terminal.
  – flat base.

  For further details contact factory.

For example, for a device with 1/2" stud base and flexible lead, reverse polarity,
V_{RRM} = 1600V, order as: R18CR16A.

Figure 7-5 *(Continued)*

Diode Rectifiers   Chap. 7

R18C, R18S, R18CR & R18SR SERIES
1800 – 1200 VOLTS RANGE

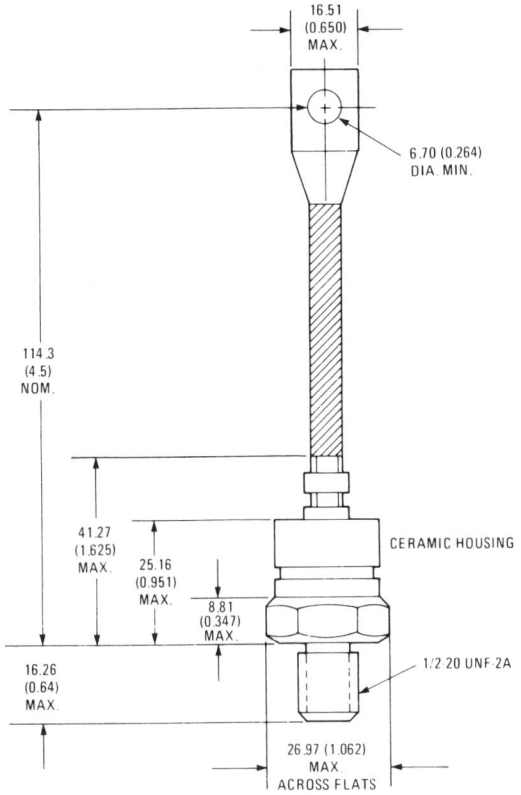

Figure 7-5 *(Continued)*

```
          VX     4    0    DC    0V   ; Voltage source to measure the output current
          D1     1    2    DMOD
          .MODEL    DMOD    D(IS=2.22E-15  BV=1200V   IBV=13E-3  CJO=2PF   TT=1US)
ANALYSIS  ■ ■ ■ .TRAN  10US  50.0MS  16.6667MS  10US       ; Transient analysis
          .FOUR   60HZ  I(D1)  V(2)  ;Fourier analysis of input current and output
          *                                                              voltage
          .PROBE                                    ; Graphics post-processor
          .OPTIONS ABSTOL = 1.0N RELTOL = .01 VNTOL = 1.0M ITL5=10000 ;
          *                                                         Convergence
       .END
■ ■ ■ ■ ■ ■ ■ ■ ■ ■ ■ ■ ■ ■ ■ ■ ■ ■ ■ ■ ■ ■ ■ ■ ■ ■ ■ ■ ■ ■ ■ ■ ■ ■ ■ ■ ■ ■ ■ ■
```

(a) The PSpice plots of the instantaneous output voltage V(2) and load current I(VX) are shown in Fig. 7-7. The output voltage becomes negative due to the inductive load, because the current has to fall to zero before the diode can cease to conduct.

(b) Fourier coefficients and THD will depend slightly on the internal time step TMAX discussed in Section 6-9.2.

Dimensions in Millimeters and (Inches)

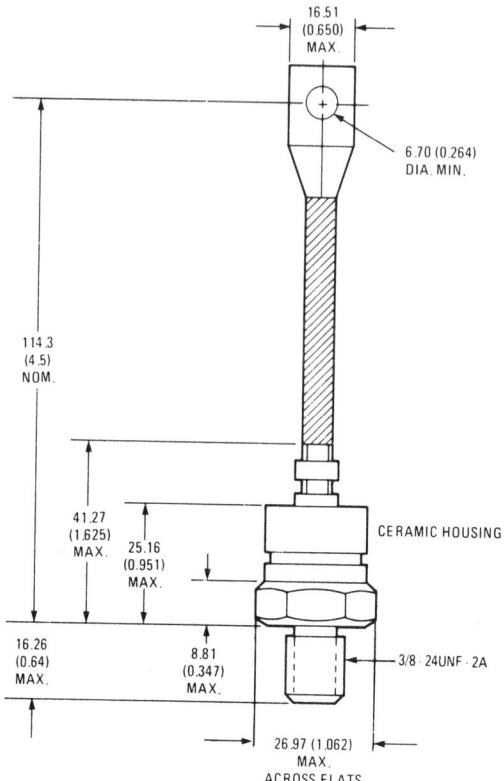

Figure 7-5 *(Continued)*

THE FOURIER COMPONENTS OF TRANSIENT RESPONSE V(2):
DC COMPONENT = 2.218909E+01

HARMONIC NO	FREQUENCY (HZ)	FOURIER COMPONENT	NORMALIZED COMPONENT	PHASE (DEG)	NORMALIZED PHASE (DEG)
1	6.000E+01	1.377E+02	1.000E+00	1.119E+01	0.000E+00
2	1.200E+02	3.646E+01	2.647E-01	-1.642E+02	-1.754E+02
3	1.800E+02	2.651E+01	1.925E-01	-1.076E+02	-1.188E+02
4	2.400E+02	1.649E+01	1.197E-01	-4.324E+01	-5.442E+01
5	3.000E+02	9.585E+00	6.958E-02	3.968E+01	2.849E+01
6	3.600E+02	8.108E+00	5.887E-02	1.366E+02	1.254E+02
7	4.200E+02	8.486E+00	6.161E-02	-1.431E+02	-1.542E+02
8	4.800E+02	7.607E+00	5.523E-02	-7.090E+01	-8.209E+01
9	5.400E+02	5.897E+00	4.281E-02	6.598E+00	-4.591E+00

TOTAL HARMONIC DISTORTION = 3.720415E+01 PERCENT

(c) To find the input power factor, we need to find the Fourier series of the input current, which is the same as the current through diode D_1.

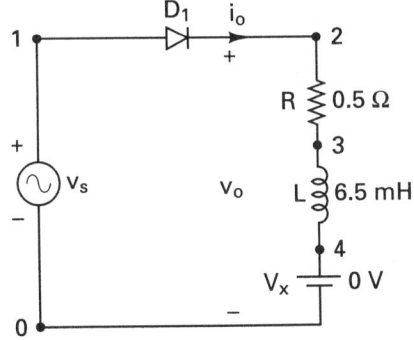

Figure 7-6 Single-phase half-wave rectifier for PSpice simulation.

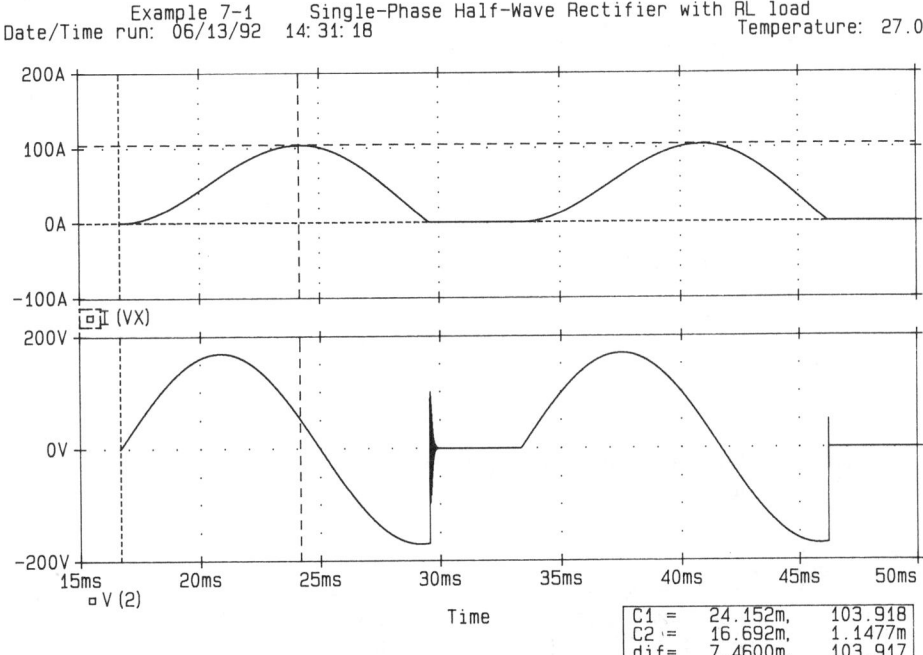

Example 7-1   Single-Phase Half-Wave Rectifier with RL load
Date/Time run: 06/13/92   14:31:18   Temperature: 27.0

C1 =	24.152m,	103.918
C2 =	16.692m,	1.1477m
dif=	7.4600m,	103.917

Figure 7-7 Plots for Example 7-1.

THE FOURIER COMPONENTS OF TRANSIENT RESPONSE I(D1)

DC COMPONENT =   4.438586E+01

HARMONIC NO	FREQUENCY (HZ)	FOURIER COMPONENT	NORMALIZED COMPONENT	PHASE (DEG)	NORMALIZED PHASE (DEG)
1	6.000E+01	5.507E+01	1.000E+00	−6.738E+01	0.000E+00
2	1.200E+02	7.448E+00	1.352E−01	1.107E+02	1.781E+02
3	1.800E+02	3.634E+00	6.598E−02	1.681E+02	2.355E+02
4	2.400E+02	1.606E+00	2.916E−02	−1.317E+02	−6.432E+01
5	3.000E+02	8.217E−01	1.492E−02	−5.059E+01	1.678E+01
6	3.600E+02	5.774E−01	1.048E−02	5.326E+01	1.206E+02

HARMONIC NO	FREQUENCY (HZ)	FOURIER COMPONENT	NORMALIZED COMPONENT	PHASE (DEG)	NORMALIZED PHASE (DEG)
7	4.200E+02	4.422E-01	8.030E-03	1.264E+02	1.938E+02
8	4.800E+02	4.177E-01	7.584E-03	-1.633E+02	-9.597E+01
9	5.400E+02	2.892E-01	5.251E-03	-7.644E+01	-9.061E+00

TOTAL HARMONIC DISTORTION = 1.548379E+01 PERCENT

Dc input current, $I_{in(dc)} = 44.39$ A

Rms fundamental input current, $I_{1(rms)} = 55.07/\sqrt{2} = 38.94$ A

Total harmonic distortion of input current, THD = 15.48% = 0.1548

Harmonic input current, $I_{h[(rms)} = I_{1(rms)}$ THD = 38.94 × 0.1548 = 6.028 A

Rms input current, $I_s = [I_{in(dc)}^2 + I_{r(rms)}^2 + I_{h(rms)}^2]^{1/2}$
$$= (44.39^2 + 38.94^2 + 6.028^2)^{1/2} = 59.36 \text{ A}$$

Displacement angle, $\phi_1 = -67.38°$

Displacement factor, DF = $\cos \phi_1 = \cos(-67.38) = 0.3846$ (lagging)

Thus the input power factor is given [1] by

$$\text{PF} = \frac{I_{1(rms)}}{I_s} \cos \phi_1 = \frac{38.94}{59.36} \times 0.3846 = 0.2523 \quad \text{(lagging)}$$

The power factor can be determined directly from the THD as follows:

$$\text{PF} = \frac{I_{1(rms)}}{I_s} \cos \phi_1 = \frac{1}{[1 + (\%\text{THD}/100)^2]^{1/2}} \cos \phi_1$$

(7-3)

$$= \frac{1}{(1 + 0.1548^2)^{1/2}} \times 0.3846 = 0.3801 \quad \text{(lagging)}$$

which gives a higher value and cannot be applied if there is a significant amount of dc component.

Example 7-2

A single-phase bridge rectifier is shown in Fig. 7-8. The sinusoidal input voltage has a peak of 169.7 V, 60 Hz. The load inductance L is 6.5 mH, and the load resistance R is 0.5 Ω. Use PSpice (a) to plot the instantaneous output voltage v_o and the load current i_o, and (b) to calculate the Fourier coefficients of the input current and the input power factor.

Solution $V_m = 169.7$ V and $f = 60$ Hz. The list of the circuit file is as follows:

■ ■

Example 7-2   Single-phase bridge rectifier with *RL* load

```
SOURCE   ■ VS    1   0   SIN (0   169.7V   60HZ)
CIRCUIT  ■ ■ R   3   5   0.5
           L     5   6   6.5MH
           VX    6   4   DC   0V  ; Voltage source to measure the output current
           VY    1   2   DC   0V  ; Voltage source to measure the input current
           D1    2   3   DMOD
           D3    0   3   DMOD
```

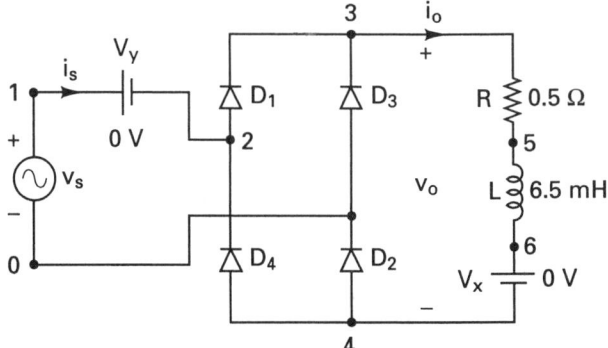

Figure 7-8 Single-phase bridge rectifier for PSpice simulation.

```
              D2    4    0    DMOD
              D4    4    2    DMOD
              .MODEL    DMOD    D(IS=2.22E-15  BV=1200V  IBV=13E-3  CJO=2PF  TT=1US)
ANALYSIS ■ ■ ■ .TRAN    10US    50MS    33.3333MS    10US  ; Transient analysis
              .FOUR    60HZ    I(VY)                      ; Fourier analysis of input
              *                                                                  current
              .PROBE                                      ; Graphics post-processor
              .OPTIONS ABSTOL = 1.0N RELTOL = .01 VNTOL = 1.0M ITL5=10000 ;
              *                                                              Convergence
       .END
```

(a) The PSpice plots of the instantaneous output voltage V(3,4) and load current I(VX) are shown in Fig. 7-9. One of the diode pairs always conducts. The load current contains ripples and has not reached steady-state conditions.

(b) The input current, which is the same as the current through the voltage source VY, is equal to I(VY).

THE FOURIER COMPONENTS OF TRANSIENT RESPONSE I(VY)
DC COMPONENT =   −2.56451E+00

HARMONIC NO	FREQUENCY (HZ)	FOURIER COMPONENT	NORMALIZED COMPONENT	PHASE (DEG)	NORMALIZED PHASE (DEG)
1	6.000E+01	2.595E+02	1.000E+00	−3.224E+00	0.000E+00
2	1.200E+02	7.374E−01	2.842E−03	1.410E+02	1.442E+02
3	1.800E+02	8.517E+01	3.282E−01	4.468E+00	7.693E+00
4	2.400E+02	5.856E−01	2.257E−03	1.199E+02	1.232E+02
5	3.000E+02	5.118E+01	1.972E−01	3.216E+00	6.440E+00
6	3.600E+02	5.526E−01	2.130E−03	1.111E+02	1.143E+02
7	4.200E+02	3.658E+01	1.410E−01	2.868E+00	6.092E+00
8	4.800E+02	5.406E−01	2.083E−03	1.065E+02	1.097E+02
9	5.400E+02	2.846E+01	1.097E−01	2.822E+00	6.047E+00

TOTAL HARMONIC DISTORTION =   4.225668E+01 PERCENT

Dc input current, $I_{in(dc)} = -2.56$ A, which should ideally be zero

Rms fundamental input current, $I_{1(rms)} = 259.5/\sqrt{2} = 183.49$ A

Total harmonic distortion of input current, THD = 42.26% = 0.4226

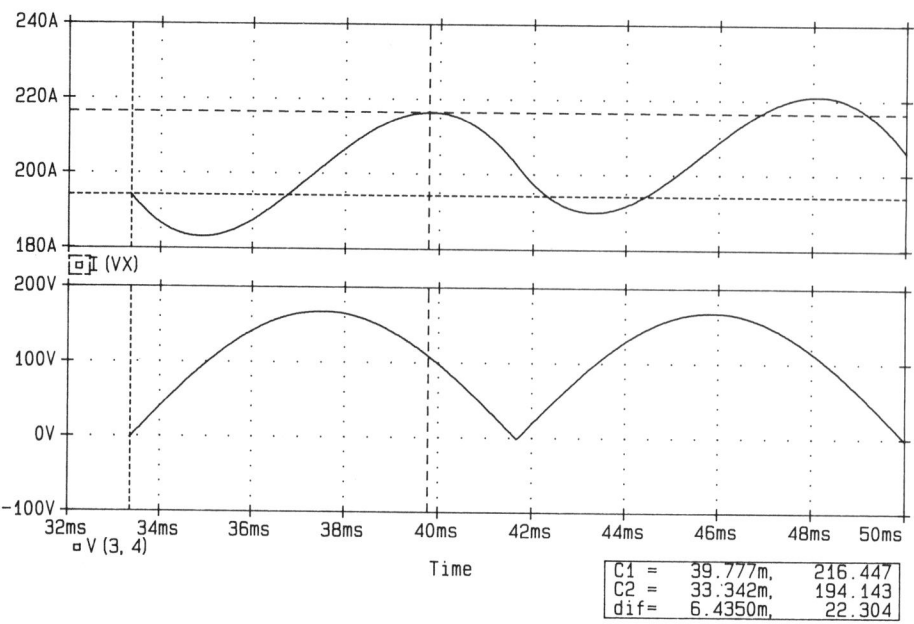

C1 = 39.777m,   216.447
C2 = 33.342m,   194.143
dif= 6.4350m,   22.304

Figure 7-9 Plots for Example 7-2.

Rms harmonic current, $I_{h(rms)} = I_{1(rms)} \times \text{THD} = 183.49 \times 0.4226 = 77.54$ A

Rms input current $I_s = [I_{in(dc)}^2 + I_{1(rms)}^2 + I_{h(rms)}^2]^{1/2}$
$$= (2.56^2 + 183.49^2 + 77.54^2)^{1/2} = 199.22 \text{ A}$$

Displacement angle, $\phi_1 = -3.22$

Displacement factor, $\text{DF} = \cos \phi_1 = \cos(-3.22) = 0.998$ (lagging)

Thus the input power factor is

$$\text{PF} = \frac{I_{1(rms)}}{I_s} \cos \phi_1 = \frac{183.49}{199.22} \times 0.998 = 0.9192 \quad \text{(lagging)}$$

Assuming that $I_{in(dc)} = 0$, Eq. (7-3) gives the power factor as

$$\text{PF} = \frac{1}{(1 + 0.4226^2)^{1/2}} \times 0.9981 = 0.9193 \quad \text{(lagging)}$$

Example 7-3

A single-phase bridge rectifier with an *LC* filter is shown in Fig. 7-10. The sinusoidal input voltage has a peak of 169.7 V, 60 Hz. The load inductance *L* is 10 mH, and the load resistance *R* is 40 Ω. The filter inductance L_e is 30.83 mH, and filter capacitance C_e is 326 μF. Use PSpice (a) to plot the instantaneous output voltage v_o and the load current i_o, (b) to calculate the Fourier coefficients of the output voltage,

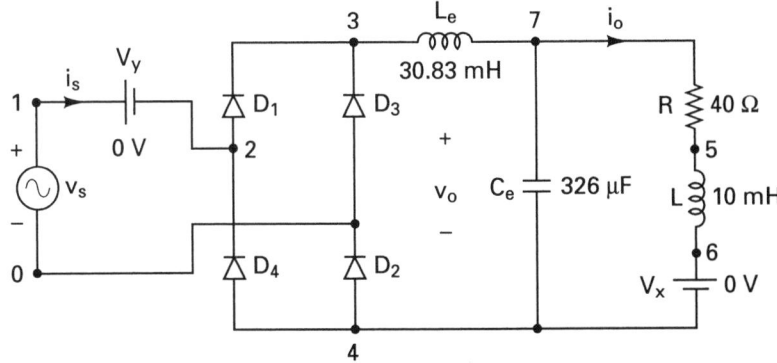

Figure 7-10 Single-phase bridge rectifier with load filter for PSpice simulation.

and (c) to calculate the Fourier coefficients of the input current and input power factor.

Solution Peak voltage V_m = 169.7 V, and f = 60 Hz. The list of the circuit file is as follows:

■ ■

Example 7-3   Single-phase bridge rectifier with *RL* load

```
SOURCE    ■ VS    1    0    SIN (0    169.7V    60HZ)
CIRCUIT   ■ ■ LE  3    7    30.83MH
            CE    7    4    326UF
            R     7    5    40
            L     5    6    10MH
            VX    6    4    DC    0V  ; Voltage source to measure the output current
            VY    1    2    DC    0V  ; Voltage source to measure the input current
            D1    2    3    DMOD
            D3    0    3    DMOD
            D2    4    0    DMOD
            D4    4    2    DMOD
            .MODEL  DMOD  D(IS=2.22E-15  BV=1200V  IBV=13E-3  CJO=2PF  TT=1US)
ANALYSIS  ■ ■ ■ .TRAN  10US  50MS  33.3333MS  10US ; Transient analysis
            .FOUR  120HZ  V(7,4)              ; Fourier analysis of output
            *                                                        voltage
            .OPTIONS ABSTOL = 1.0N RELTOL = .01 VNTOL = 1.0M ITL5=10000 ;
            *                                                     Convergence
          .END
```

■ ■

(a) The PSpice plots of the instantaneous output voltage V(7,4) and load current I(VX) are shown in Fig. 7-11. The *LC* filter smooths the load voltage and reduces ripples.

(b) The Fourier coefficients of the output voltage are:

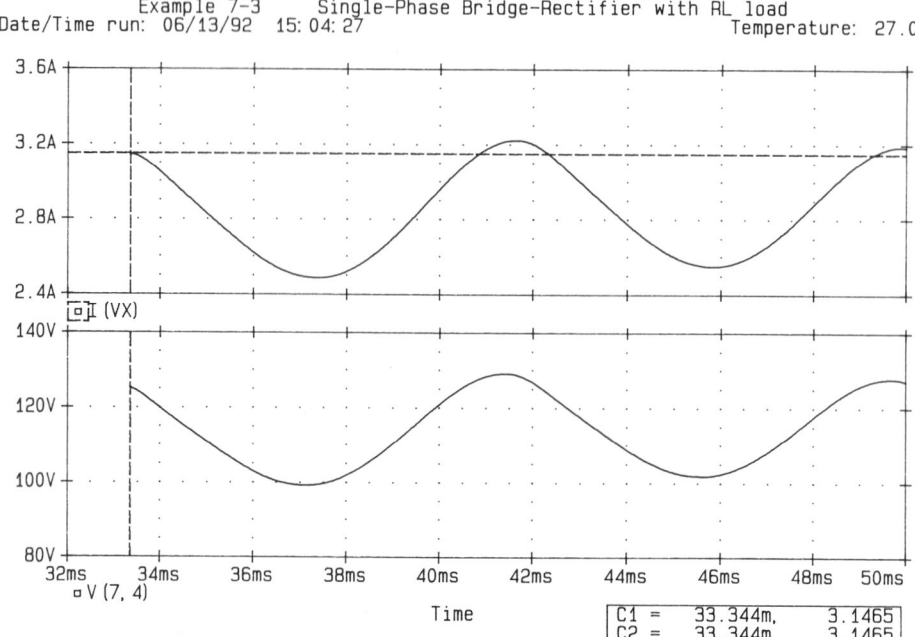

Example 7-3   Single-Phase Bridge-Rectifier with RL load
Date/Time run: 06/13/92 15:04:27   Temperature: 27.0

Figure 7-11 Plots for Example 7-3.

FOURIER COMPONENTS OF TRANSIENT RESPONSE V(7,4)

DC COMPONENT =   1.143072E+02

HARMONIC NO	FREQUENCY (HZ)	FOURIER COMPONENT	NORMALIZED COMPONENT	PHASE (DEG)	NORMALIZED PHASE (DEG)
1	1.200E+02	1.306E+01	1.000E+00	1.034E+02	0.000E+00
2	2.400E+02	6.509E-01	4.983E-02	1.225E+02	1.907E+01
3	3.600E+02	2.315E-01	1.772E-02	9.039E+01	-1.305E+01
4	4.800E+02	1.617E-01	1.238E-02	4.774E+01	-5.570E+01
5	6.000E+02	1.316E-01	1.007E-02	2.218E+01	-8.126E+01
6	7.200E+02	1.050E-01	8.039E-03	8.698E+00	-9.474E+01
7	8.400E+02	8.482E-02	6.494E-03	2.760E+00	-1.007E+02
8	9.600E+02	7.149E-02	5.473E-03	5.647E-02	-1.034E+02
9	1.080E+03	6.137E-02	4.699E-03	-2.062E+00	-1.055E+02

TOTAL HARMONIC DISTORTION =   5.666466E+00 PERCENT

(c) The input current, which is the same as the current through voltage source VY, is equal to I(VY). After running PSpice to obtain the Fourier series of the input current, using the command

```
.FOUR   60HZ   I(VY)                ; Fourier analysis of input current
```

we get

FOURIER COMPONENTS OF TRANSIENT RESPONSE I(VY)

DC COMPONENT = -2.229026E+00

HARMONIC NO	FREQUENCY (HZ)	FOURIER COMPONENT	NORMALIZED COMPONENT	PHASE (DEG)	NORMALIZED PHASE (DEG)
1	6.000E+01	2.555E+02	1.000E+00	-3.492E+00	0.000E+00
2	1.200E+02	1.146E+00	4.486E-03	1.192E+02	1.227E+02
3	1.800E+02	8.372E+01	3.277E-01	3.125E+00	6.617E+00
4	2.400E+02	1.092E+00	4.273E-03	1.044E+02	1.079E+02
5	3.000E+02	5.024E+01	1.967E-01	1.036E+00	4.528E+00
6	3.600E+02	1.082E+00	4.237E-03	9.819E+01	1.017E+02
7	4.200E+02	3.586E+01	1.404E-01	-1.011E-01	3.391E+00
8	4.800E+02	1.070E+00	4.187E-03	9.446E+01	9.795E+01
9	5.400E+02	2.788E+01	1.091E-01	-9.085E-01	2.583E+00

TOTAL HARMONIC DISTORTION = 4.216000E+01 PERCENT

Total harmonic distortion of input current, THD = 42.16% = 0.4216

Displacement angle, $\phi_1 = -3.492°$

Displacement factor, DF $= \cos \phi_1 = \cos(-3.492) = 0.9981$ (lagging)

Neglecting the dc input current $I_{in(dc)} = -2.229$ A, which is small relative to the fundamental component, we can find power factor from Eq. (7-3) as

$$PF = \frac{1}{(1 + 0.4216^2)^{1/12}} \times 0.9981 = 0.9197 \quad \text{(lagging)}$$

Example 7-4

A three-phase bridge rectifier is shown in Fig. 7-12. The rectifier is supplied from a balanced three-phase balanced supply whose per phase voltage has a peak of 169.7 V, 60 Hz. The load inductance L is 6.5 mH, and the load resistance R is 0.5 Ω. Use PSpice (a) to plot the instantaneous output voltage v_o and line (phase) current i_a, (b) to plot the rms and average currents of diode D_1, (c) to plot the average output power, and (d) to calculate the Fourier coefficients of the input current and the input power factor.

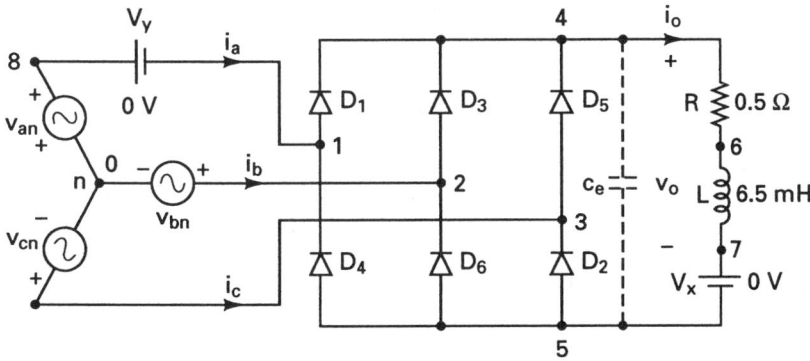

Figure 7-12   Three-phase bridge rectifier for PSpice simulation.

Solution Peak voltage per phase $V_m = 169.7$ V, and $f = 60$ Hz. The list of the circuit file is as follows:

■■■

Example 7-4   Three-phase bridge rectifier

```
SOURCE   ■ Van    8   0   SIN(0   169.7V   60HZ)
           Vbn    2   0   SIN(0   169.7V   60HZ   0   0   120DEG)
           Vcn    3   0   SIN(0   169.7V   60HZ   0   0   240DEG)
CIRCUIT  ■■ CE    4   5   1UF         ; Small capacitance to aid convergence
           R      4   6   0.5
           L      6   7   6.5MH
           VX     7   5   DC     0V   ; Voltage source to measure the output current
           VY     8   1   DC     0V   ; Voltage source to measure the input current
           D1     1   4   DMOD
           D3     2   4   DMOD
           D5     3   4   DMOD
           D2     5   3   DMOD
           D6     5   2   DMOD
           D4     5   1   DMOD
           .MODEL  DMOD  D(IS=2.22E-15  BV=1200V  IBV=13E-3  CJO=2PF  TT=1US)
ANALYSIS ■■■ .TRAN  10US  33.3333MS  0  10US ; Transient analysis
           .FOUR   60HZ   I(VY)              ; Fourier analysis of line current
           .PROBE                            ; Graphics post-processor
           .OPTIONS ABSTOL = 1.0N RELTOL = 1.0M VNTOL = 1.0M ITL5=10000 ;
           *                                                      Convergence
         .END
```

■■■

(a) The PSpice plots of the instantaneous output voltage V(4,5) and line current I(VY) are shown in Fig. 7-13. As expected, there are six output pulses over the period of the input voltage. The input current is rectangular.

(b) The plots of the instantaneous rms and average currents of diode D_1 are shown in Fig. 7-14. Averaging over a small time at the very beginning yields a large value. But after a sufficiently long time, it gives the true average or rms values.

(c) The plot of the instantaneous average output power is shown in Fig. 7-15. The plots of average current, rms current, and average power will come to a steady-state fixed value if the transient analysis is continued for a longer period.

(d) The Fourier coefficients of the input current are:

```
FOURIER COMPONENTS OF TRANSIENT RESPONSE I(VY)
DC COMPONENT =   2.066274E-01
```

HARMONIC NO	FREQUENCY (HZ)	FOURIER COMPONENT	NORMALIZED COMPONENT	PHASE (DEG)	NORMALIZED PHASE (DEG)
1	6.000E+01	6.161E+02	1.000E+00	-8.420E-03	0.000E+00
2	1.200E+02	1.182E+00	1.919E-03	-1.692E+02	-1.692E+02
3	1.800E+02	9.265E-01	1.504E-03	-6.353E+00	-6.345E+00
4	2.400E+02	1.219E+00	1.979E-03	-1.767E+02	-1.767E+02
5	3.000E+02	1.227E+02	1.991E-01	1.797E+02	1.797E+02
6	3.600E+02	6.153E-02	9.987E-05	1.145E+02	1.145E+02
7	4.200E+02	8.839E+01	1.435E-01	-1.797E+02	-1.797E+02

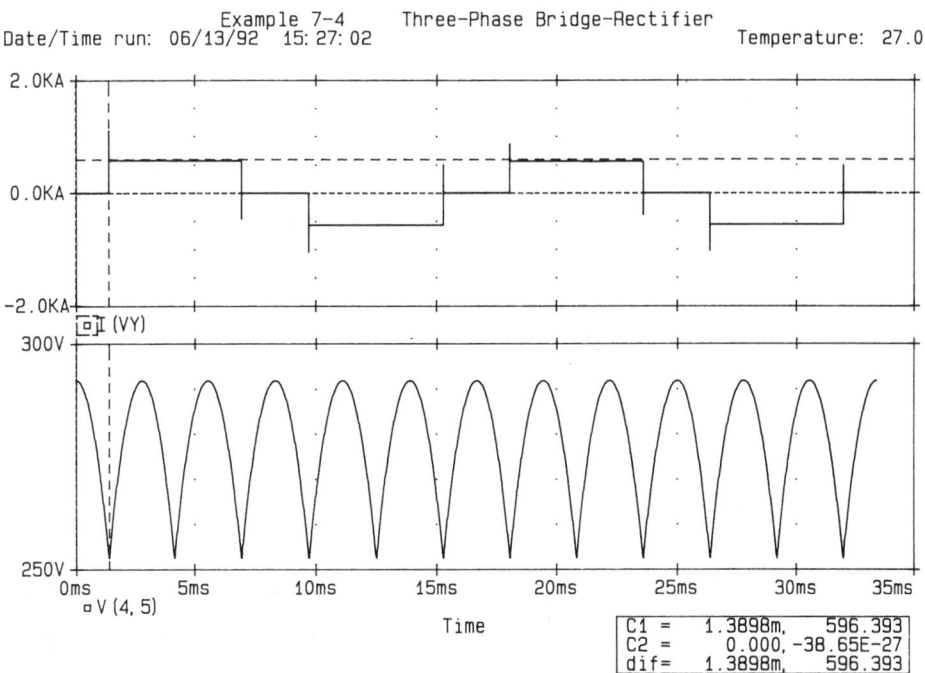

Figure 7-13 Plots of output voltage V(4.5) and line current I(VY).

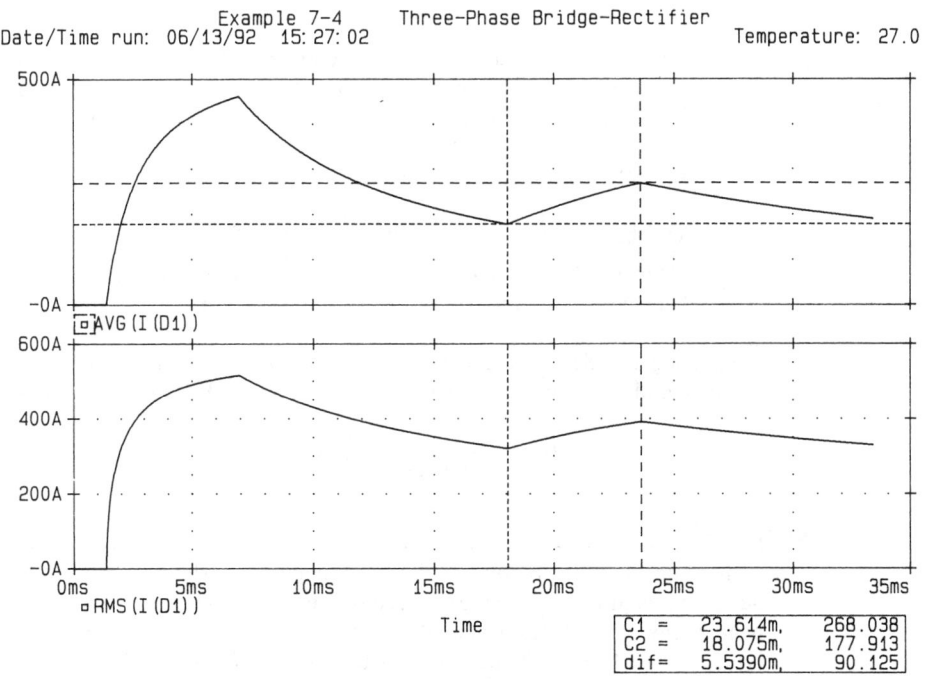

Figure 7-14 Plots of rms and average currents of diode D_1.

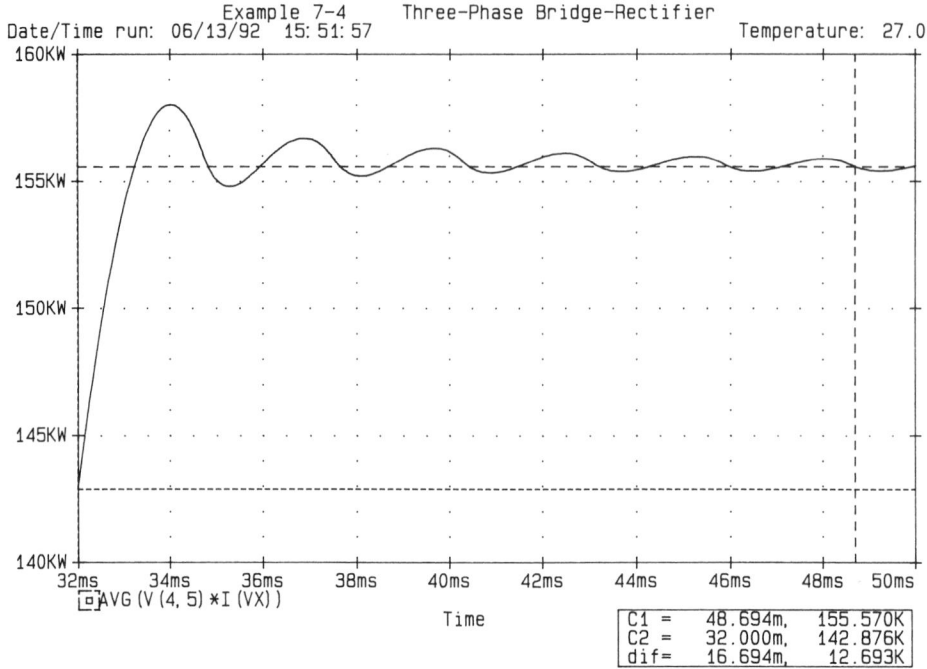

Figure 7-15 Plot of average output power.

HARMONIC NO	FREQUENCY (HZ)	FOURIER COMPONENT	NORMALIZED COMPONENT	PHASE (DEG)	NORMALIZED PHASE (DEG)
8	4.800E+02	1.196E+00	1.941E-03	3.666E+00	3.675E+00
9	5.400E+02	9.152E-01	1.485E-03	1.779E+02	1.779E+02

TOTAL HARMONIC DISTORTION = 2.454718E+01 PERCENT

Total harmonic distortion of input current, THD = 24.55% = 0.2455

Displacement angle, $\phi_1 \approx 0°$

Displacement factor, DF = $\cos \phi_1 = \cos(0) \approx 1$

Neglecting the dc input current $I_{in(dc)} = 0.207$ A, which is small relative to the fundamental component, we can find power factor PF from Eq. (7-3) as

$$\text{PF} = \frac{1}{(1 + 0.2455^2)^{1/2}} \times 1 = 0.971 \quad \text{(lagging)}$$

Example 7-5

A three-phase bridge rectifier with line inductances is shown in Fig. 7-16. The rectifier is supplied from a balanced three-phase supply whose per phase voltage has a peak of 169.7 V, 60 Hz. The load inductance L is 6.5 mH, and the load resistance R is 0.5 Ω. The line inductances are equal $L_1 = L_2 = L_3 = 0.5$ mH. Use PSpice to plot the instantaneous line voltages v_{ac} and v_{bc}, and the instantaneous currents through diodes D_1, D_3, and D_5.

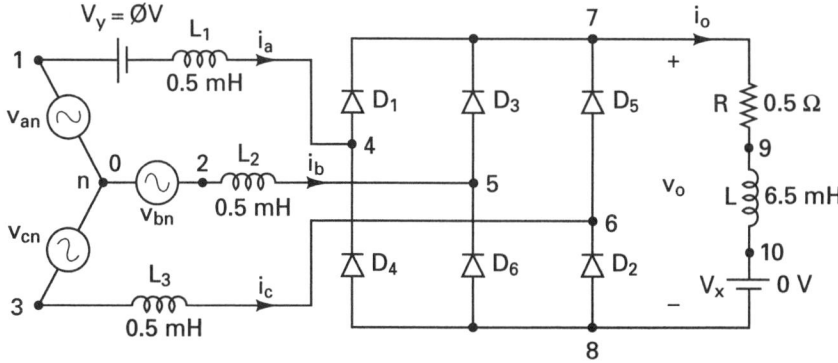

Figure 7-16 Three-phase bridge rectifier with source inductance for PSpice simulation.

Solution Peak phase voltages $V_m = 169.7$ V, and $f = 60$ Hz. The list of the circuit file is as follows:

■ ■

Example 7-5   Three-phase bridge rectifier with source inductances

```
          Van   1    0    SIN(0    169.7V    60HZ)
          L1    1    4    0.5MH
          Vbn   2    0    SIN(0    169.7V    60HZ    0    0    120DEG)
          L2    2    5    0.5MH
          Vcn   3    0    SIN(0    169.7V    60HZ    0    0    240DEG)
          L3    3    6    0.5MH
          R     7    9    0.5
          L     9    10   6.5MH
          VX    10   8    DC    0V    ; Voltage source to measure the output current
          D1    4    7    DMOD
          D3    5    7    DMOD
          D5    6    7    DMOD
          D2    8    6    DMOD
          D6    8    5    DMOD
          D4    8    4    DMOD
          .MODEL    DMOD    D(IS=2.22E-15   BV=1200V   IBV=13E-3   CJO=2PF   TT=1US)
          .TRAN    10US    50MS    33.3333MS    10US    ; Transient analysis
          .PROBE                                       ; Graphics post-processor
          .OPTIONS ABSTOL = 1.0N RELTOL = 0.01 VNTOL = 1.0M ITL5=10000 ; Convergence
          .END
```

SOURCE
CIRCUIT

ANALYSIS

■ ■

The PSpice plots of the instantaneous currents through diode D_1, I(D1), through diode D_3, I(D3), and through diode D_5, I(D5), and the line voltages V(1,3) and V(2,3) are shown in Fig. 7-17. Due to the source inductances, a commutation interval exists. During this interval, the current through the incoming diode rises and that through the outgoing diode falls. The sum of these currents must equal the load current.

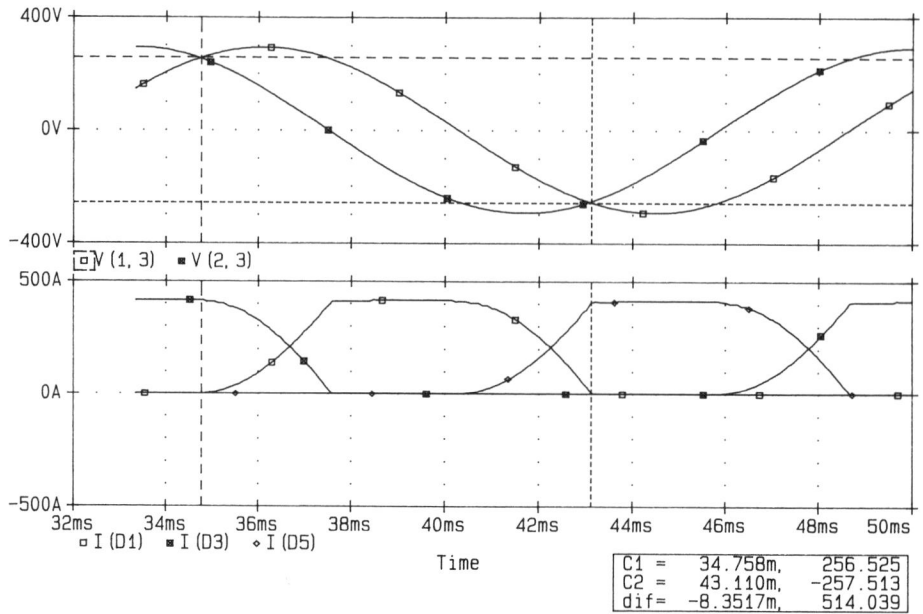

Example 7-5   Three-Phase Bridge-Rectifier with Source Inductances
Date/Time run: 06/13/92 15:59:05   Temperature: 27.0

□◇ V (1, 3)   ■ V (2, 3)

□ I (D1)   ■ I (D3)   ◇ I (D5)

Time

C1 =	34.758m,	256.525
C2 =	43.110m,	-257.513
dif=	-8.3517m,	514.039

Figure 7-17   Plots for Example 7-5.

7-7 LABORATORY EXPERIMENTS

It is possible to develop many experiments for demonstrating the operation and characteristics of diode rectifiers. The following three experiments are suggested:

Single-phase full-wave center-tapped rectifier
Single-phase bridge rectifier
Three-phase bridge rectifier

7-7.1 Experiment DR-1

SINGLE-PHASE FULL-WAVE CENTER-TAPPED RECTIFIER

Objective   The objective is to study the operation and characteristics of a single-phase full-wave rectifier under various load conditions.

Applications   A single-phase full-wave rectifier is used as an input stage in many applications (e.g., power supplies).

Textbook   See Ref. 1, Secs. 3-8 and 3-9.

Apparatus   1. Two diodes with ratings of at least 50 A and 400 V, mounted on heat sinks
2. One center-tapped (step-down) transformer
3. An RL load
4. One dual-beam oscilloscope with floating or isolating probes
5. Ac and dc voltmeters and ammeters and one noninductive shunt

Warning   Before making any circuit connection, switch the ac power OFF. **Do not** switch the power ON unless the circuit is checked and approved by your lab instructor. **Do not** touch the diode heat sinks, which are connected to live terminals.

Experimental procedure   1. Set up the circuit as shown in Fig. 7-18. Use the load resistance R only.
2. Connect the measuring instruments as required.
3. Observe and record the waveforms of the load voltage v_o and the load current i_o.
4. Measure the average load voltage $V_{o(dc)}$, the rms load voltage $V_{o(rms)}$, the average load current $I_{o(dc)}$, the rms load current $I_{o(rms)}$, the rms input current $I_{s(rms)}$, the rms input voltage $V_{s(rms)}$, and the average load power P_L.
5. Repeat steps 2 to 4 with the load inductance L only.
6. Repeat steps 2 to 4 with both load resistance R and load inductance L.

Report   1. Present all recorded waveforms and discuss all significant points.
2. Compare the waveforms generated by SPICE with the experimental results, and comment.

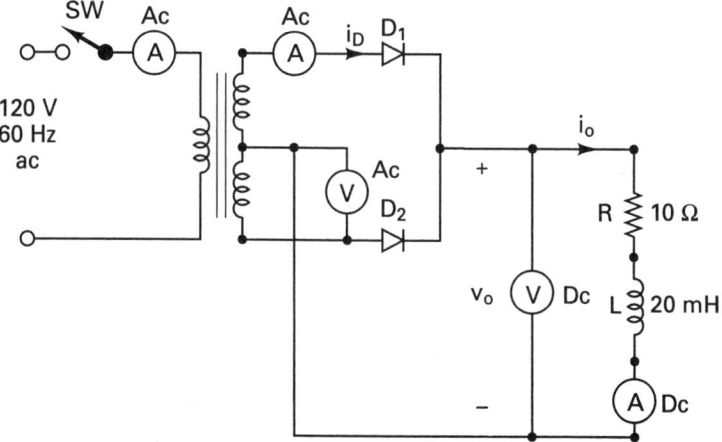

Figure 7-18   Single-phase full-wave rectifier.

3. Compare the experimental results with the results predicted.
4. Discuss the advantages and disadvantages of this type of rectifier.

7-7.2 Experiment DR-2

SINGLE-PHASE BRIDGE RECTIFIER

Objective The objective is to study the operation and characteristics of single-phase bridge-rectifier under various load conditions.

Applications A single-phase bridge rectifier is used as an input stage in many applications (e.g., power supplies and variable-speed ac/dc motor drives).

Textbook See Ref. 1, Secs. 3-8 and 3-9.

Apparatus Same as Experiment DR-1, except that four diodes are required.

Warning See Experiment DR-1.

Experimental procedure Set up the circuit as shown in Fig. 7-19 and follow the steps for Experiment DR-1.

Report Repeat the steps in Experiment DR-1.

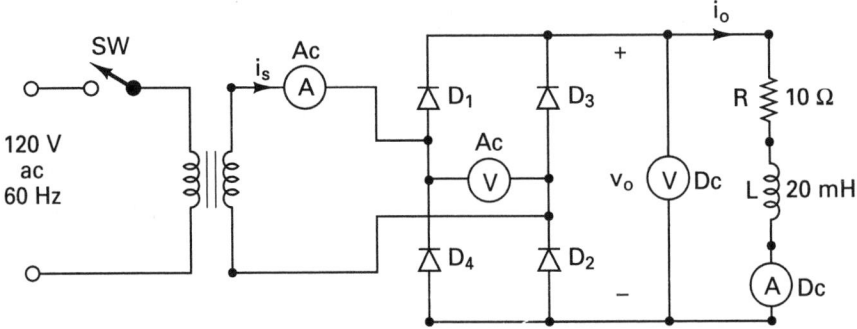

Figure 7-19 Single-phase bridge rectifier.

7-7.3 Experiment DR-3

THREE-PHASE BRIDGE RECTIFIER

Objective   The objective is to study the operation and characteristics of a three-phase bridge rectifier under various load conditions.

Applications   A three-phase bridge rectifier is used an an input stage in many applications (e.g., variable-speed ac motor drives).

Textbook   See Ref. 1, Secs. 3-11 and 3-12.

Apparatus   Same as Experiment DR-1, except that six diodes are required.

Warning   See Experiment DR-1.

Experimental procedure   Set up the circuit as shown in Fig. 7-20 and follow the steps for Experiment DR-1.

Report   Repeat the steps in Experiment DR-1.

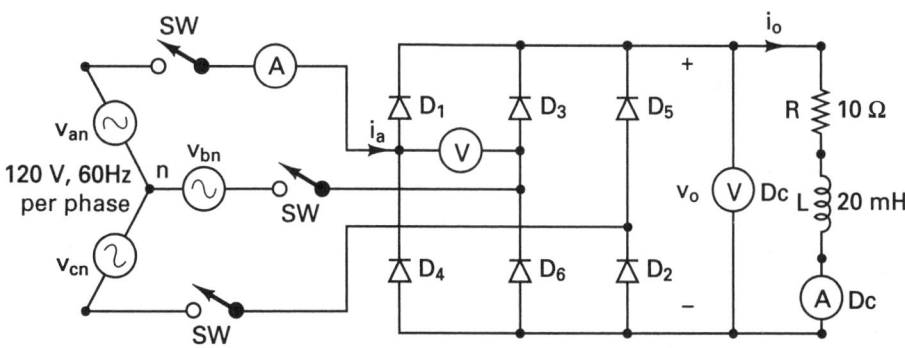

Figure 7-20   Three-phase bridge rectifier.

SUMMARY

The statements for diodes are:

```
D⟨name⟩   NA  NK  DNAME   [(area) value]
.MODEL   DNAME   TYPE  (P1=V1  P2=V2  P3=V3 ... PN=VN)
```

I(D1)	Instantaneous current through diode D_1
RMS(I(D1))	Rms current of diode D_1
AVG(I(D1))	Average current of diode D_1
V(D1)	Instantaneous anode to cathode voltage of diode D_1

SUGGESTED READING

1. M. H. Rashid, *Power Electronics: Circuits, Devices, and Applications,* 2nd ed. Englewood Cliffs, N.J.: Prentice Hall, 1993.

2. M. H. Rashid, *SPICE for Circuits and Electronics Using PSpice.* Englewood Cliffs, N.J.: Prentice Hall, 1990.

3. N. Roussy et al., *Basics of Power Electronics.* Ste-Foy, Quebec, Canada: Lab-Volt Ltd., 1990.

4. *PSpice Manual.* Irvine, Calif.: MicroSim Corporation, 1992.

5. P. Antognetti, *Power Integrated Circuits.* New York: McGraw-Hill, 1986.

DESIGN PROBLEMS

7-1. Design the single-phase full-wave rectifier of Fig. 7-18 with the following specifications:

Ac supply voltage V_s = 120 V (rms), 60 Hz

Load resistance R = 5 Ω

Load inductance L = 15 mH

Dc output voltage $V_{o(dc)}$ = 24 V

(a) Determine the ratings of all components and devices.
(b) Use SPICE to verify your design.
(c) Provide a cost estimate of the circuit.

7-2. (a) Design an output C filter for the single-phase full-wave rectifier of Problem 7-1. The harmonic content of the load current should be less less than 5% of the value without the filter.
(b) Use SPICE to verify your design in part (a).

7-3. Design the single-phase bridge rectifier of Fig. 7-19 with the following specifications:

Ac supply voltage V_s = 120 V (rms), 60 Hz

Load resistance R = 5 Ω

Load inductance L = 15 mH

Dc output voltage $V_{o(dc)}$ = 48 V

(a) Determine the ratings of all components and devices.
(b) Use SPICE to verify your design.
(c) Provide a cost estimate of the circuit.

7-4. (a) Design an output C filter for the single-phase bridge rectifier of Problem 7-3. The harmonic content of the load current should be less than 5% of the value without the filter.
(b) Use SPICE to verify your design in part (a).

7-5. (a) Design an output C filter for the single-phase bridge rectifier of Problem 7-3. The harmonic content of the load voltage should be less less than 5% of the value without the filter.
(b) Use SPICE to verify your design in part (a).

7-6. The rms input voltage to the single-phase bridge rectifier of Fig. 7-19 is 120 V, 60 Hz, and it has an output LC filter. If the dc output voltage is V_{dc} = 48 V at I_{dc} = 25 A, determine the value of filter inductance L.

7-7. It is required to design the three-phase bridge rectifier of Fig. 7-20 with the following specifications:

Ac supply voltage per phase V_s
$$= 120 \text{ V (rms)}, 60 \text{ Hz}$$

Load resistance $R = 5 \; \Omega$

Load inductance $L = 15 \text{ mH}$

Dc output voltage $V_{o(dc)}$
$$= \text{maximum possible value}$$

(a) Determine the ratings of all components and devices.
(b) Use SPICE to verify your design.
(c) Provide a cost estimate of the circuit.

7-8. (a) Design an output C filter for the three-phase bridge rectifier of Problem 7-7. The harmonic content of the load current should be less than 5% of the value without the filter.
(b) Use SPICE to verify your design in part (a).

7-9. (a) Design an output C filter for the three-phase bridge rectifier of Problem 7-7. The harmonic content of the load voltage should be less than 5% of the value without the filter.
(b) Use SPICE to verify your design in part (a).

Chapter 8

■■■■■■■■■

Controlled Rectifiers

8-1 INTRODUCTION

A thyristor can be turned on by applying a pulse of short duration. Once the thyristor is on, the gate pulse has no effect, and it remains on until its current is reduced to zero. It is a latching device.

8-2 AC THYRISTOR MODEL

There are a number of published ac thyristor models [3–7]. Lauretzen [7] summarizes the various power semiconductor device models for use in circuit simulation. We shall use a very simple model that can be used to obtain the various waveforms of controlled rectifiers. Let us assume that the thyristor shown in Fig. 8-1(a) is operated from an ac supply. This thyristor should exhibit the following characteristics:

1. It should switch to the on-state with the application of a small positive gate voltage, provided that the anode-to-cathode voltage is positive.
2. It should remain in the on-state as long as the anode current flows.
3. It should switch to the off-state when the anode current goes through zero in the negative direction.

The switching action of the thyristor can be modeled by a voltage-controlled switch and a polynomial current source [3]. This is shown in Fig. 8-1(b). The turn-on process can be explained by the following steps:

1. For a positive gate voltage V_g between nodes 3 and 2, the gate current is $I_g =$ I(VX) = V_g/R_G.

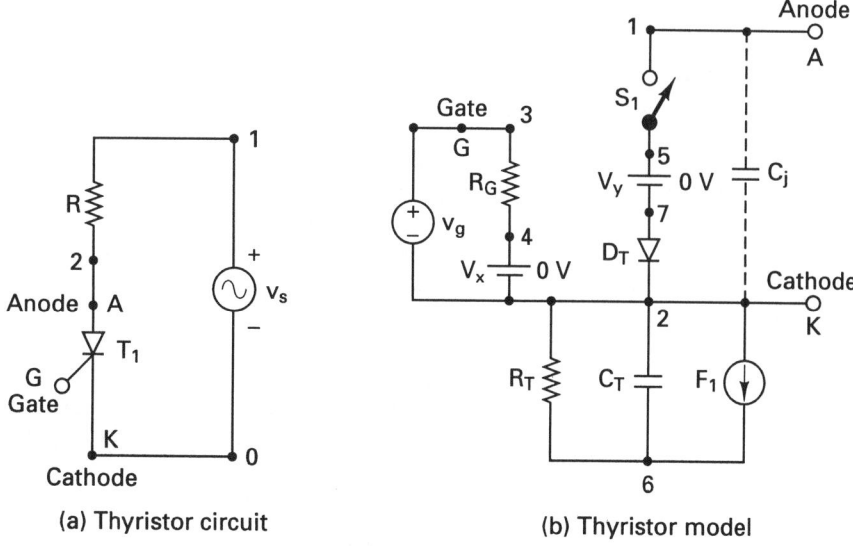

Figure 8-1 Ac thyristor model.

(a) Thyristor circuit

(b) Thyristor model

2. The gate current I_g activates the current-controlled current source F_1 and produces a current of value $F_g = P_1I_g = P_1 \times \text{I(VX)}$, such that $F_1 = F_g + F_a$.

3. The current source F_g produces a rapidly rising voltage V_R across resistance R_T.

4. As the voltage V_R increases above zero, the resistance R_S of the voltage-controlled switch S_1 decreases from R_{OFF} toward R_{ON}.

5. As the switch resistance R_S decreases, the anode current $I_a = \text{I(VY)}$ increases, provided that the anode-to-cathode voltage is positive. This increasing anode current I_a produces a current $F_a = P_2I_a = P_2 \times \text{I(VY)}$. This causes an increased value of voltage V_R.

6. This then produces a regenerative condition with the switch rapidly being driven into low resistance (the on-state). The switch remains on if the gate voltage V_g is removed.

7. The anode current I_a continues to flow as long as it is positive and the switch remains in the on-state.

During turn-off, the gate current is off and $I_g = 0$. That is, $I_g = 0$ and $F_g = 0$, $F_1 = F_g + F_a = F_a$. The turn-off operation can be explained by the following steps:

1. As the anode current I_a goes negative, the current F_1 reverses provided that the gate voltage V_g is no longer present.

2. With a negative F_1, the capacitor C_T discharges through the current source F_1 and the resistance R_T.

---

3. With the fall of voltage V_R to a low level, the resistance R_S of switch S_1 increases from a low (R_{ON}) value to a high (R_{OFF}) value.

4. This is, again, a regenerative condition with the switch resistance being driven rapidly to an R_{OFF} value as the voltage V_R becomes zero.

This model works well with a converter circuit in which the thyristor current falls to zero itself: for example, in half-wave controlled rectifiers and ac voltage controllers. But in full-wave converters with a continuous load current, the current of a thyristor is diverted to another thyristor, and this model may not give the true output. This problem can be remedied by adding diode D_T as shown in Fig. 8-1(b). The diode prevents reverse current flow through the thyristor resulting from the firing of another thyristor in the circuit.

Let us consider the thyristor whose data sheets are shown in Fig. 8-2. Suitable values of the model parameters can be chosen to satisfy the characteristics of a particular thyristor by the following steps:

1. The switch parameters V_{ON} and V_{OFF} can be chosen arbitrarily. Let $V_{ON} = 1$ V and $V_{OFF} = 0$ V.

2. R_T can also be chosen arbitrarily. Let $R_T = 1\ \Omega$.

3. R_{ON} should be chosen to model the on-state resistance of the thyristor so that $R_{ON} = V_{TM}/I_{TM}$ at 25°C. From the on-state characteristic of the data sheet, $V_{TM} = 2$ V at $I_{TM} = 190$ A. Thus $R_{ON} = 2/190 = 0.0105\ \Omega$.

4. R_{OFF} should be chosen to model the off-state resistance of the thyristor so that $R_{OFF} = V_{RRM}/I_{DRM}$ at 25°C. From the data sheet, $V_{RRM} = 1200$ V at $I_{DRM} = 10$ mA. Thus $R_{OFF} = 1200/(10 \times 10^{-3}) = 120$ kΩ.

5. The switch resistance R_s can be found from

$$R_S = R_{ON} \times \frac{R_{OFF}}{R_{ON}} \left[0.5 + 2\left(V_R - 0.5\,\frac{V_{ON} + V_{OFF}}{V_{ON} - V_{OFF}}\right)^3 \right.$$
$$\left. - 1.5\left(V_R - 0.5\,\frac{V_{ON} + V_{OFF}}{V_{ON} - V_{OFF}}\right) \right] \tag{8-1}$$

For $V_R = 0$, $R_S = R_{OFF}$, and for $V_R = 1$, $R_S = R_{ON}$. As V_R varies from 0 to 1 V, the switch resistance R_S changes from R_{OFF} to R_{ON}.

6. The value of P_2 must be such that the device turns off (with zero gate current $I_g = 0$) at the maximum anode-to-cathode voltage V_{DRM}. If V_{R1} is 15% of V_R, that is, $V_{R1} = 0.15 V_R$, the switch will essentially be off. The voltage V_{R1} due to the anode current only is given by

$$V_{R1} < P_2 I_a = P_2\,\frac{V_{DRM}}{R_{S1}}$$
$$P_2 > \frac{V_{R1} R_{S1}}{V_{DRM}} \tag{8-2}$$

At the latching current I_L, the switch must be fully on, $V_R = 1$ V. That is,

S18CF SERIES
1200-1000 VOLTS RANGE
STANDARD TURN-OFF TIME 16 μs
110 AMP RMS, CENTER AMPLIFYING GATE
INVERTER TYPE STUD MOUNTED SCRs

VOLTAGE RATINGS

VOLTAGE CODE [1]	V_{RRM}, V_{DRM} – (V) Max. rep. peak reverse and off-state voltage	V_{RSM} – (V) Max. non-rep. peak reverse voltage $t_p \leq 5$ms	NOTES
	T_J = -40° to 125°C	T_J = 25° to 125°C	
12	1200	1300	Gate open
10	1000	1100	

MAXIMUM ALLOWABLE RATINGS

PARAMETER		VALUE	UNITS	NOTES
T_J	Junction temperature	-40 to 125	°C	
T_{stg}	Storage temperature	-40 to 150	°C	
$I_{T(AV)}$	Max. av. current	70	A	180° half sine wave
	@ Max. T_C	85	°C	
$I_{T(RMS)}$	Max. RMS current	110	A	
I_{TSM}	Max. peak non-repetitive surge current	1910		50Hz half cycle sine wave — Initial T_J = 125°C, rated V_{RRM} applied after surge.
		2000		60Hz half cycle sine wave
		2270	A	50Hz half cycle sine wave — Initial T_J = 125°C, no voltage applied after surge.
		2380		60Hz half cycle sine wave
I^2t	Max. I^2t capability	18		t = 10ms Initial T_J = 125°C, rated V_{RRM} applied after surge.
		17		t = 8.3ms
		26	kA²s	t = 10ms Initial T_J = 125°C, no voltage applied after surge.
		24		t = 8.3ms
$I^2\sqrt{t}$	Max. $I^2\sqrt{t}$ capability	258	kA²√s	Initial T_J = 125°C, no voltage applied after surge. I^2t for time t_x = $I^2\sqrt{t} \cdot \sqrt{t_x}$, 0.1 $\leq t_x \leq$ 10ms.
di/dt	Max. non-repetitive rate-of-rise of current	800	A/μs	T_J = 125°C, V_D = V_{DRM}, I_{TH} = 1600A. Gate pulse: 20V, 20Ω, 10μs, 0.5μs rise time. Max. repetitive di/dt is approximately 40% of non-repetitive value.
P_{GM}	Max. peak gate power	10	W	$t_p \leq 5$ms
$P_{G(AV)}$	Max. av. gate power	2	W	
$+I_{GM}$	Max. peak gate current	3	A	$t_p \leq 5$ms
$-V_{GM}$	Max. peak negative gate voltage	15	V	
T	Mounting torque	15.5(137) ± 10%	N•m	Non-lubricated threads
		14(120) ± 10%	(lbf-in)	Lubricated threads

[1] To complete the part number, refer to the Ordering Information table.

Figure 8-2 Data sheets for IR thyristors type S18. (Courtesy of International Rectifier.)

CHARACTERISTICS

PARAMETER		MIN.	TYP.	MAX.	UNITS	TEST CONDITIONS
V_{TM}	Peak on-state voltage	—	1.95	2.06	V	Initial T_J = 25°C, 50-60Hz half sine, I_{peak} = 220A.
$V_{T(TO)1}$	Low-level threshold	—	—	1.28	V	T_J = 125°C
$V_{T(TO)2}$	High-level threshold	—	—	1.61		Av. power = $V_{T(TO)} \cdot I_{T(AV)} + r_T \cdot [I_{T(RMS)}]^2$
r_{T1}	Low-level resistance	—	—	3.54	mΩ	Use low level values for
r_{T2}	High-level resistance	—	—	2.27		$I_{TM} \leq \pi$ rated $I_{T(AV)}$
I_L	Latching current	—	270	—	mA	T_C = 25°C, 12V anode. Gate pulse: 10V, 20Ω, 100μs.
I_H	Holding current	—	90	500	mA	T_C = 25°C, 12V anode. Initial I_T = 3A.
t_d	Delay time	—	0.5	1.5	μs	T_C = 25°C, V_D = rated V_{DRM}, 50A resistive load. Gate pulse: 10V, 20Ω, 10μs, 1μs rise time.
t_q	Turn-off time					
	"A" suffix	—	---	16	μs	T_J = 125°C. I_{TM} = 200A, di_R/dt = 10A/μs, V_R = 50V,
	"B" suffix	---	—	20		dv/dt = 200V/μs lin. to 80% rated V_{DRM}. Gate: 0V, 100Ω.
$t_{q(diode)}$	Turn-off time with feedback diode					
	"A" suffix	—	---	20	μs	T_J = 125°C. I_{TM} = 200A, di_R/dt = 10A/μs, V_R = 1V,
	"B" suffix	---	—	25		dv/dt = 600V/μs lin. to 40% rated V_{DRM}. Gate: 0V, 100Ω.
$I_{RM(REC)}$	Recovery current	—	57	—	A	T_J = 125°C, I_{TM} = 400A, di_R/dt = 50A/μs.
Q_{RR}	Recovered charge	—	58	—	μC	
dv/dt	Critical rate-of-rise of off-state voltage	500	700	---	V/μs	T_J = 125°C. Exp. to 100% or lin. Higher dv/dt values to 80% V_{DRM}, gate open. available.
		1000	—	—		T_J = 125°C. Exp. to 67% V_{DRM}, gate open.
I_{RM}, I_{DM}	Peak reverse and off-state current	—	10	20	mA	T_J = 125°C. Rated V_{RRM} and V_{DRM}, gate open.
I_{GT}	DC gate current to trigger	—	---	300	mA	T_C = -40°C. +12V anode-to-cathode. For recommended gate drive see "Gate Characteristics" figure.
		25	50	150		T_C = 25°C
V_{GT}	DC gate voltage to trigger	---	—	3.3	V	T_C = -40°C
		—	1.2	2.5		T_C = 25°C
V_{GD}	DC gate voltage not to trigger	—	---	0.3	V	T_C = 125°C. Max. value which will not trigger with rated V_{DRM} anode-to-cathode.
R_{thJC}	Thermal resistance, junction-to-case	—	—	0.250	°C/W	DC operation
		—	---	0.291	°C/W	180° sine wave
		---	—	0.302	°C/W	120° rectangular wave
R_{thCS}	Thermal resistance, case-to-sink	—	---	0.100	°C/W	Mtg. surface smooth, flat and greased.
wt	Weight	—	100(3.5)	—	g(oz.)	
	Case Style		TO-209AC (TO-94)		JEDEC	

Figure 8-2 *(Continued)*

$P_2 > V_{R1}/I_L$. Thus the value of P_2 must satisfy the condition

$$\frac{V_R}{I_L} < P_2 < \frac{V_{R1}R_{S1}}{V_{DRM}} \tag{8-3}$$

From the data sheet, I_L = 270 mA. That is, (1/270 mA) < P_2 > (0.15 × 120 kΩ/1200) or 3.7 < P_2 > 15. Let us choose P_2 = 11.

S18CF SERIES
1200–1000 VOLTS RANGE

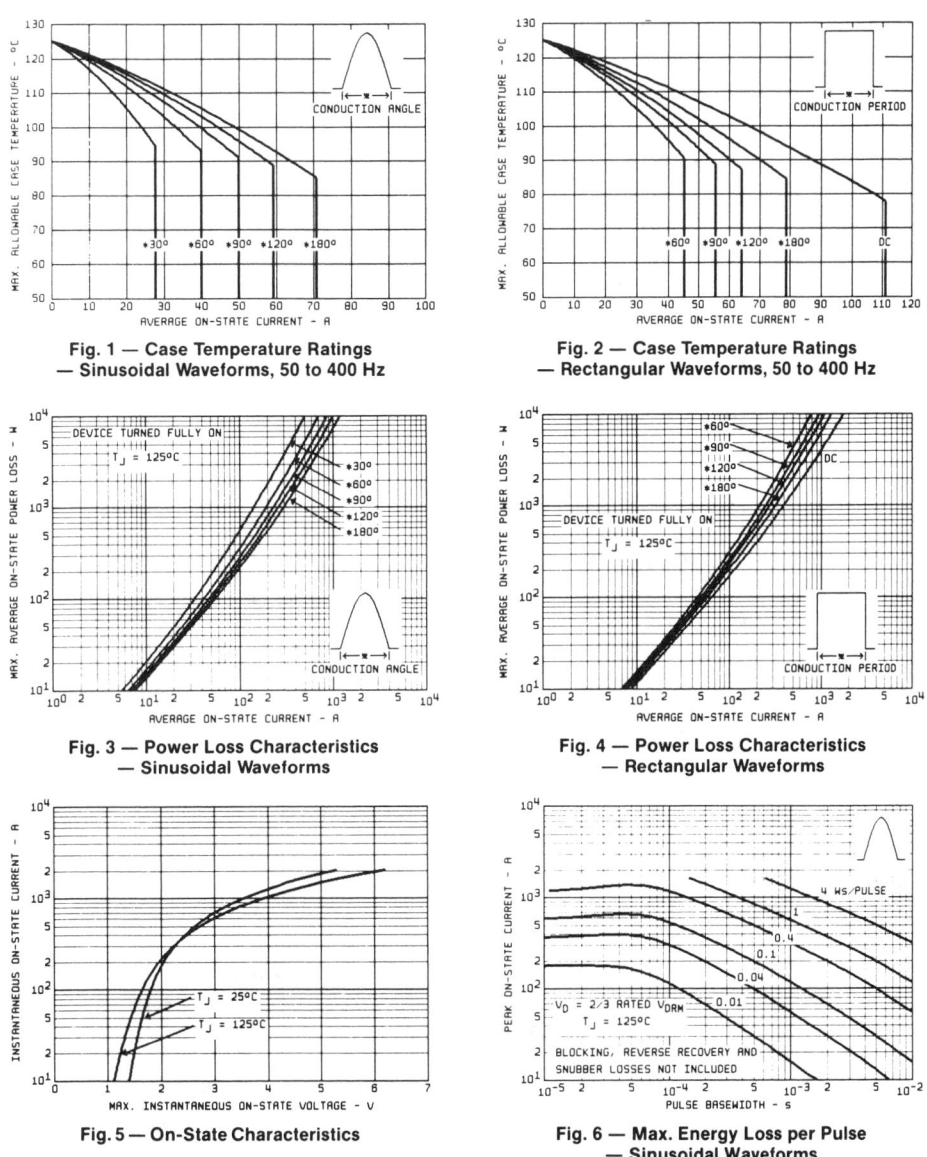

Fig. 1 — Case Temperature Ratings
— Sinusoidal Waveforms, 50 to 400 Hz

Fig. 2 — Case Temperature Ratings
— Rectangular Waveforms, 50 to 400 Hz

Fig. 3 — Power Loss Characteristics
— Sinusoidal Waveforms

Fig. 4 — Power Loss Characteristics
— Rectangular Waveforms

Fig. 5 — On-State Characteristics

Fig. 6 — Max. Energy Loss per Pulse
— Sinusoidal Waveforms

Figure 8-2 (*Continued*)

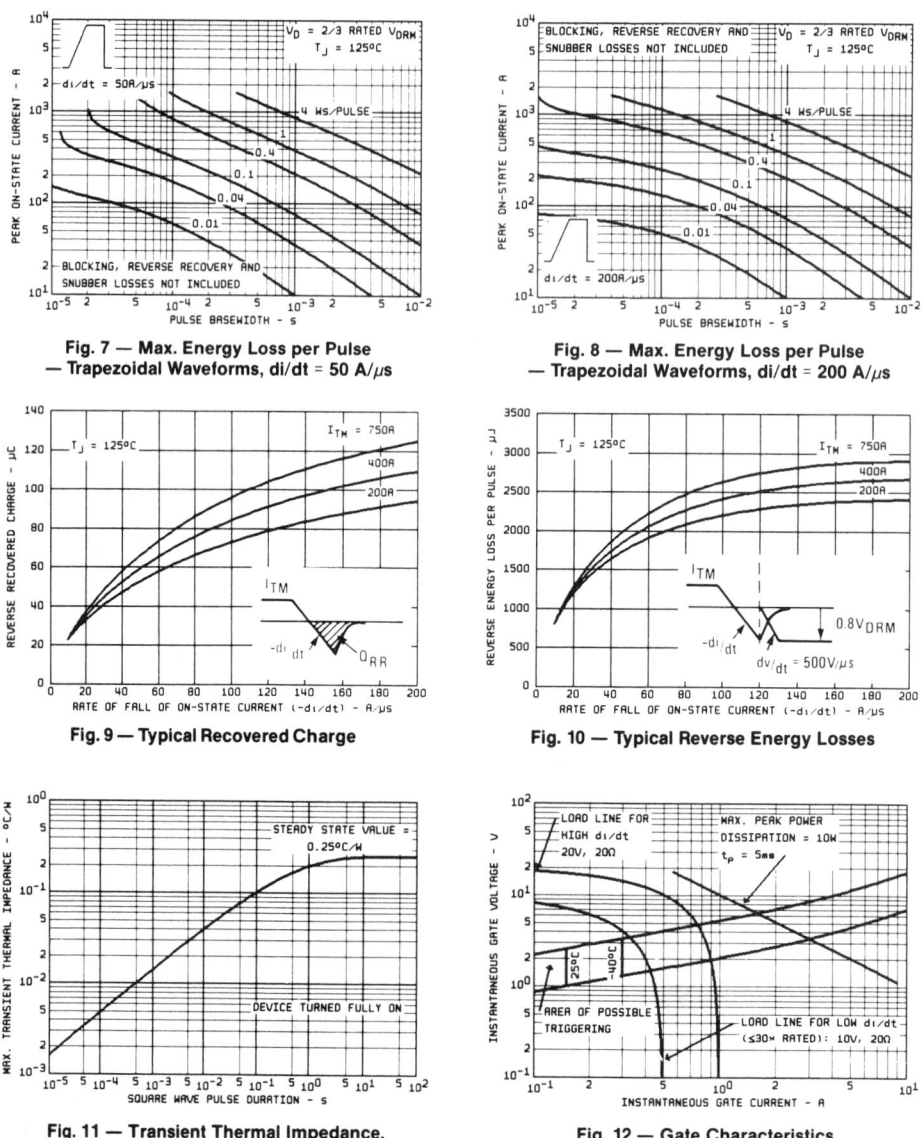

Fig. 7 — Max. Energy Loss per Pulse
— Trapezoidal Waveforms, di/dt = 50 A/μs

Fig. 8 — Max. Energy Loss per Pulse
— Trapezoidal Waveforms, di/dt = 200 A/μs

Fig. 9 — Typical Recovered Charge

Fig. 10 — Typical Reverse Energy Losses

Fig. 11 — Transient Thermal Impedance,
Junction-to-Case

Fig. 12 — Gate Characteristics

Figure 8-2 *(Continued)*

S18CF SERIES
1200-1000 VOLTS RANGE

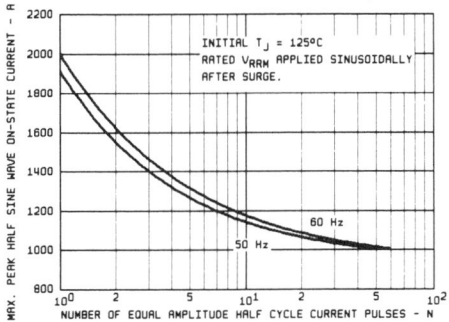

Fig. 13 — Non-Repetitive Surge Current Ratings

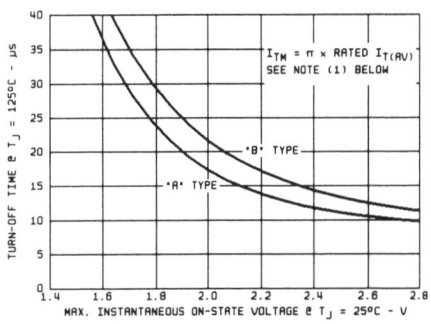

Fig. 14 — Trend for Turn-Off
Time vs. On-State Voltage

(1) These curves are intended as a guideline. To specify
non-standard t_q/V_{TM} contact factory.

ORDERING INFORMATION

TYPE	PACKAGE [1]		FAST	TEMPERATURE		VOLTAGE		TURN-OFF		LEADS & TERMINALS	
	CODE	DESCRIPTION		CODE	MAX. T_J	CODE	V_{DRM}	CODE	MAX. t_q	CODE	DESCRIPTION
S18	C	1/2" stud, ceramic housing.	F	—	125°C	12	1200V	A	16µs	0	Flexible leads, eyelet terminals. Standard in USA. (Fig. 1)
						10	1000V	B	20µs		
										1	Flexible leads, fast-on terminals. Standard in Europe. (Fig. 2)

[1] Other packages are also available:
 - Supplied with flag terminals.
For further details contact factory.

For a device with standard USA case, max T_J = 125°C, V_{DRM} = 1200V, max. t_q = 16µs, order as: S18CF12A0.

Figure 8-2 *(Continued)*

7. The capacitor C_T is introduced primarily to facilitate SPICE convergence. If C_T is too small, the thyristor will go off before the anode current becomes zero. If C_T has a sufficiently large value, the thyristor will continue to conduct beyond the zero crossing of supply voltage, and C_R can be chosen large enough to model turn-off time t_{off} of the thyristor. The time constant $C_T R_T$ should be much smaller than the period T of the supply voltage. This condition is generally satisfied by the relation $C_T R_T \le 0.01 T$. Let us choose $C_T =$ 10 μF.

S18CF SERIES
1200-1000 VOLTS RANGE

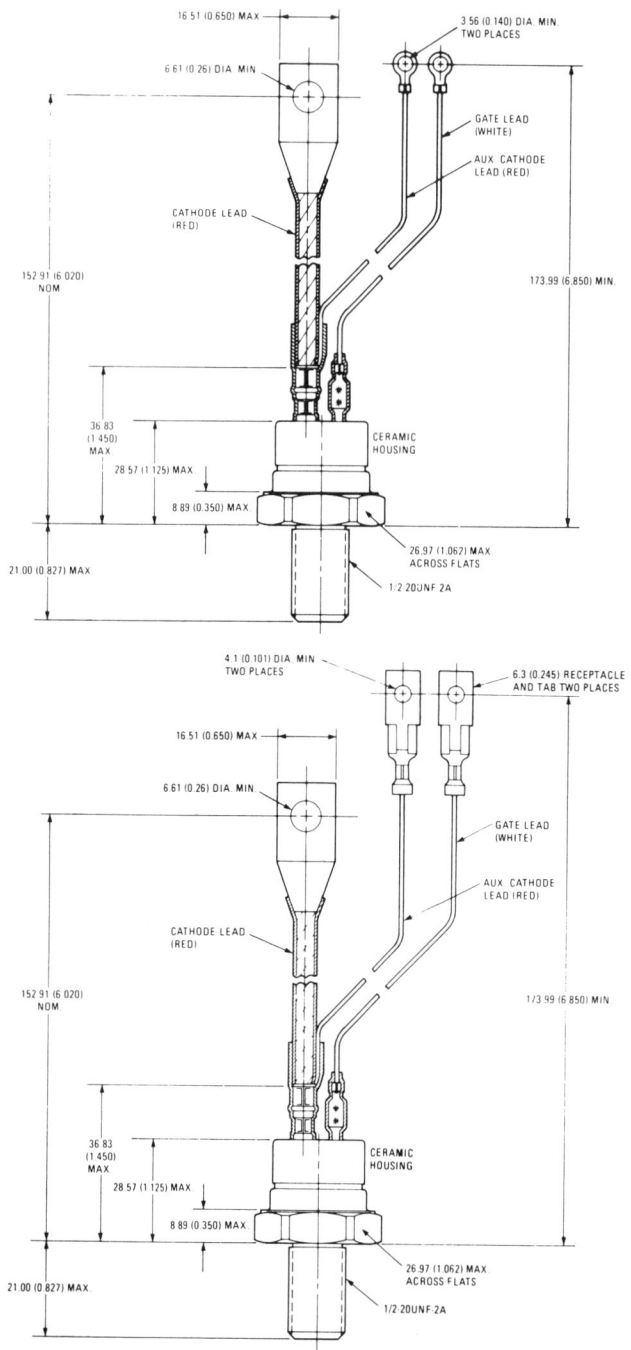

Figure 8-2 *(Continued)*

8. As far as the model is concerned, R_G can be chosen arbitrarily. From the data sheet, $V_g = 5$ V at $I_g = 100$ mA. Therefore, we get $R_G = V_g/I_g = 5/0.1 = 50 \ \Omega$.

9. For a known value of R_G, the multiplier P_1 can be determined from $V_{ON} = R_T P_1 I_g$, which gives

$$P_1 = \frac{V_{ON}}{R_T I_g} = \frac{V_{ON}}{R_T V_{g(peak)}/R_G} = \frac{V_{ON} R_G}{R_T V_{g(peak)}}$$

$$= \frac{1 \times 50}{1 \times 5} = 10 \tag{8-4}$$

where $V_{g(peak)}$ is the peak voltage of the gate pulse. At the leading edge of the gate pulse, most of the controlling current F_1 flows through C_T. Thus the gate pulse must be applied for a sufficient time to cause on-triggering. The voltage v_{R2} due to gate triggering only (with zero anode current) is given by

$$v_{R2} = R_T P_1 I_g (1 - e^{-t/R_T C_T}) \tag{8-5}$$

The time $t = t_n$ at which $v_{R2} = V_{ON}$ gives the turn-on delay of the thyristor. Due to the presence of C_T, the voltage v_{R2} will be delayed and will not be equal to $V_{ON} = R_T P_1 I_g$ instantly. A higher value of P_1 is required to turn on the thyristor. Taking five times more than the value given by Eq. (8-4), let $P_1 = 50$.

This thyristor model can be used as a subcircuit. The switch S_1 is controlled by the controlling voltage V_R connected between nodes 6 and 2. The switch and/or diode parameters can be adjusted to yield the desired on-state drop of the thyristor. In the following examples, we shall use a superdiode with parameters IS=2.2E−15, BV=1200V, TT=0, and CJO=0 and the switch parameters RON=0.0105, ROFF=10E+5, VON=0.5V, and VOFF=0V. The subcircuit definition for the thyristor model *SCR* can be described as follows:

```
*     Subcircuit for ac thyristor model
.SUBCKT      SCR     1        2         3          2
*            model   anode    cathode   +control   −control
*            name                       voltage    voltage
S1    1      5       6    2   SMOD          ; Switch
RG    3      4       50
VX    4      2       DC       0V
VY    5      7       DC       0V
DT    7      2       DMOD                   ; Switch diode
RT    6      2       1
CT    6      2       10UF
F1    2      6       POLY(2)  VX   VY  0    50    11
.MODEL SMOD VSWITCH (RON=0.0105 ROFF=10E+5 VON=0.5V VOFF=0V) ; Switch model
.MODEL  DMOD  D(IS=2.2E−15 BV=1200V TT=0 CJO=0)  ; Diode model parameters
.ENDS  SCR                        ; Ends subcircuit definition
```

8-3 CONTROLLED RECTIFIERS

A controlled rectifier converts a fixed ac voltage to a variable dc voltage and uses thyristors as switching devices. The output voltage of an ideal rectifier should be pure dc and contain no harmonics or ripples. Similarly, the input current should be pure sine wave and contain no harmonics. That is, the total harmonic distortion (THD) of the input current and output voltage should be zero, and the input power factor should be unity.

8-4 EXAMPLES OF CONTROLLED RECTIFIERS

Let us apply the thyristor model of Fig. 8-1(b) to the circuit of Fig. 8-3(a). Thyristor T_1 is turned on by the voltage V_g connected between the gate and cathode voltage. The gate voltage V_g is shown in Fig. 8-3(b). The list of the PSpice circuit file for determining the transient response is as follows:

```
■■■■■■■■■■■■■■■■■■■■■■■■■■■■■■■■■■■■■■■■■■■■■■■■■■■■

         Ac thyristor circuit
SOURCE   ■ VS   1   0   SIN (0   169.7V   60HZ)
           Vg   4   0   PULSE (0V   10V   2777.8US   1NS   1NS   100US   16666.7US)
CIRCUIT  ■ ■ R    1   2   2.5
           VX   2   3   DC   0V   ; Voltage source to measure the load current
           *   Subcircuit call for thyristor model:
           XT1  3   0   4   0   SCR        ; Thyristor T1
ANALYSIS ■ ■ ■ .TRAN   1US   5MS           ; Transient analysis from 0 to 5 ms
             .PROBE                        ; Graphics post-processor
         .END
■■■■■■■■■■■■■■■■■■■■■■■■■■■■■■■■■■■■■■■■■■■■■■■■■■■■
```

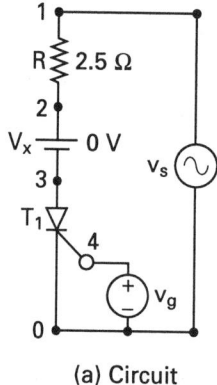

(a) Circuit

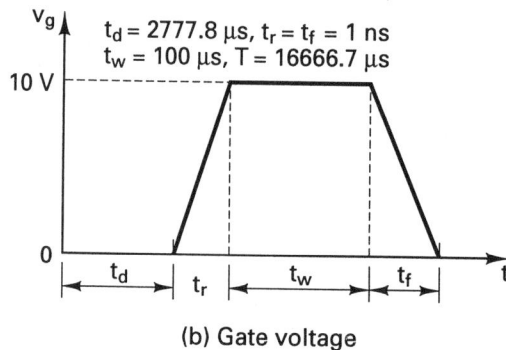

$t_d = 2777.8$ µs, $t_r = t_f = 1$ ns
$t_w = 100$ µs, $T = 16666.7$ µs

(b) Gate voltage

Figure 8-3 Ac thyristor circuit.

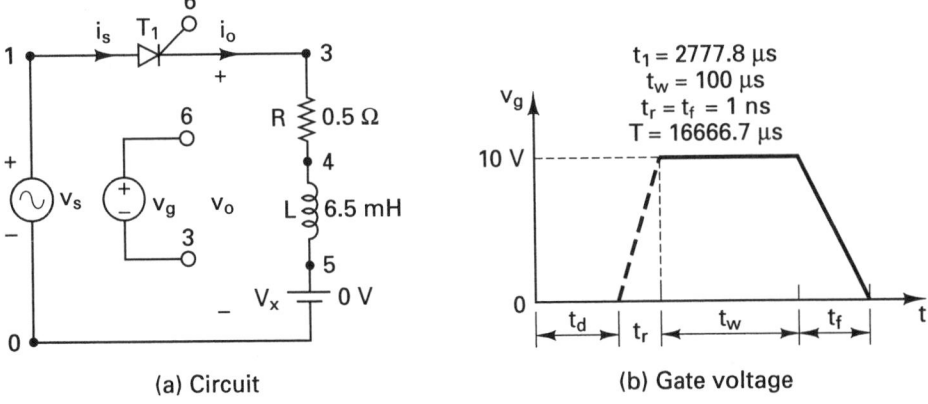

(a) Circuit   (b) Gate voltage

Figure 8-4 Single-phase half-wave controlled rectifier for PSpice simulation.

Example 8-1

A single-phase half-wave rectifier is shown in Fig. 8-4(a). The input has a peak voltage of 169.7 V, 60 Hz. The load inductance L is 6.5 mH, and the load resistance R is 0.5 Ω. The delay angle is $\alpha = 60°$. The gate voltage is shown in Fig. 8-4(b). Use PSpice (a) to plot the instantaneous output voltage v_o and the load current i_o, and (b) to calculate the Fourier coefficients of the input current i_s and the input power factor PF.

Solution The peak supply voltage $V_m = 169.7$ V. For $\alpha = 60°$, time delay $t_1 = (60/360) \times (1000/60 \text{ Hz}) \times 1000 = 2777.78 \ \mu s$. The list of the circuit file is as follows:

■ ■

Example 8-1   Single-phase half-wave controlled rectifier

SOURCE ■	VS	1	0	SIN (0	169.7V	60HZ)		
	Vg	6	3	PULSE (0V	10V	2777.8US	1NS 1NS 100US 16666.7US)
CIRCUIT ■ ■	R	3	4	0.5				
	L	4	5	6.5MH				
	VX	5	0	DC	0V	; Voltage source to measure the load current
	*	Subcircuit call for thyristor model:						
	XT1	1	3	6	3	SCR	; Thyristor T1	
	*	Subcircuit SCR, which is missing, <u>must</u> be inserted.						
ANALYSIS ■ ■ ■	.TRAN	20US	50.0MS	33.33MS	20US	; Transient analysis		
	.PROBE					; Graphics post-processor		
	.OPTIONS ABSTOL=1.0N RELTOL=1.0M VNTOL=1.0M ITL5=10000 ; Convergence
	.FOUR	60HZ	I(VX)			; Fourier analysis		
.END								

■ ■

(a) The PSpice plots of the instantaneous output voltage V(3) and the load current I(VX) are shown in Fig. 8-5. The thyristor T_1 turns off when its current falls to zero but not when the input voltage becomes zero.

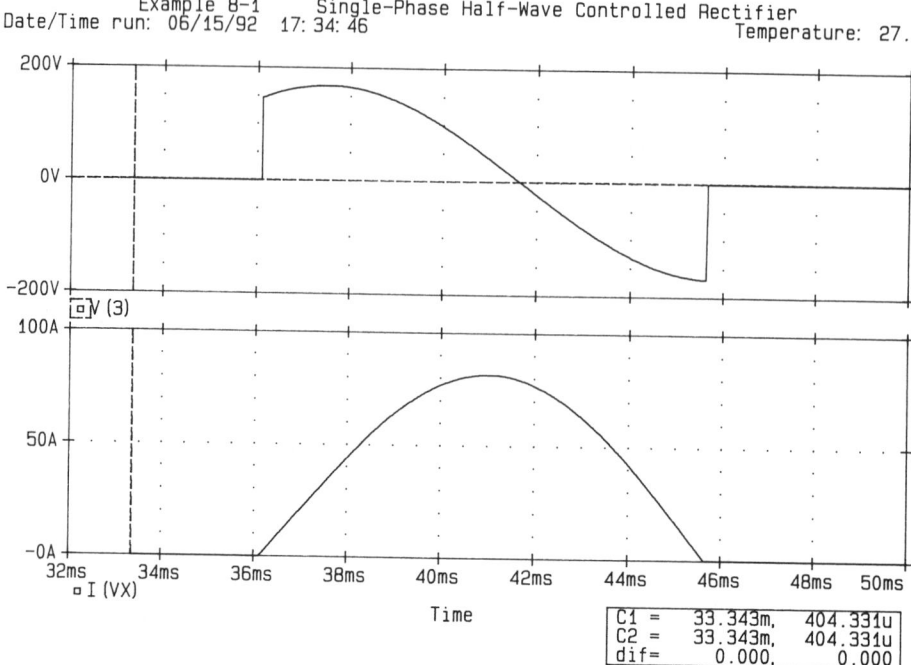

Figure 8-5 Plots for Example 8-1.

(b) To find the input power factor, we need to find the Fourier series of the input current, which is the same as the current through source VX.

FOURIER COMPONENTS OF TRANSIENT RESPONSE I (VX)

DC COMPONENT =   2.912470E+01

HARMONIC NO	FREQUENCY (HZ)	FOURIER COMPONENT	NORMALIZED COMPONENT	PHASE (DEG)	NORMALIZED PHASE (DEG)
1	6.000E+01	4.266E+01	1.000E+00	−7.433E+01	0.000E+00
2	1.200E+02	1.315E+01	3.082E−01	1.200E+02	1.943E+02
3	1.800E+02	2.897E+00	6.792E−02	1.448E+02	2.191E+02
4	2.400E+02	1.845E+00	4.325E−02	−2.775E+01	4.657E+01
5	3.000E+02	1.510E+00	3.540E−02	−3.585E+00	7.074E+01
6	3.600E+02	3.159E−01	7.406E−03	1.761E+02	2.504E+02
7	4.200E+02	8.563E−01	2.007E−02	−1.511E+02	−7.681E+01
8	4.800E+02	1.373E−01	3.218E−03	−1.137E+02	−3.936E+01
9	5.400E+02	4.678E−01	1.097E−02	6.119E+01	1.355E+02

TOTAL HARMONIC DISTORTION =   3.214001E+01 PERCENT

Dc input current, $I_{in(dc)} = 29.12$ A

Rms fundamental input current, $I_{1(rms)} = 42.66/\sqrt{2} = 30.17$ A

Total harmonic distortion of input current, THD = 32.14% = 0.3214

Rms harmonic current, $I_{h(rms)} = I_{1(rms)} \times \text{THD} = 30.17 \times 0.3214 = 9.69$

Rms input current, $I_s = [I_{in(dc)}^2 + I_{1(rms)}^2 + I_{h(rms)}^2]^{1/2}$
$$= (29.12^2 + 30.17^2 + 9.69^2)^{1/2} = 43.4 \text{ A}$$

Displacement angle, $\phi_1 = -74.33°$

Displacement factor, $DF = \cos \phi_1 = \cos(-74.33) = 0.27$ (lagging)

The input power factor is

$$PF = \frac{I_{1(rms)}}{I_s} \cos \phi_1 = \frac{30.17}{43.4} \times 0.27 = 0.1877 \quad \text{(lagging)}$$

Note. For a .FOUR command, TSTART and TSTOP values should be a multiple of the period of input voltage (e.g., for 60 Hz, a multiple of 16.667 ms). The Fourier coefficients will vary slightly with TMAX, because it sets the number of samples in a period.

Example 8-2

A single-phase semiconverter is shown in Fig. 8-6(a). The input voltage has a peak of 169.7 V, 60 Hz. The load inductance L is 6.5 mH, and the load resistance R is 0.5 Ω. The load battery voltage is $V_x = 10$ V. The delay angle is $\alpha = 60°$. The gate voltages are shown in Fig. 8-6(b). Use PSpice (a) to plot the instantaneous output voltage, the input current i_s and the load current i_o, and (b) to calculate the Fourier coefficients of the input current i_s and the input power factor PF.

Solution The peak supply voltage $V_m = 169.7$ V. For $\alpha = 60°$,

$$\text{time delay } t_1 = \frac{60}{360} \times \frac{1000}{60 \text{ Hz}} \times 1000 = 2777.78 \ \mu s$$

$$\text{time delay } t_2 = \frac{240}{360} \times \frac{1000}{60 \text{ Hz}} \times 1000 = 11{,}111.1 \ \mu s$$

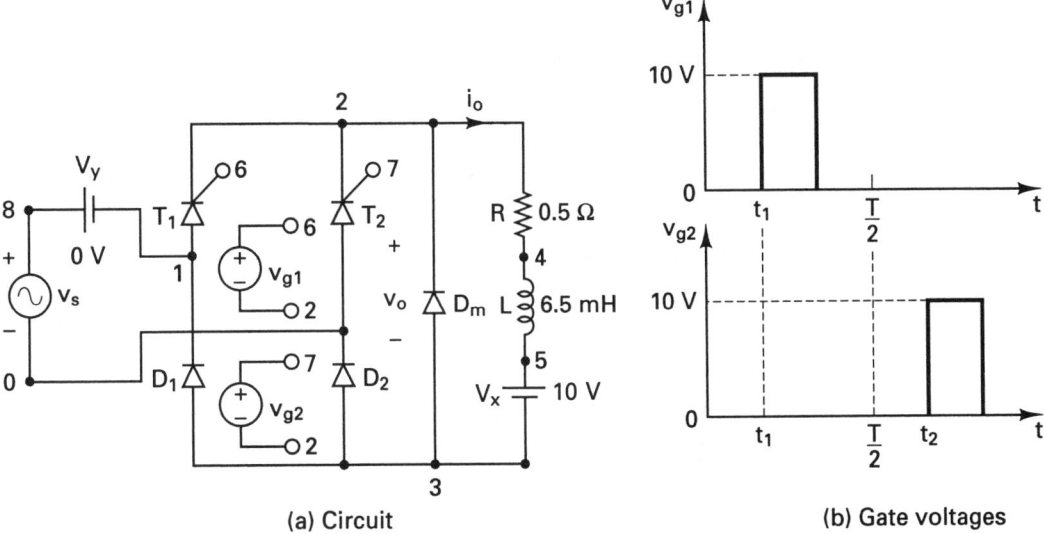

(a) Circuit　　　　　　　　　　　　　　　　(b) Gate voltages

Figure 8-6 Single-phase semiconverter for PSpice simulation.

The list of the circuit file is as follows:

■ ■

Example 8-2   Single-phase semiconverter

SOURCE ■ VS   8   0   SIN (0   169.7V   60HZ)

Vg1   6   2   PULSE (0V   10V   2777.8US   1NS   1NS   100US   16666.7US)

Vg2   7   2   PULSE (0V   10V   11111.1US   1NS   1NS   100US   16666.7US)

CIRCUIT ■ ■ R   2   4   0.5

L   4   5   6.5MH

VX   5   3   DC   10V   ; Load battery voltage

VY   8   1   DC   0V   ; Voltage source to supply current

D1   3   1   DMOD

D2   3   0   DMOD

DM   3   2   DMOD

.MODEL   DMOD   D(IS=2.2E-15 BV=1200V TT=0 CJO=0)   ; Diode model parameters

* Subcircuit calls for thyristor model:

XT1   1   2   6   2   SCR   ; Thyristor T1

XT2   0   2   7   2   SCR   ; Thyristor T2

* Subcircuit SCR, which is missing, <u>must</u> be inserted.

ANALYSIS ■ ■ ■ .TRAN   20US   50.0MS   33.33MS   20US   ; Transient analysis

.PROBE   ; Graphics post-processor

.OPTIONS ABSTOL=1.00N RELTOL=1.0M VNTOL=1.0M ITL5=10000 ; Convergence

.FOUR   60HZ   I(VY)

.END

■ ■

(a) The PSpice plots of the instantaneous output voltage V(2,3), the input current I(VY), and the load current I(VX) are shown in Fig. 8-7. The load current is continuous as expected, and the effects of load current ripples can be noticed on the input current.

(b) To find the input power factor, we need to find the Fourier series of the input current, which is the same as the current through source VY.

FOURIER COMPONENTS OF TRANSIENT RESPONSE I(VY)

DC COMPONENT =   4.587839E+00

HARMONIC NO	FREQUENCY (HZ)	FOURIER COMPONENT	NORMALIZED COMPONENT	PHASE (DEG)	NORMALIZED PHASE (DEG)
1	6.000E+01	1.436E+02	1.000E+00	-2.960E+01	0.000E+00
2	1.200E+02	1.144E+01	7.966E-02	1.008E+02	1.303E+02
3	1.800E+02	1.730E+01	1.204E-01	7.189E+01	1.015E+02
4	2.400E+02	1.035E+01	7.204E-02	1.200E+02	1.496E+02
5	3.000E+02	2.854E+01	1.987E-01	4.806E+01	7.766E+01
6	3.600E+02	1.037E+01	7.218E-02	1.353E+02	1.649E+02
7	4.200E+02	9.400E+00	6.545E-02	-7.113E+00	2.249E+01
8	4.800E+02	9.306E+00	6.480E-02	1.495E+02	1.791E+02
9	5.400E+02	7.692E+00	5.356E-02	1.414E+02	1.710E+02

TOTAL HARMONIC DISTORTION =   2.865256E+01 PERCENT

Total harmonic distortion of input current, THD = 28.65% = 0.2865

Displacement angle, $\phi_1 = -29.6°$

Displacement factor, DF = cos ϕ_1 = cos(-29.6) = 0.87   (lagging)

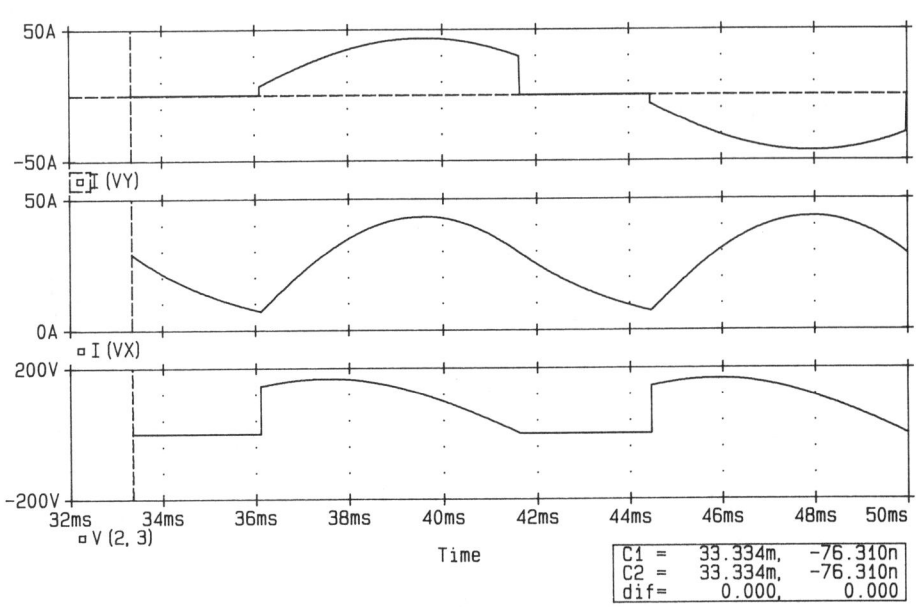

Figure 8-7 Plots for Example 8-2.

From Eq. (7-3), the input power factor

$$PF = \frac{1}{(1 + THD^2)^{1/2}} \cos \phi_1 = \frac{1}{(1 + 0.2865^2)^{1/2}} \times 0.87 = 0.836 \quad \text{(lagging)}$$

Example 8-3

A single-phase full converter is shown in Fig. 8-8(a). The input voltage has a peak of 169.7 V, 60 Hz. The load inductance L is 6.5 mH, and the load resistance R is 0.5 Ω. The load battery voltage is $V_x = 10$ V. The delay angle is $\alpha = 60°$. The gate voltages are shown in Fig. 8-8(b). Use PSpice (a) to plot the instantaneous output voltage v_o, the input current i_s, and the load current i_o, and (b) to calculate the Fourier coefficients of the input current i_s and the input power factor PF.

Solution The peak supply voltage $V_m = 169.7$ V. For $\alpha_1 = 60°$,

$$\text{time delay } t_1 = \frac{60}{360} \times \frac{1000}{60 \text{ Hz}} \times 1000 = 2777.78 \ \mu s$$

$$\text{time delay } t_2 = \frac{240}{360} \times \frac{1000}{60 \text{ Hz}} \times 1000 = 11,111.1 \ \mu s$$

The list of the circuit file is as follows:

▪ ▪

Example 8-3   Single-phase full-bridge converter

SOURCE	▪ VS	10	0	SIN (0	169.7V	60HZ)			
	Vg1	6	2	PULSE (0V	10V	2777.8US	1NS	1NS	100US	16666.7US)
	Vg2	7	0	PULSE (0V	10V	2777.8US	1NS	1NS	100US	16666.7US)

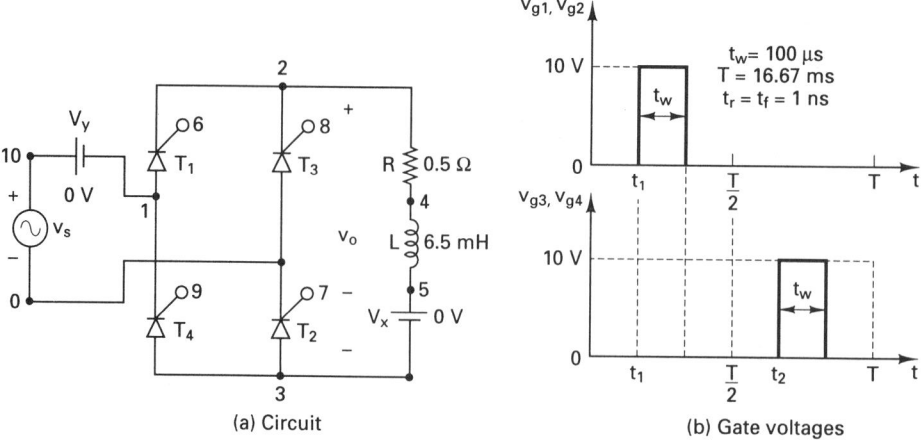

Figure 8-8 Single-phase full converter for PSpice simulation.

```
          Vg3   8    2    PULSE (0V    10V    11111.1US    1NS    1NS    100US    16666.7US)
          Vg4   9    1    PULSE (0V    10V    11111.1US    1NS    1NS    100US    16666.7US)
CIRCUIT ■ ■ R    2    4    0.5
          L     4    5    6.5MH
          VX    5    3    DC      10V  ; Load battery voltage
          VY    10   1    DC      0V   ; Voltage source to measure supply current
          *     Subcircuit calls for thyristor model:
          XT1   1    2    6    2    SCR        ; Thyristor T₁
          XT3   0    2    8    2    SCR        ; Thyristor T₃
          XT2   3    0    7    0    SCR        ; Thyristor T₂
          XT4   3    1    9    1    SCR        ; Thyristor T₄
          *     Subcircuit SCR, which is missing, must be inserted.
ANALYSIS ■ ■ ■ .TRAN   50US  50MS  33.33MS  50US   ; Transient analysis
          .PROBE                                    ; Graphics post-processor
          .OPTIONS ABSTOL = 1.00N  RELTOL = 1.0M  VNTOL = 0.01  ITL5=20000
          .FOUR    60HZ    I(VY)
     .END
```

(a) The PSpice plots of the instantaneous output voltage V(2,3), the input current I(VY), and the load current I(VX) are shown in Fig. 8-9. The instantaneous output voltage can be negative, but the load current is always positive.

(b) To find the input power factor, we need to find the Fourier series of the current through the voltage source VY.

FOURIER COMPONENTS OF TRANSIENT RESPONSE I(VY)

DC COMPONENT =   1.200052E−02

HARMONIC NO	FREQUENCY (HZ)	FOURIER COMPONENT	NORMALIZED COMPONENT	PHASE (DEG)	NORMALIZED PHASE (DEG)
1	6.000E+01	1.113E+02	1.000E+00	−6.211E+01	0.000E+00
2	1.200E+02	8.525E−01	7.662E−03	2.244E+01	8.456E+01
3	1.800E+02	1.633E+01	1.468E−01	−1.744E+02	−1.123E+02

HARMONIC NO	FREQUENCY (HZ)	FOURIER COMPONENT	NORMALIZED COMPONENT	PHASE (DEG)	NORMALIZED PHASE (DEG)
4	2.400E+02	3.016E−01	2.711E−03	−6.781E+01	−5.696E+00
5	3.000E+02	1.007E+01	9.046E−02	6.171E+01	1.238E+02
6	3.600E+02	2.470E−01	2.220E−03	9.033E+01	1.524E+02
7	4.200E+02	7.179E+00	6.452E−02	−5.831E+01	3.806E+00
8	4.800E+02	3.580E−01	3.218E−03	−3.425E+00	5.869E+01
9	5.400E+02	5.454E+00	4.902E−02	1.796E+02	2.417E+02

TOTAL HARMONIC DISTORTION =   1.907363E+01 PERCENT

Total harmonic distortion of input current, THD = 19.07% = 0.1907

Displacement angle, $\phi_1 = -62.1°$

Displacement factor, DF = $\cos \phi_1 = \cos(-62.1) = 0.468$ (lagging)

From Eq. (7-3), the input power factor is

$$PF = \frac{1}{(1 + THD^2)^{1/2}} \cos \phi_1 = \frac{1}{(1 + 0.1907^2)^{1/2}} \times 0.468 = 0.46 \quad \text{(lagging)}$$

Example 8-4

A three-phase half-wave converter is shown in Fig. 8-10(a). The input voltage per phase has a peak of 169.7 V, 60 Hz. The load inductance L is 6.5 mH, and the load resistance R is 0.5 Ω. The load battery voltage is $V_x = 10$ V. The delay angle is $\alpha = 60°$. The gate voltages are shown in Fig. 8-10(b). Use PSpice (a) to plot the instanta-

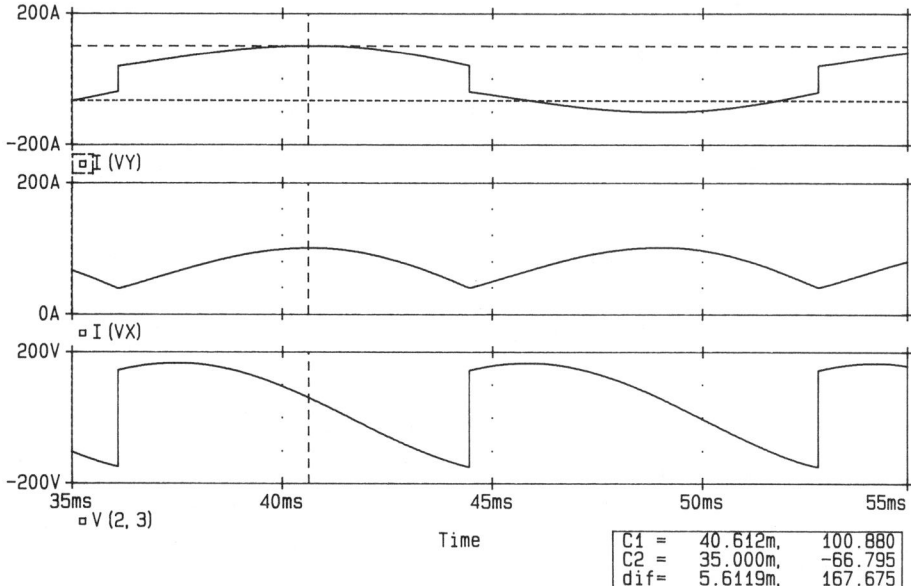

Figure 8-9   Plots for Example 8-3.

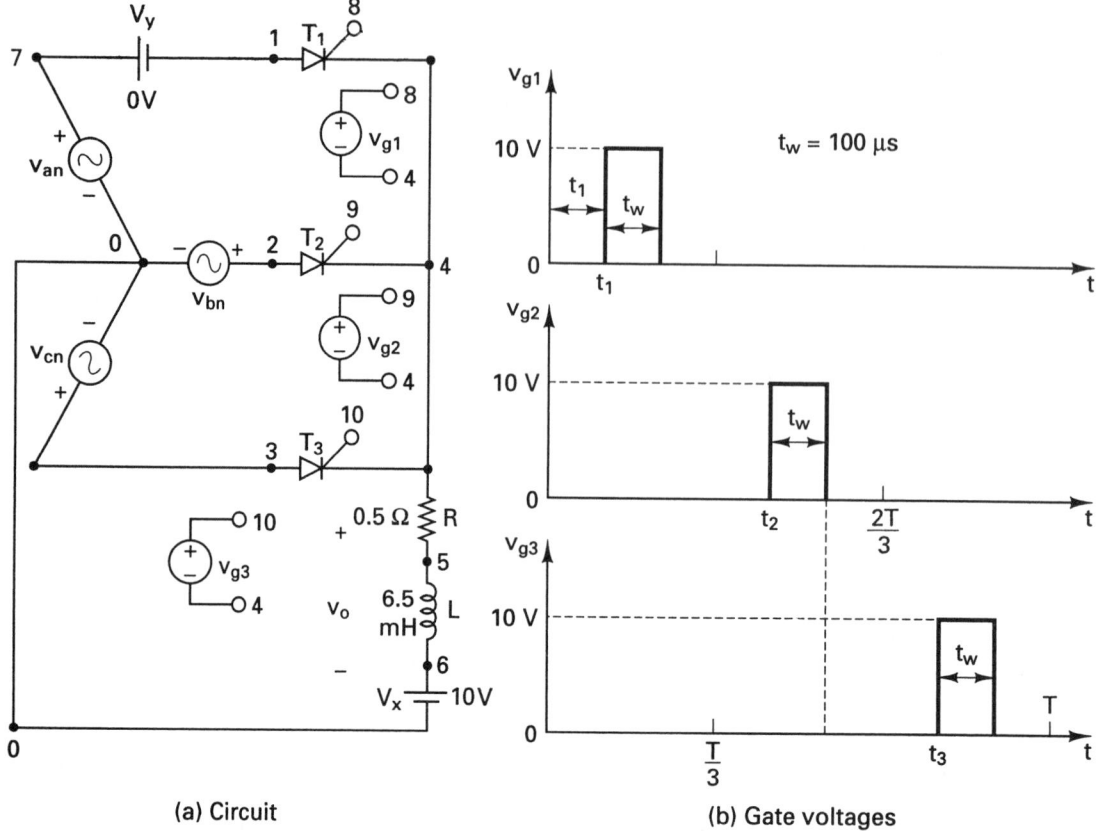

(a) Circuit (b) Gate voltages

Figure 8-10 Three-phase half-wave converter for PSpice simulation.

neous output voltage v_o and the load current i_o, and $(\dot{v})$ to calculate the Fourier coefficients of the input current i_s and the input power factor PF.
Solution The peak supply voltage $V_m = \sqrt{2}\ V_s = \sqrt{2} \times 120 = 169.7$ V. For $\alpha = 60°$,

$$\text{time delay } t_1 = \frac{90}{360} \times \frac{1000}{60\ \text{Hz}} \times 1000 = 4166.7\ \mu s$$

$$\text{time delay } t_2 = \frac{210}{360} \times \frac{1000}{60\ \text{Hz}} \times 1000 = 9722.2\ \mu s$$

$$\text{time delay } t_3 = \frac{330}{360} \times \frac{1000}{60\ \text{Hz}} \times 1000 = 15,277.8\ \mu s$$

The list of the circuit file is as follows:

■ ■

Example 8-4 Three-phase half-wave converter

SOURCE ■ Van   7   0   SIN(0   169.7V   60HZ)
Vbn   2   0   SIN(0   169.7V   60HZ   0   0   −120DEG)

```
Vcn   3   0   SIN(0    169.7V   60HZ    0    0    -240DEG)
Vg1   8   4   PULSE (0V   10V     4166.7US    1NS    1NS    100US    16666.7US)
Vg2   9   4   PULSE (0V   10V     9722.2US    1NS    1NS    100US    16666.7US)
Vg3  10   4   PULSE (0V   10V    15277.8US    1NS    1NS    100US    16666.7US)
```

CIRCUIT ■ ■
```
R     4   5   0.5
L     5   6   6.5MH
VX    6   0   DC    10V   ; Load battery voltage
VY    7   1   DC     0V   ; Voltage source to measure supply current
*     Subcircuit calls for thyristor model:
XT1   1   4    8   4   SCR        ; Thyristor T1
XT2   2   4    9   4   SCR        ; Thyristor T3
XT3   3   4   10   4   SCR        ; Thyristor T2
*     Subcircuit SCR, which is missing, must be inserted.
```

ANALYSIS ■ ■ ■
```
.TRAN    20US   50MS   33.33MS   20US   ; Transient analysis
.PROBE                                  ; Graphics post-processor
.OPTIONS ABSTOL=1.00N RELTOL=1.0M VNTOL=1.0M ITL5=20000 ; Convergence
.FOUR    60HZ   I(VY)
.END
```
■ ■

(a) The PSpice plots of the instantaneous output voltage V(4) and the load
current I(VX) are shown in Fig. 8-11. The average load current is greater than that of
a single-phase converter.

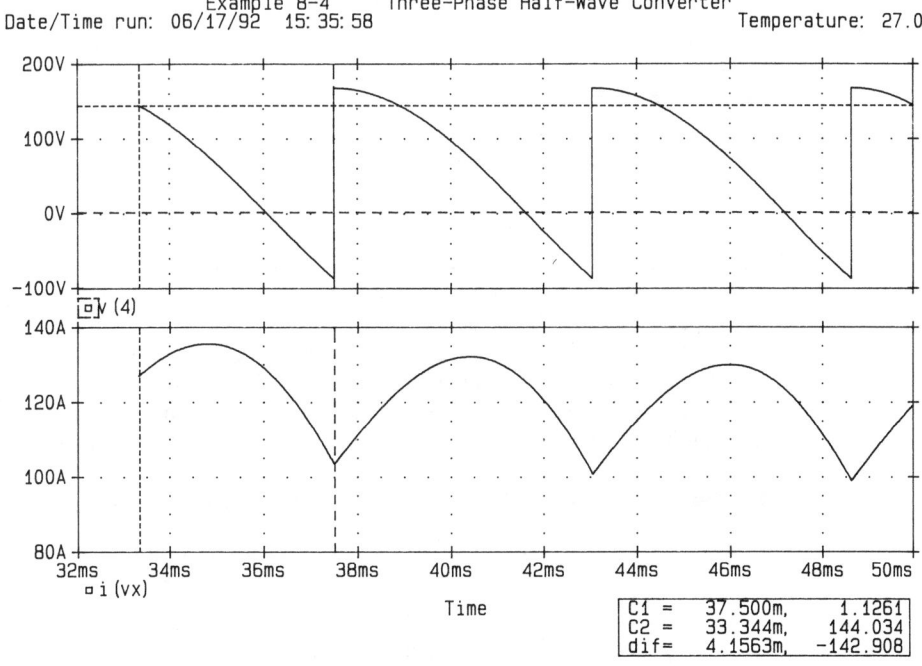

Figure 8-11 Plots for Example 8-4.

(b) The Fourier series of the current through voltage source VY is

FOURIER COMPONENTS OF TRANSIENT RESPONSE I(VY)
DC COMPONENT =   4.048497E+01

HARMONIC NO	FREQUENCY (HZ)	FOURIER COMPONENT	NORMALIZED COMPONENT	PHASE (DEG)	NORMALIZED PHASE (DEG)
1	6.000E+01	6.793E+01	1.000E+00	−6.001E+01	0.000E+00
2	1.200E+02	3.650E+01	5.373E−01	1.499E+02	2.099E+02
3	1.800E+02	4.317E+00	6.354E−02	−2.866E+00	5.715E+01
4	2.400E+02	1.315E+01	1.936E−01	3.128E+01	9.129E+01
5	3.000E+02	1.216E+01	1.790E−01	−1.188E+02	−5.882E+01
6	3.600E+02	1.108E+00	1.631E−02	9.453E+01	1.545E+02
7	4.200E+02	7.661E+00	1.128E−01	1.206E+02	1.806E+02
8	4.800E+02	7.409E+00	1.091E−01	−2.886E+01	3.116E+01
9	5.400E+02	5.800E−01	8.538E−03	−1.720E+02	−1.120E+02

TOTAL HARMONIC DISTORTION =   6.222461E+01 PERCENT

Dc input current, $I_{in(dc)}$ = 40.48 A

Rms fundamental input current, $I_{1(rms)}$ = 67.93/$\sqrt{2}$ = 48.03 A

Total harmonic distortion of input current, THD = 62.22% = 0.6222

Rms harmonic current, $I_{h(rms)}$ = $I_{1(rms)}$ × THD = 48.03 × 0.6222 = 29.88 A

Rms input current, I_s = $[I_{in(dc)}^2 + I_{1(rms)}^2 + I_{h(rms)}^2]^{1/2}$
$$= (40.48^2 + 48.03^2 + 29.88^2)^{1/2} = 69.56 \text{ A}$$

Displacement angle, ϕ_1 = −60.01°

Displacement factor, DF = cos ϕ_1 = cos(−60.01) = 0.5   (lagging)

Thus the input power factor is

$$\text{PF} = \frac{I_{1(rms)}}{I_s} \cos \phi_1 = \frac{48.03}{69.56} \times 0.5 = 0.3452 \quad \text{(lagging)}$$

Example 8-5

A three-phase semiconverter is shown in Fig. 8-12(a). The input voltage per phase has a peak of 169.7 V, 60 Hz. The load inductance L is 6.5 mH, and the load resistance R is 0.5 Ω. The load battery voltage is V_x = 10 V. The delay angle is α = 60°. The gate voltages are shown in Fig. 8-12(b). Use PSpice (a) to plot the instantaneous output voltage v_o, the load current i_o, and the input current i_s, and (b) to calculate the Fourier coefficients of the input current i_s and the input power factor PF.

Solution The peak supply voltage V_m = 169.7 V. For α = 60°,

$$\text{time delay } t_1 = \frac{90}{360} \times \frac{1000}{60 \text{ Hz}} \times 1000 = 4166.7 \ \mu\text{s}$$

$$\text{time delay } t_2 = \frac{210}{360} \times \frac{1000}{60 \text{ Hz}} \times 1000 = 9722.2 \ \mu\text{s}$$

$$\text{time delay } t_3 = \frac{330}{360} \times \frac{1000}{60 \text{ Hz}} \times 1000 = 15,277.8 \ \mu\text{s}$$

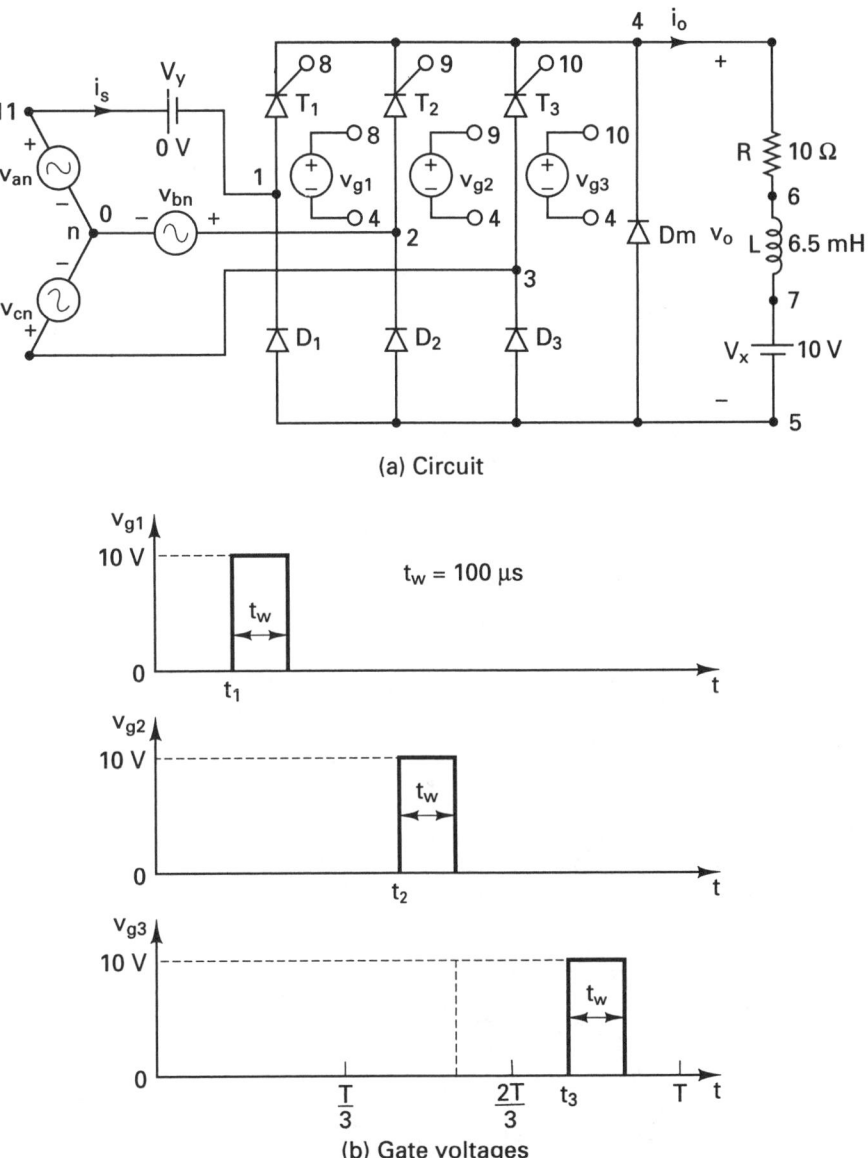

(a) Circuit

(b) Gate voltages

Figure 8-12 Three-phase semiconverter for PSpice simulation.

The list of the circuit file is as follows:

■■■

Example 8-5 Three-phase semiconverter

SOURCE ■ Van 11 0   SIN(0   169.7V   60HZ)
          Vbn   2   0   SIN(0   169.7V   60HZ   0   0   −120DEG)
          Vcn   3   0   SIN(0   169.7V   60HZ   0   0   −240DEG)
          Vg1   8   4   PULSE (0V   10V   4166.7US   1NS   1NS   100US   16666.7US)

```
                 Vg2    9    4    PULSE (0V    10V    9722.2US    1NS    1NS    100US    16666.7US)
                 Vg3   10    4    PULSE (0V    10V    15277.8US   1NS    1NS    100US    16666.7US)
CIRCUIT   ■ ■ ■  R      4    6    0.5
                 L      6    7    6.5MH
                 VX     7    5    DC     10V   ; Load battery voltage
                 VY    11    1    DC      0V   ; Voltage source to measure supply current
                 D1     5    1    DMOD
                 D2     5    2    DMOD
                 D3     5    3    DMOD
                 DM     5    4    DMOD
                 .MODEL   DMOD   D(IS=2.2E-15 BV=1200V TT=0 CJO=1PF)  ; Diode model parameters
                 *    Subcircuit calls for thyristor model:
                 XT1    1    4    8    4    SCR        ; Thyristor T1
                 XT2    2    4    9    4    SCR        ; Thyristor T2
                 XT3    3    4   10    4    SCR        ; Thyristor T3
                 *    Subcircuit SCR, which is missing, must be inserted.
ANALYSIS  ■ ■ ■  .TRAN   50US   50MS   33.33MS   50US   ; Transient analysis
                 .PROBE                                 ; Graphics post-processor
                 .OPTIONS ABSTOL = 100.U  RELTOL = 0.01  VNTOL = 0.01   ITL5=20000
                 .FOUR    60HZ   I(VY)
                 .END
■ ■ ■ ■ ■ ■ ■ ■ ■ ■ ■ ■ ■ ■ ■ ■ ■ ■ ■ ■ ■ ■ ■ ■ ■ ■ ■ ■ ■ ■ ■ ■ ■ ■ ■ ■ ■ ■ ■ ■ ■ ■ ■ ■ ■ ■
```

(a) The PSpice plots of the instantaneous output voltage V(4,5), the load current I(VX), and the input current I(VY) are shown in Fig. 8-13. The ripple contents

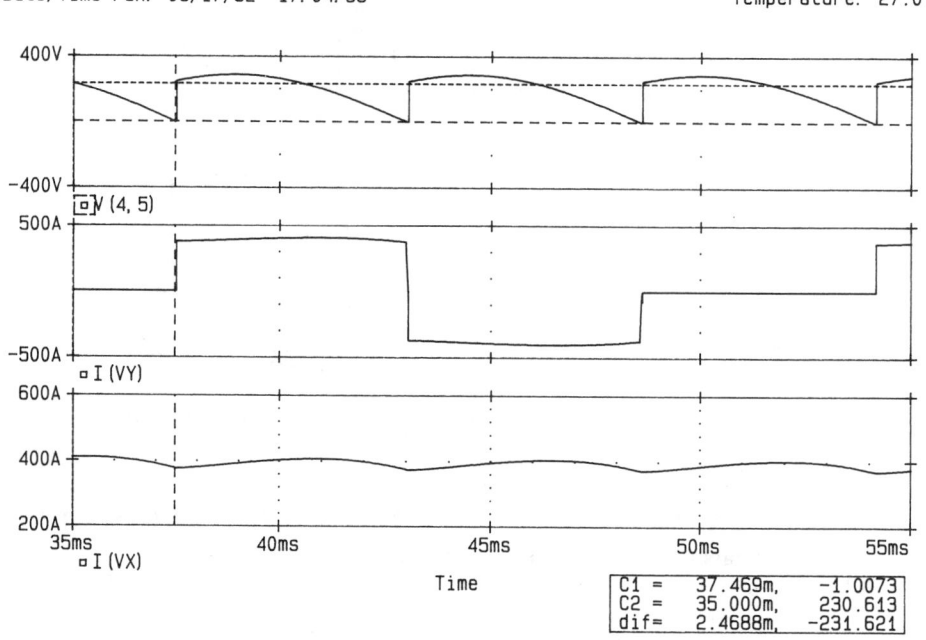

Figure 8-13 Plots for Example 8-5.

on the output voltage and output current are lower than those of a single-phase converter.

(b) The Fourier series of the input current is as follows:

```
FOURIER COMPONENTS OF TRANSIENT RESPONSE I (VY)
DC COMPONENT =    8.208448E-01
```

HARMONIC NO	FREQUENCY (HZ)	FOURIER COMPONENT	NORMALIZED COMPONENT	PHASE (DEG)	NORMALIZED PHASE (DEG)
1	6.000E+01	3.753E+02	1.000E+00	-3.022E+01	0.000E+00
2	1.200E+02	1.943E+02	5.177E-01	1.195E+02	1.498E+02
3	1.800E+02	1.153E-01	3.073E-04	8.665E+01	1.169E+02
4	2.400E+02	8.585E+01	2.287E-01	6.213E+01	9.235E+01
5	3.000E+02	7.424E+01	1.978E-01	-1.483E+02	-1.180E+02
6	3.600E+02	5.945E-02	1.584E-04	1.676E+02	1.978E+02
7	4.200E+02	4.864E+01	1.296E-01	1.519E+02	1.822E+02
8	4.800E+02	4.660E+01	1.242E-01	-5.781E+01	-2.759E+01
9	5.400E+02	4.144E-02	1.104E-04	-1.095E+02	-7.932E+01

```
TOTAL HARMONIC DISTORTION =    6.258354E+01 PERCENT
```

Total harmonic distortion of input current, THD = 62.58% = 0.6258

Displacement angle, $\phi_1 = -30.2°$

Displacement factor, DF = $\cos \phi_1 = \cos(-30.2) = 0.864$ (lagging)

The input power factor is

$$\text{PF} = \frac{1}{(1 + \text{THD}^2)^{1/2}} \cos \phi_1 = \frac{1}{(1 + 0.6258^2)^{1/2}} \times 0.864 = 0.732 \quad \text{(lagging)}$$

Example 8-6

A three-phase full-bridge converter is shown in Fig. 8-14(a). The input voltage per phase has a peak of 169.7 V, 60 Hz. The load inductance L is 6.5 mH, and the load resistance R is 0.5 Ω. The load battery voltage is $V_x = 10$ V. The delay angle is $\alpha = 60°$. The gate voltages are shown in Fig. 8-14(b). Use PSpice (a) to plot the instantaneous output voltage v_o, the load current i_o, and the input current i_s, and (b) to calculate the Fourier coefficients of the input current i_s and the input power factor PF.

Solution For $\alpha = 60°$,

$$\text{time delay } t_1 = \frac{90}{360} \times \frac{1000}{60 \text{ Hz}} \times 1000 = 4166.7 \ \mu\text{s}$$

$$\text{time delay } t_3 = \frac{210}{360} \times \frac{1000}{60 \text{ Hz}} \times 1000 = 9722.2 \ \mu\text{s}$$

$$\text{time delay } t_5 = \frac{330}{360} \times \frac{1000}{60 \text{ Hz}} \times 1000 = 15,277.8 \ \mu\text{s}$$

$$\text{time delay } t_2 = \frac{150}{360} \times \frac{1000}{60 \text{ Hz}} \times 1000 = 6944.4 \ \mu\text{s}$$

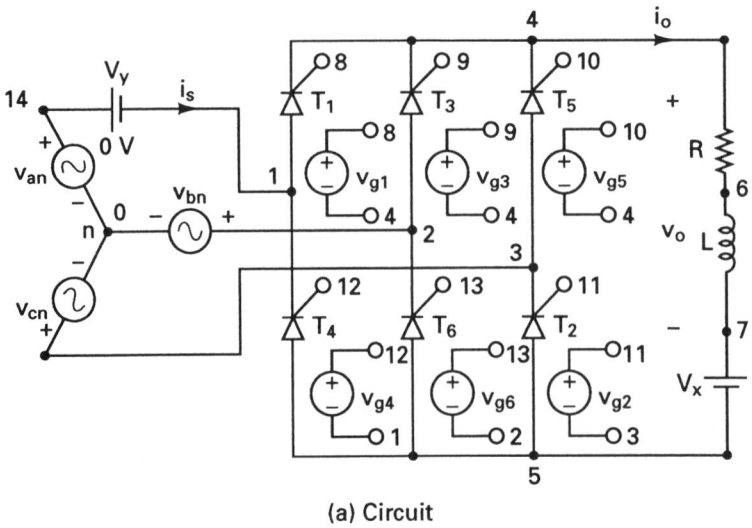

(a) Circuit

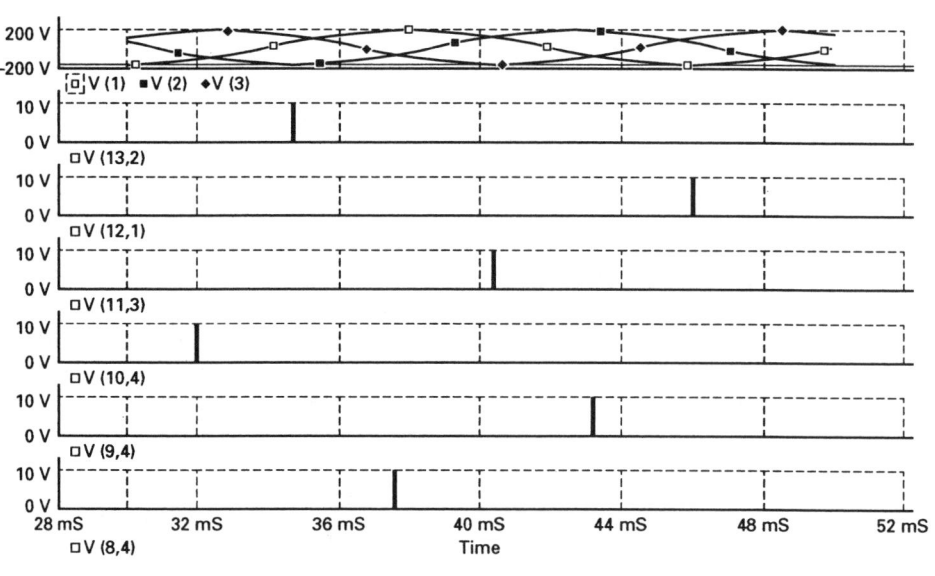

(b) Gate voltages

Figure 8-14 Three-phase full-bridge converter for PSpice simulation.

$$\text{time delay } t_4 = \frac{270}{360} \times \frac{1000}{60 \text{ Hz}} \times 1000 = 12{,}500.0 \ \mu s$$

$$\text{time delay } t_6 = \frac{30}{360} \times \frac{1000}{60 \text{ Hz}} \times 1000 = 1388.9 \ \mu s$$

The list of the circuit file is as follows:

■■■ ■■■■

Example 8-6 Three-phase full-bridge converter

```
SOURCE    ■ Van   14   0    SIN(0    169.7V    60HZ)
            Vbn    2   0    SIN(0    169.7V    60HZ   0    0   -120DEG)
            Vcn    3   0    SIN(0    169.7V    60HZ   0    0   -240DEG)
            Vg1    8   4    PULSE (0V    10V     4166.7US    1NS   1NS   100US   16666.7US)
            Vg3    9   4    PULSE (0V    10V     9722.2US    1NS   1NS   100US   16666.7US)
            Vg5   10   4    PULSE (0V    10V    15277.8US    1NS   1NS   100US   16666.7US)
            Vg2   11   3    PULSE (0V    10V     6944.4US    1NS   1NS   100US   16666.7US)
            Vg4   12   1    PULSE (0V    10V    12500.0US    1NS   1NS   100US   16666.7US)
            Vg6   13   2    PULSE (0V    10V     1388.9US    1NS   1NS   100US   16666.7US)
CIRCUIT   ■■ R     4   6     0.5
            L      6   7     6.5MH
            VX     7   5     DC     10V   ; Load battery voltage
            VY    14   1     DC      0V   ; Voltage source to measure supply current
            *   Subcircuit calls for thyristor model:
            XT1    1   4    8    4    SCR        ; Thyristor T1
            XT3    2   4    9    4    SCR        ; Thyristor T3
            XT5    3   4   10    4    SCR        ; Thyristor T5
            XT2    5   3   11    3    SCR        ; Thyristor T2
            XT4    5   1   12    1    SCR        ; Thyristor T4
            XT6    5   2   13    2    SCR        ; Thyristor T6
            *   Subcircuit SCR, which is missing, must be inserted.
ANALYSIS  ■■■ .TRAN   50US   50MS   33.33MS   50US    ; Transient analysis
            .PROBE                                    ; Graphics post-processor
            .FOUR    60HZ    I(VY)
            .OPTIONS   ABSTOL = 1.00N RELTOL = 0.01   VNTOL = 0.01 ITL5=20000 ;
            *                                                               Convergence
          .END
```

■■■

(a) The PSpice plots of the instantaneous output voltage V(4,5), the load current I(VX), and the input current I(VY) are shown in Fig. 8-15. The load current has not reached the steady-state condition.

(b) The Fourier series of the input current is as follows:

FOURIER COMPONENTS OF TRANSIENT RESPONSE I(VY)
DC COMPONENT = 7.862956E−01

HARMONIC NO	FREQUENCY (HZ)	FOURIER COMPONENT	NORMALIZED COMPONENT	PHASE (DEG)	NORMALIZED PHASE (DEG)
1	6.000E+01	2.823E+02	1.000E+00	−6.021E+01	0.000E+00
2	1.200E+02	2.387E+00	8.453E−03	−1.651E+02	−1.048E+02
3	1.800E+02	1.885E+00	6.676E−03	−1.540E+02	−9.383E+01
4	2.400E+02	1.952E+00	6.915E−03	−1.020E+02	−4.179E+01

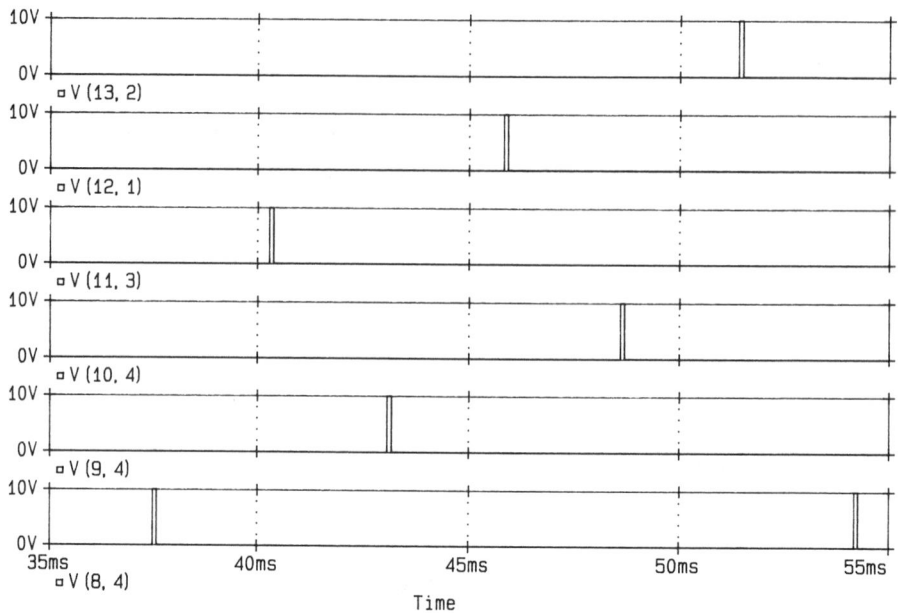

Figure 8-15 Plots for Example 8-6.

HARMONIC NO	FREQUENCY (HZ)	FOURIER COMPONENT	NORMALIZED COMPONENT	PHASE (DEG)	NORMALIZED PHASE (DEG)
5	3.000E+02	6.049E+01	2.143E−01	−1.211E+02	−6.088E+01
6	3.600E+02	1.836E+00	6.503E−03	−1.761E+02	−1.159E+02
7	4.200E+02	3.774E+01	1.337E−01	1.216E+02	1.818E+02
8	4.800E+02	1.473E+00	5.216E−03	−1.451E+02	−8.488E+01
9	5.400E+02	6.411E−01	2.271E−03	1.662E+02	2.264E+02

TOTAL HARMONIC DISTORTION =   2.530035E+01 PERCENT

Total harmonic distortion of input current, THD = 25.3% = 0.253

Displacement angle, $\phi_1 = -60.21°$

Displacement factor, DF = $\cos \phi_1 = \cos(-60.21) = 0.497$ (lagging)

The input power factor

$$PF = \frac{1}{(1 + THD^2)^{1/2}} \cos \phi_1 = \frac{1}{(1 + 0.253^2)^{1/2}} \times 0.497 = 0.482 \quad (lagging)$$

8-5 SWITCHED THYRISTOR DC MODEL

Forced commutated converters are being used increasingly to improve the input power factor. These converters operate power devices as switches. The switches are turned on or off at a specified time. We do not need latching characteristics.

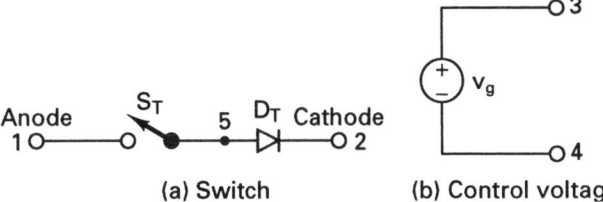

(a) Switch   (b) Control voltage

Figure 8-16 Switched dc thyristor model.

Rather, we need to operate them as on and off switches, like gate turn-off (GTO) thyristors. We shall model a power device as the voltage-controlled switch shown in Fig. 8-16. This model, called a *switched thyristor dc model*, is also used in Chapter 10. The subcircuit definition SSCR can be described as follows:

```
*     Subcircuit for switched thyristor model:
.SUBCKT    SSCR      1         2          3           4
*          model    anode    cathode    +control    -control
*          name                         voltage     voltage
DT    5    2    DMOD                    ; Switch diode
ST    1    5    3    4    SMOD          ; Switch
.MODEL    DMOD    D(IS=2.2E-15 BV=1200V TT=0 CJO=0)   ; Diode model parameters
.MODEL    SMOD    VSWITCH (RON=0.01 ROFF=10E+6 VON=10V VOFF=5V)
.ENDS    SSCR                          ; Ends subcircuit definition
```

8-6 GATE-TURN-OFF THYRISTOR MODEL

The ac thyristor model of Fig. 8-1(b) can be used as a gate-turn-off (GTO). The turn-on is similar to a normal SCR. However, the turn-off gate voltage must be negative with appropriate magnitude and be capable of turning off at the maximum possible current. Therefore, the turn-off gate voltage v_{gn} must satisfy the condition

$$P_1 I_g + P_2 I_{a(max)} \leq 0 \quad \text{or} \quad P_1 \frac{v_{gn}}{R_G} + P_2 I_{a(max)} \leq 0$$

which gives the magnitude of v_{gn} as

$$|v_{gn}| \geq \frac{P_2 I_{a(max)}}{P_1} R_G \tag{8-6}$$

For $I_{T(RMS)} = 110$ A, $I_{a(max)} = \sqrt{2} I_{T(RMS)} = \sqrt{2} \times 110 = 156$ V. Using $P_1 = 50$, $P_2 = 11$, and $R_G = 50\ \Omega$, we get $|v_{gn}| \geq (11 \times 156 \times 50/50) = 1716$ V. Thus subcircuit model SCR must be gated with a positive pulse voltage v_g to turn on and a negative pulse voltage v_{gn} to turn-off.

The conduction angles of one or more switches can be generated by using a comparator with a reference signal v_r and a carrier signal v_c as shown in Fig. 8-17(a). The pulse width δ can be varied by varying the carrier voltage v_c. This technique, known as a *pulse-width modulator* (PWM), can be implemented by an op-amp circuit as shown in Fig. 8-17(b). The inputs to the amplifier are v_r and v_c. Its output is the conduction angle δ, during which a switch remains on.

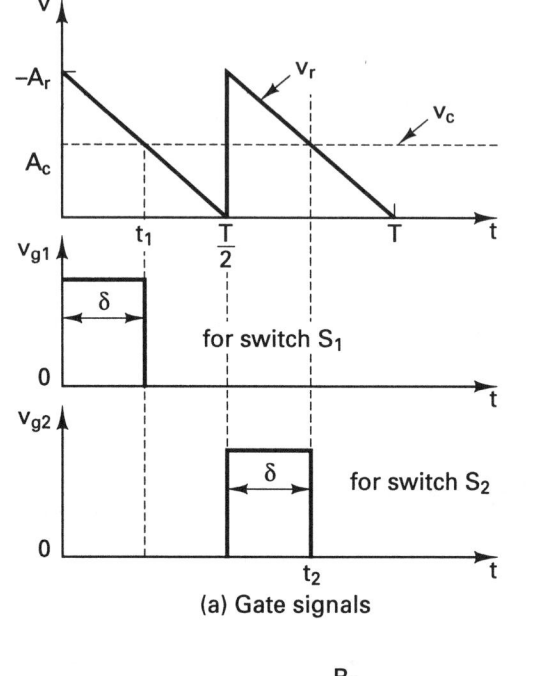

(a) Gate signals

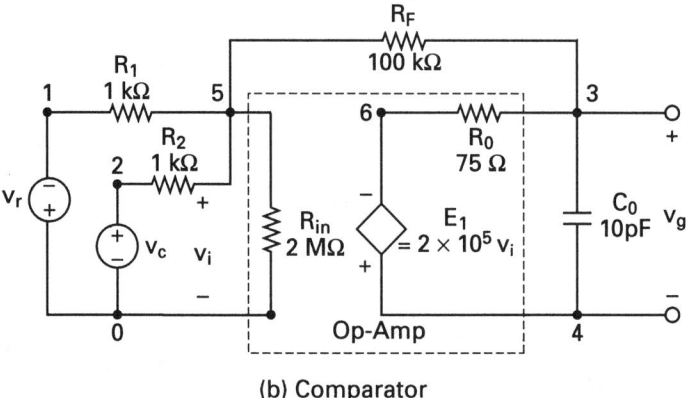

(b) Comparator

Figure 8-17   Pulse-width modulation (PWM) control.

The modulation index M is defined by

$$M = \frac{A_c}{A_r} \tag{8-7}$$

where A_r is the peak value of a reference signal v_r and A_c is the peak value of a carrier signal v_c.

The PWM modulator can be used as a subcircuit to generate control signals for a triangular reference voltage of one or more pulses per half-cycle and a dc (or sinusoidal) carrier signal. The subcircuit definition for the modulator model PWM can be described as follows:

```
*    Subcircuit for PWM control:
.SUBCKT    PWM       1        2        3         4
*          model    ref.    carrier  +control  -control
*          name     input    input   voltage   voltage
R1      1    5     1K
R2      2    5     1K
RIN     5    0     2MEG
RF      5    6     100K
RO      6    3     75
CO      3    4     10PF
E1      6    4     0    5    2E+5      ; Voltage-controlled voltage source
  .ENDS    PWM                        ; Ends subcircuit definition
```

Example 8-7

A single-phase converter with an extinction angle control is shown in Fig. 8-18(a). The input voltage has a peak of 169.7 V, 60 Hz. The load inductance L is 6.5 mH, and the load resistance R is 0.5 Ω. The load battery voltage is $V_x = 10$ V. The extinction angle is $\beta = 60°$. Use PSpice (a) to plot the instantaneous output voltage v_o, the load current i_o, and the input current i_s, and (b) to calculate the Fourier coefficients of the input current i_s and the input power factor PF.

Solution The conduction angles are generated with two carrier voltages as shown in Fig. 8-18(b). Let us assume that $A_r = 10$ V. For $\beta = 60$, $A_c = 10 \times 60/360 = 3.33$ V, and $M = 3.33/10 = 0.333$. v_c is generated by a PWL waveform, and v_r by a PULSE waveform. The list of the circuit file for the converter is as follows:

■ ■

Example 8-7   Single-phase semiconverter with extinction angle control

```
SOURCE   ■ VS    6    0    SIN (0    169.7V    60HZ)
           Vr1   9    0    PULSE (0 -10V  0   1NS  8333.3US  1NS  16666.67US)
           Vc    10   0    PWL (0  0  1NS 3.33V  50MS 3.33V)
           Rg1   7    0    10MEG
           Vr2   11   0    PULSE (0 -10V 8333.33US 1NS  8333.33US 1NS 16666.67US)
           Rg2   8    0    10MEG
```

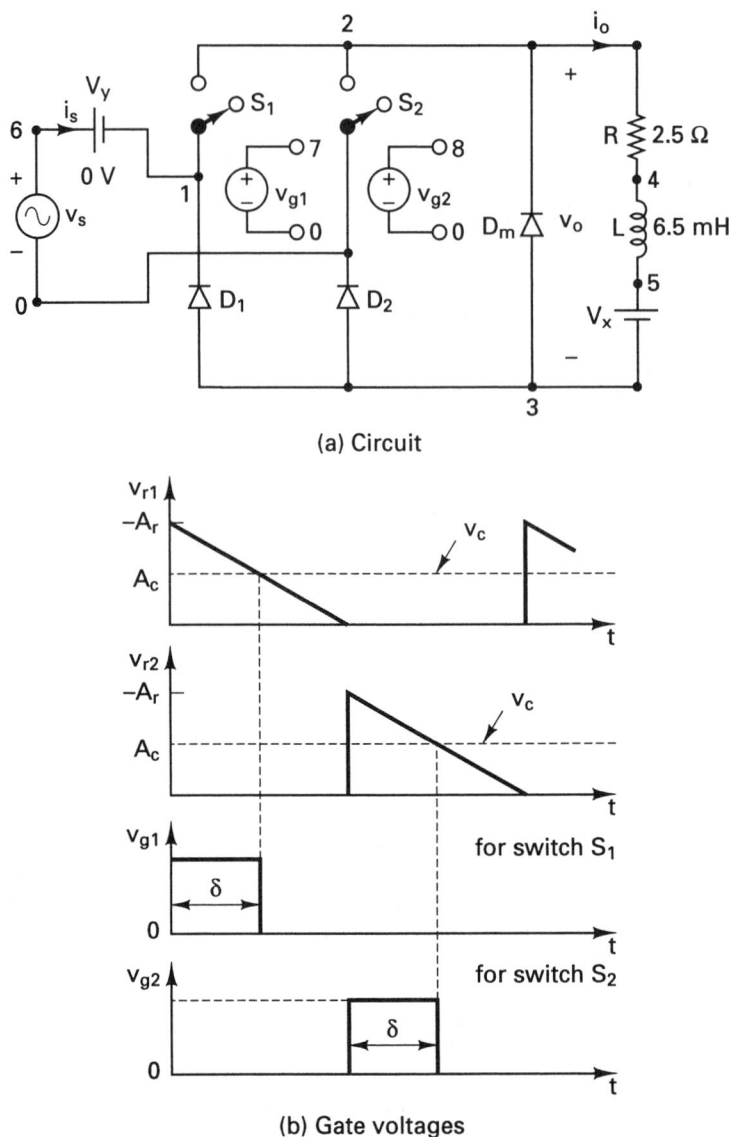

(a) Circuit

(b) Gate voltages

Figure 8-18 Single-phase converter with extinction angle control.

```
CIRCUIT  ■ ■  R    2   4    2.5
            L    4   5    6.5MH
            VX   5   3    DC    10V  ; Load battery voltage
            VY   6   1    DC    0V   ; Voltage source to measure supply current
            D1   3   1    DMOD              ; Diode
```

```
D2     3    0    DMOD                  ; Diode
DM     3    2    DMOD                  ; Freewheeling diode
.MODEL  DMOD  D(IS=2.2E-15 BV=1200V TT=0 CJO=2PF)  ; Diode model parameters
*    Subcircuit calls for switched thyristor model:
XT1    1    2    7    0    SSCR        ; Thyristor T1
XT2    0    2    8    0    SSCR        ; Thyristor T2
*    Subcircuit calls for PWM control:
XPW1   9   10    7    0    PWM         ; Control voltage for thyristor T1
XPW2  11   10    8    0    PWM         ; Control voltage for thyristor T2
*    Subcircuit SSCR, which is missing, must be inserted.
*    Subcircuit PWM, which is missing, must be inserted.
```

ANALYSIS ■ ■ ■

```
.TRAN    50US  50MS  33.333MS  50US    ; Transient analysis
.PROBE                                 ; Graphics post-processor
.OPTIONS ABSTOL = 1.00U RELTOL = 0.01 VNTOL = 0.01 ITL5=20000 ;
*                                                      Convergence
.FOUR    60HZ   I(VY)                  ; Fourier analysis
.END
```
■ ■

(a) The PSpice plots of the instantaneous output voltage V(2,3), the load current I(VX), and the input current I(VY) are shown in Fig. 8-19. The input current is not symmetrical about the 0-axis, and its shape depends on the load current.

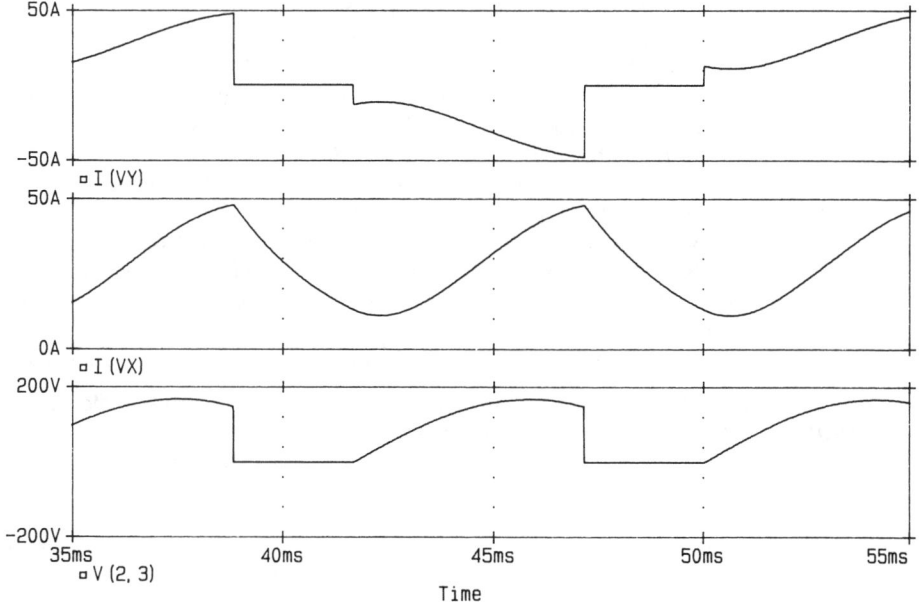

Example 8-7   Single-Phase Semiconverter with Extinction Angle Control
Date/Time run: 06/17/92 18:41:21   Temperature: 27.0

Figure 8-19 Plots for Example 8-7.

(b) The Fourier series of the input current is as follows:

```
FOURIER COMPONENTS OF TRANSIENT RESPONSE I(VY)
DC COMPONENT =   -8.990786E-02
```

HARMONIC NO	FREQUENCY (HZ)	FOURIER COMPONENT	NORMALIZED COMPONENT	PHASE (DEG)	NORMALIZED PHASE (DEG)
1	6.000E+01	3.078E+01	1.000E+00	1.338E+01	0.000E+00
2	1.200E+02	1.281E-01	4.163E-03	1.912E+01	5.742E+00
3	1.800E+02	1.057E+01	3.434E-01	1.779E+02	1.645E+02
4	2.400E+02	1.335E-01	4.338E-03	1.697E+02	1.564E+02
5	3.000E+02	6.870E+00	2.232E-01	-4.495E+01	-5.833E+01
6	3.600E+02	1.798E-01	5.841E-03	-8.377E+01	-9.714E+01
7	4.200E+02	5.066E+00	1.646E-01	5.811E+01	4.473E+01
8	4.800E+02	1.225E-01	3.980E-03	2.827E+01	1.490E+01
9	5.400E+02	2.582E+00	8.390E-02	-1.671E+02	-1.805E+02

```
TOTAL HARMONIC DISTORTION =   4.493854E+01 PERCENT
```

Total harmonic distortion of input current, THD = 44.93% = 0.4493

Displacement angle, ϕ_1 = 13.38

Displacement factor, DF = cos ϕ_1 = cos(13.38) = 0.9728 (leading)

The input power factor is

$$\text{PF} = \frac{1}{(1 + \text{THD}^2)^{1/2}} \cos \phi_1 = \frac{1}{(1 + 0.4493^2)^{1/2}} \times 0.9728 = 0.8873 \quad \text{(leading)}$$

Example 8-8

A symmetrical angle control is applied to the converter of Fig. 8-18(a). The input voltage has a peak of 169.7 V, 60 Hz. The load inductance L is 6.5 mH, and the load resistance R is 2.5 Ω. The load battery voltage is V_x = 10 V. The conduction angle is $\beta = 60°$. Use PSpice (a) to plot the instantaneous output voltage v_o, the load current i_o, and the input current i_s, and (b) to calculate the Fourier coefficients of the input current i_s and the input power factor PF.

Solution The conduction angles can be generated with two carrier voltages as shown in Fig. 8-20. Let us assume that A_r = 10 V. For $\beta = 60°$, $A_c = 10 \times 60/360 = 3.33$ V, and $M = 3.33/10 = 0.333$. The subcircuit PWM is used to generate control signals. v_r is generated by a PULSE generator with a very small pulse width, say, TW = 1NS. v_c is generated by a PWL generator. The list of the circuit file is as follows:

■ ■

Example 8-8 Single-phase semiconverter with symmetrical angle control

```
SOURCE  ■ VS    6   0   SIN (0   169.7V   60HZ)
          Vr1   9   0   PULSE (0 -10V 0  4166.67US  4166.67US  1NS  16666.67US)
          Vc   10   0   PWL (0 0  1NS 3.33V  50MS 3.33V)
          Rg1   7   0   10MEG
          Vr2  11   0   PULSE (0 -10V 8333.33US 4166.67US 4166.67US 1NS 16666.67US)
          Rg2   8   0   2MEG
CIRCUIT ■ ■ R    2   4   2.5
          L     4   5   6.5MH
          VX    5   3   DC   10V  ; Load battery voltage
```

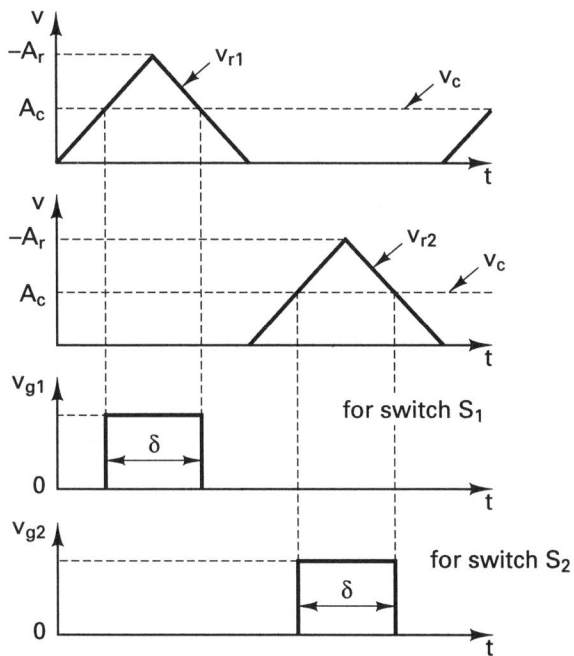

Figure 8-20 Gate voltages with symmetrical angle control.

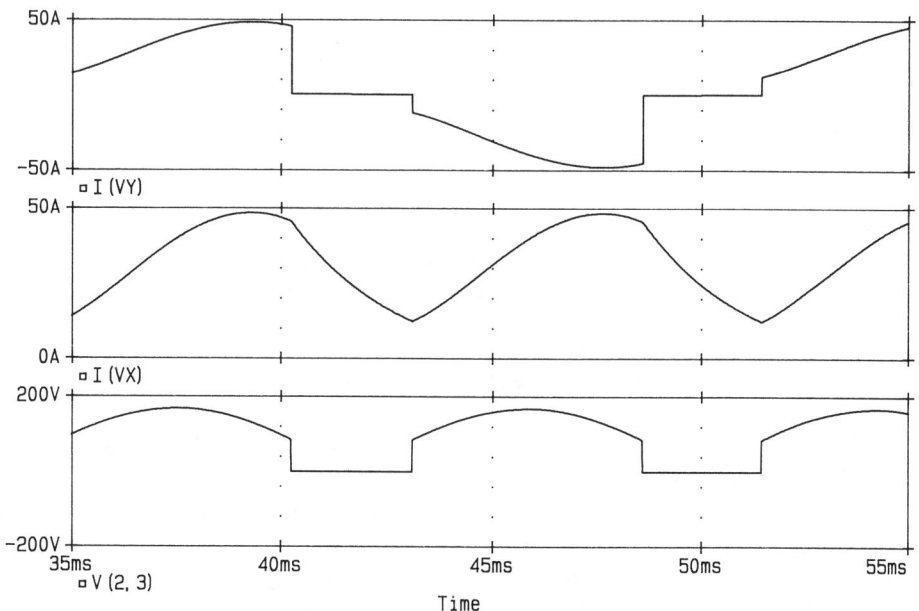

Figure 8-21 Plots for Example 8-8.

```
VY     6   1    DC    0V    ; Voltage source to measure supply current
D1     3   1    DMOD              ; Diode
D2     3   0    DMOD              ; Diode
DM     3   2    DMOD              ; Freewheeling diode
.MODEL  DMOD  D(IS=2.2E-15 BV=1200V TT=0 CJO=0)  ; Diode model parameters
*    Subcircuit calls for switched thyristor model:
XT1    1   2    7    0    SSCR    ; Thyristor T1
XT2    0   2    8    0    SSCR    ; Thyristor T2
*    Subcircuit calls for PWM control:
XPW1   9   10   7    0    PWM     ; Control voltage for thyristor T1
XPW2   11  10   8    0    PWM     ; Control voltage for thyristor T2
*    Subcircuit SSCR, which is missing, must be inserted.
*    Subcircuit PWM, which is missing, must be inserted.
```
■ ■ ■
```
.TRAN   20US  50MS  33.33MS  20US    ; Transient analysis
.PROBE                               ; Graphics post-processor
.OPTIONS ABSTOL = 1.00N RELTOL = 0.01 VNTOL = 0.1 ITL5=20000 ;
*                                                      Convergence
.FOUR   60HZ   I(VY)                 ; Fourier analysis
.END
```

(a) The PSpice plots of the instantaneous output voltage V(2,3), the load current I(VX), and the input current I(VY) are shown in Fig. 8-21. The output voltage is symmetrical, as expected.

(b) The Fourier series of the input current is as follows:

FOURIER COMPONENTS OF TRANSIENT RESPONSE I(VY)

DC COMPONENT =   1.888986E-02

HARMONIC NO	FREQUENCY (HZ)	FOURIER COMPONENT	NORMALIZED COMPONENT	PHASE (DEG)	NORMALIZED PHASE (DEG)
1	6.000E+01	4.063E+01	1.000E+00	-1.143E+01	0.000E+00
2	1.200E+02	4.407E-02	1.085E-03	1.524E+02	1.639E+02
3	1.800E+02	9.826E+00	2.419E-01	1.115E+02	1.230E+02
4	2.400E+02	4.161E-02	1.024E-03	-1.490E+02	-1.376E+02
5	3.000E+02	7.352E+00	1.810E-01	1.630E+02	1.744E+02
6	3.600E+02	4.074E-02	1.003E-03	-8.388E+01	-7.245E+01
7	4.200E+02	4.407E+00	1.085E-01	-1.558E+02	-1.444E+02
8	4.800E+02	4.312E-02	1.061E-03	-2.285E+01	-1.142E+01
9	5.400E+02	2.517E+00	6.195E-02	-7.257E+01	-6.113E+01

TOTAL HARMONIC DISTORTION =   3.269051E+01 PERCENT

Total harmonic distortion of input current, THD = 32.69% = 0.3269

Displacement angle, $\phi_1 = -11.43°$

Displacement factor, DF = $\cos \phi_1 = \cos(-11.434) = 0.98$ (lagging)

From Eq. (7-3), the input power factor

$$PF = \frac{1}{(1 + THD^2)^{1/2}} \cos \phi_1 = \frac{1}{(1 + 0.3269^2)^{1/2}} \times 0.98 = 0.9315 \quad \text{(lagging)}$$

Example 8-9

A pulse-width modulation (PWM) control with four pulses half-cycle is applied to the converter of Fig. 8-18(a). The input voltage has a peak of 169.7 V, 60 Hz. The load inductance L is 6.5 mH, and the load resistance R is 0.5 Ω. The load battery voltage is $V_x = 10$ V. Use PSpice (a) to plot the instantaneous output voltage v_o, the load current i_o, and the input current i_s, and (b) to calculate the Fourier coefficients of the input current i_s and the input power factor PF. Assume four pulses per half-cycle and a modulation index of 4.

Solution The conduction angles are generated with two carrier voltages as shown in Fig. 8-22. Assume that reference voltage $V_r = 10$ V. For $M = 0.4$, the carrier voltage $V_c = MV_r = 4$ V. We can use the same subcircuit PWM for generation of control signals. Note that the carrier voltages are generated with a PWL generator. Instead, we could use PULSE generator with a very small pulse width, say, TW = 1 ns. The list of the circuit file is as follows:

∎∎

Example 8-9   Single-phase semiconverter with PWM control

SOURCE	∎ VS	6	0	SIN (0	169.7V	60HZ)			
	Vr1	9	0	PULSE (0	−6V	0	1NS	1NS	8333.33US	16666.67US)
	Rg1	7	0	10MEG					
	Vr2	11	0	PULSE (0	−6V	8333.33US	1NS	1NS	8333.33US	16666.67US)
	Rg2	8	0	10MEG					
	Vc	10	0	PULSE (10V	0	0	1041.67US	1041.67US	1NS	2083.34US)

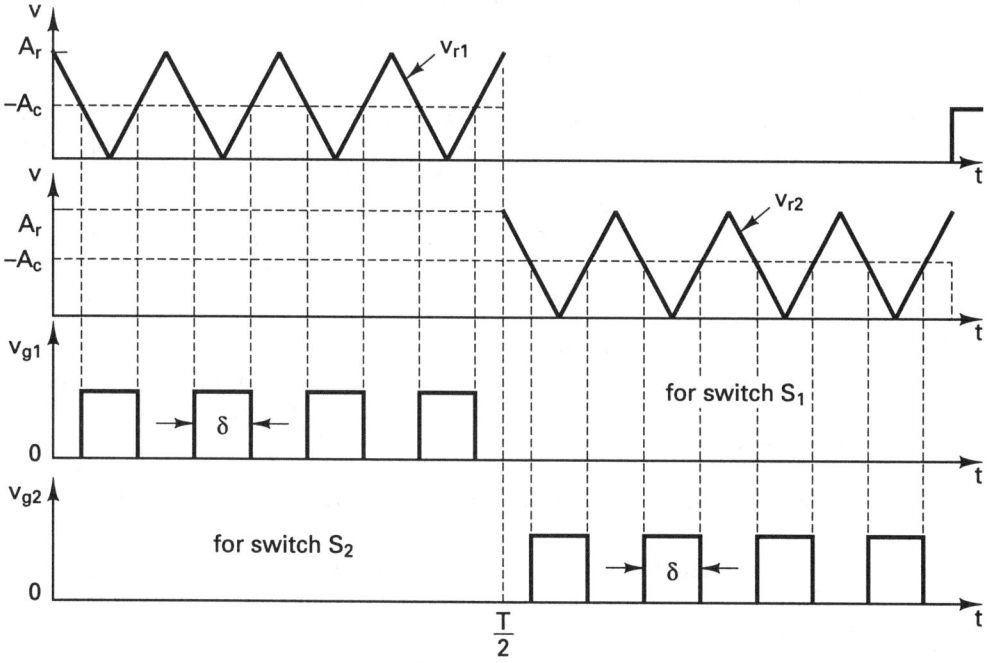

Figure 8-22 Gate voltages with PWM control.

CIRCUIT	■ ■	R	2	4	2.5		
		L	4	5	6.5MH		
		VX	5	3	DC	10V	; Load battery voltage
		VY	6	1	DC	0V	; Voltage source to measure supply current
		D1	3	1	DMOD		; Diode
		D2	3	0	DMOD		; Diode
		DM	3	2	DMOD		; Freewheeling diode

```
               .MODEL   DMOD   D(IS=2.2E-15 BV=1200V TT=0 CJO=1PF) ; Diode model parameters
               *    Subcircuit calls for switched thyristor model:
               XT1  1   2   7   0   SSCR     ; Thyristor T1
               XT2  0   2   8   0   SSCR     ; Thyristor T2
               *    Subcircuit calls for PWM controls:
               XPW1  9  10   7   0   PWM      ; Control voltage for thyristor T1
               XPW2 11  10   8   0   PWM      ; Control voltage for thyristor T2
               *    Subcircuit SSCR, which is missing, must be inserted.
               *    Subcircuit PWM, which is missing, must be inserted.
ANALYSIS  ■ ■ ■ .TRAN   50US 50MS   33.33MS   50US   ; Transient analysis
               .PROBE                                ; Graphics post-processor
               .OPTIONS ABSTOL = 1.00N  RELTOL = 0.01  VNTOL = 0.1  ITL5=20000
               .FOUR    60HZ   I(VY)                 ; Fourier analysis
          .END
```

(a) The PSpice plots of the instantaneous output voltage V(2,3), the load current I(VX), and the input current I(VY) are shown in Fig. 8-23. The input current consists of four pulses per half-cycle.

(b) The Fourier series of the input current is as follows:

FOURIER COMPONENTS OF TRANSIENT RESPONSE I(VY)

DC COMPONENT = 5.261076E-02

HARMONIC NO	FREQUENCY (HZ)	FOURIER COMPONENT	NORMALIZED COMPONENT	PHASE (DEG)	NORMALIZED PHASE (DEG)
1	6.000E+01	1.768E+01	1.000E+00	-1.163E+01	0.000E+00
2	1.200E+02	1.070E-01	6.051E-03	1.324E+02	1.441E+02
3	1.800E+02	4.811E+00	2.721E-01	1.433E+01	2.596E+01
4	2.400E+02	2.142E-01	1.212E-02	1.416E+00	1.305E+01
5	3.000E+02	4.184E+00	2.367E-01	-4.615E+00	7.015E+00
6	3.600E+02	2.012E-01	1.138E-02	1.386E+02	1.502E+02
7	4.200E+02	1.039E+01	5.876E-01	1.230E+00	1.286E+01
8	4.800E+02	1.529E-01	8.651E-03	1.689E+02	1.805E+02
9	5.400E+02	8.397E+00	4.750E-01	1.564E+02	1.681E+02

TOTAL HARMONIC DISTORTION = 8.374184E+01 PERCENT

Total harmonic distortion of input current, THD = 83.7% = 0.837

Displacement angle, $\phi_1 = -11.63°$

Displacement factor, DF = cos ϕ_1 = cos(−11.63°) = 0.979 (lagging)

From Eq. (7-3), the input power factor

$$\text{PF} = \frac{1}{(1 + \text{THD}^2)^{1/2}} \cos \phi_1 = \frac{1}{(1 + 0.837^2)^{1/2}} \times 0.979 = 0.75 \quad \text{(lagging)}$$

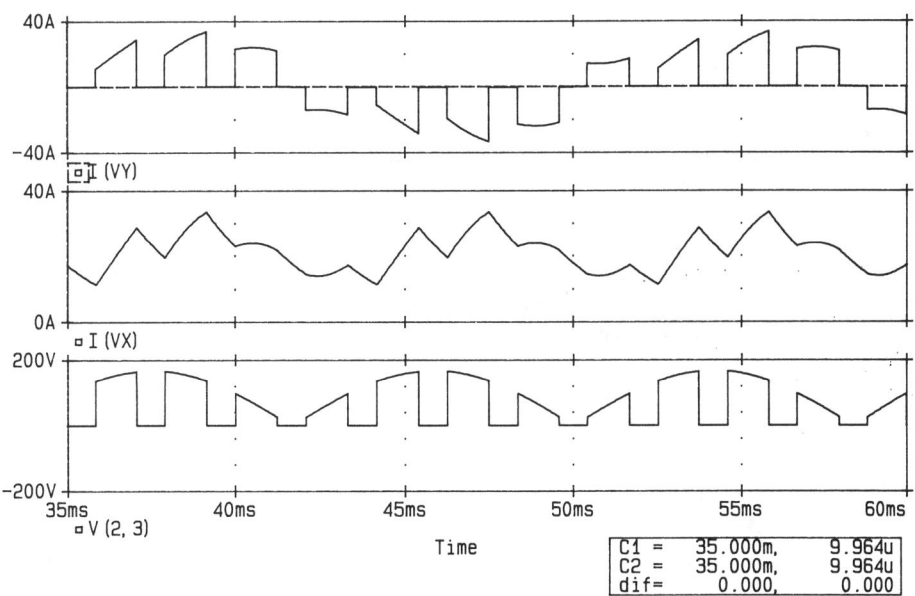

Figure 8-23 Plots for Example 8-9.

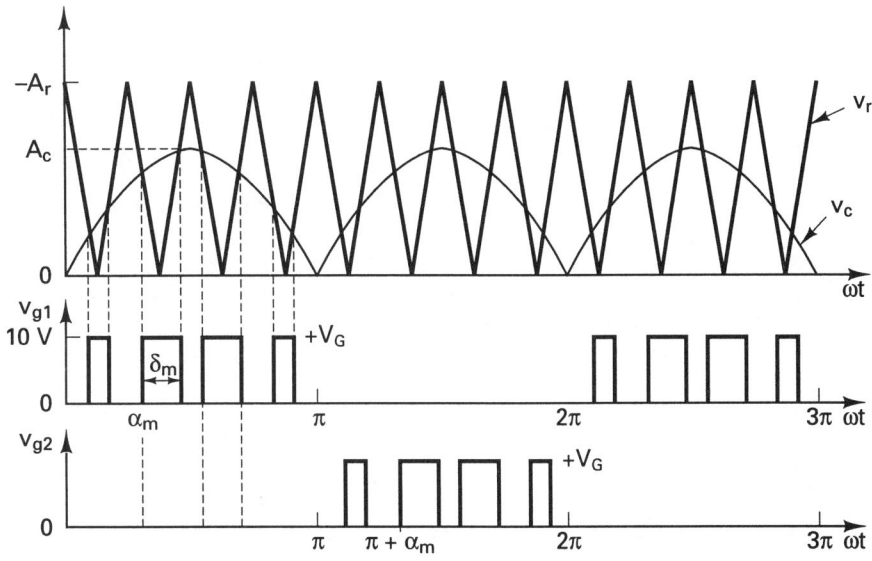

Figure 8-24 Gate voltages with a sinusoidal PWM control.

Example 8-10

A sinusoidal pulse-width modulation (SPWM) control with four pulses per half-cycle is applied to the converter of Fig. 8-18(a). The input voltage has a peak of 169.7 V, 60 Hz. The load inductance L is 6.5 mH, and the load resistance R is 2.5 Ω. The load battery voltage is $V_x = 10$ V. Use PSpice (a) to plot the instantaneous output voltage v_o, the load current i_o, and the input current i_s, and (b) to calculate the Fourier coefficients of the input current i_s and the input power factor PF. Assume four pulses per half-cycle and a modulation index of 0.4.

Solution The conduction angles are generated with two carrier voltages as shown in Fig. 8-24. In this type of control, the carrier signal must be a rectified sine wave. PSpice generates only a sine wave. Thus we can use a precision rectifier to convert a sine-wave input signal to two sine-wave pulses, and use a comparator to generate a PWM waveform. This is implemented in Fig. 8-25. We can use the subcircuit SPWM for generation of the control signals.

The subcircuit definition for the modulator model SPWM can be described as follows:

```
*     Subcircuit for sinusoidal PWM control
.SUBCKT    SPWM       1        2        3        4        8
*          model    ref.    carrier  +control  -control  rectified
*          name     input    input    voltage   voltage  carrier sine wave
R1     1    9    1K
R2     8    9    1K
RF     9    3    100K
R3     2    10   50K
R4     6    7    50K
R5     10   6    50K
R6     2    7    100K
R7     7    8    100K
CO     3    4    10PF              ; Capacitor to aid convergence
D1     6    5    DMD
D2     5    10   DMD
.MODEL    DMD      D               ; Default model parameters
X1     10   0    5    0    OPAMP   ; Call subcircuit for op-amp A1
X2     7    0    8    0    OPAMP   ; Call subcircuit for op-amp A2
X3     9    0    3    4    OPAMP   ; Call subcircuit for op-amp A3
.SUBCKT   OPAMP     1     5     2     3
*          name    -vi   +vi   +vo   -vo
RIN    1    5    2MEG
RO     4    2    75
E1     3    4    1     5    2E+5     ; Voltage-controlled voltage source
.ENDS    OPAMP                      ; Ends subcircuit OPAMP definition
.ENDS    SPWM                       ; Ends subcircuit PWM definition
```

Let us assume a reference voltage $V_r = 10$ V. For $M = 0.4$, the carrier voltage $V_c = MV_r = 4$ V. The list of the circuit file for the converter is as follows:

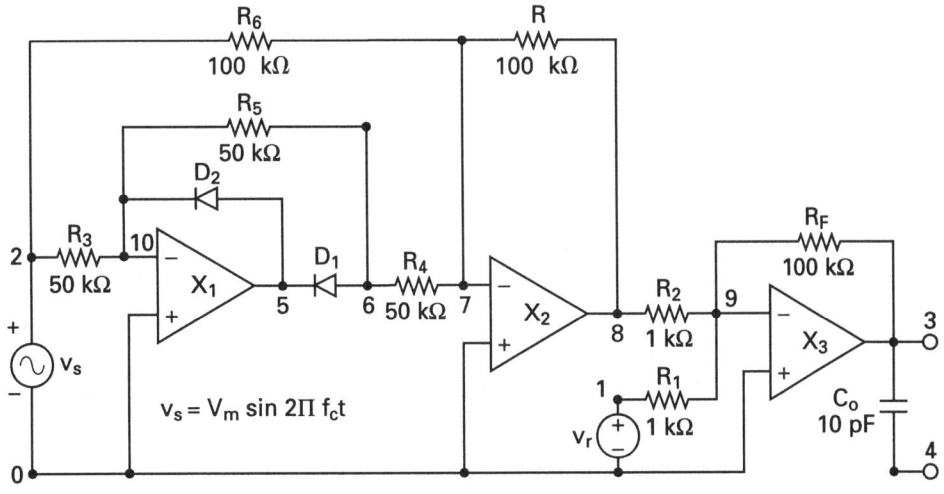

(a) Precision rectifier and comparator

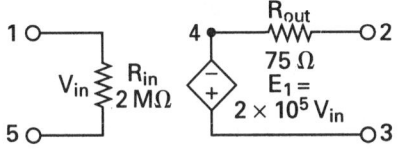

(b) Op-amp model

Figure 8-25   Sinusoidal PWM generator.

■ ■

Example 8-10   Single-phase semiconverter with sinusoidal PWM control

```
SOURCE   ■ VS     6    0    SIN (0    169.7V    60HZ)
           Vr1    9    0    PULSE (−10V 0  0  1041.67US 1041.67US  1NS  2083.34US)
           Rg1    7    0    10MEG
           Vr2    11   0    PULSE (−10V 0 8333.33US 1041.67US 1041.67US 1NS  2083.34US)
           Rg2    8    0    10MEG
           Vc     10   0    SIN (0    4V    60HZ)
CIRCUIT  ■ ■ Rc   12   0    10MEG        ; Rectified carrier sine wave
           R      2    4    2.5
           L      4    5    6.5MH
           VX     5    3    DC    10V  ; Load battery voltage
           VY     6    1    DC    0V   ; Voltage source to measure supply current
           D1     3    1    DMOD              ; Diode
           D2     3    0    DMOD              ; Diode
           DM     3    2    DMOD              ; Freewheeling diode
          .MODEL  DMOD  D(IS=2.2E−15 BV=1200V TT=0)   ; Diode model parameters
           *   Subcircuit calls for switched thyristor model:
           XT1    1    2    7    0    SSCR    ; Thyristor T1
```

```
XT2    0    2    8    0    SSCR      ; Thyristor T2
*    Subcircuit calls for SPWM control:
XPW1   9   10    7    0    12    SPWM  ; Control voltage for thyristor T1
XPW2  11   10    8    0    12    SPWM  ; Control voltage for thyristor T2
*    Subcircuit SSCR, which is missing, must be inserted.
*    Subcircuit SPWM, which is missing, must be inserted.
```

`.TRAN    50US  50MS  33.33MS  50US     ; Transient analysis`
`.PROBE                                  ; Graphics post-processor`
`.OPTIONS ABSTOL = 1.00N  RELTOL = 0.01  VNTOL = 0.1  ITL5=20000`
`.FOUR    60HZ   I(VY)                   ; Fourier analysis`
`.END`

(a) The PSpice plots of the instantaneous output voltage V(2,3), the load current I(VX), and the input current I(VY) are shown in Fig. 8-26. The input current pulses follow a sinusoidal pattern.

(b) The Fourier series of the input current is as follows:

FOURIER COMPONENTS OF TRANSIENT RESPONSE I(VY)

DC COMPONENT = −3.111144E−02

HARMONIC NO	FREQUENCY (HZ)	FOURIER COMPONENT	NORMALIZED COMPONENT	PHASE (DEG)	NORMALIZED PHASE (DEG)
1	6.000E+01	2.255E+01	1.000E+00	−1.127E+01	0.000E+00
2	1.200E+02	2.386E−01	1.058E−02	9.025E+00	2.029E+01
3	1.800E+02	8.748E+00	3.880E−01	6.041E+00	1.731E+01
4	2.400E+02	2.635E−01	1.169E−02	−1.781E+02	−1.668E+02

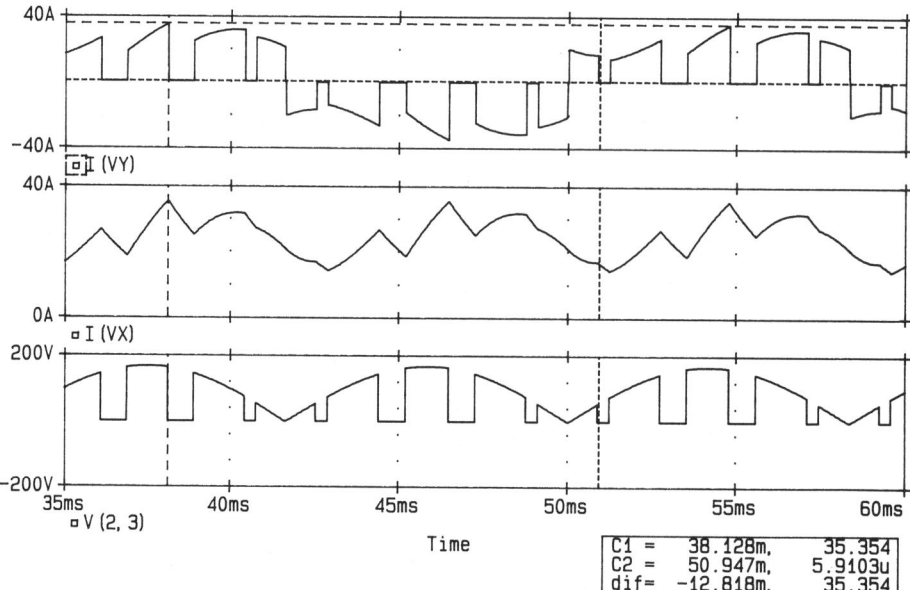

Example 8-10   Single-Phase Semiconverter with Sinusoidal PWM Control
Date/Time run: 06/18/92 16:08:03   Temperature: 27.0

Figure 8-26 Plots for Example 8-10.

5	3.000E+02	4.986E+00	2.211E-01	1.760E+01	2.887E+01
6	3.600E+02	9.425E-02	4.180E-03	3.283E+01	4.410E+01
7	4.200E+02	5.257E+00	2.332E-01	1.662E+02	1.775E+02
8	4.800E+02	1.603E-01	7.111E-03	1.688E+02	1.801E+02
9	5.400E+02	1.184E+01	5.252E-01	-1.120E+01	7.249E-02

TOTAL HARMONIC DISTORTION =   7.279670E+01 PERCENT

Total harmonic distortion of input current, THD = 72.8% = 0.728

Displacement angle, $\phi_1 = -11.27°$

Displacement factor, DF = $\cos \phi_1 = \cos(-11.27) = 0.9807$ (lagging)

From Eq. (7-3), the input power factor

$$PF = \frac{1}{(1 + THD^2)^{1/2}} \cos \phi_1 = \frac{1}{(1 + 0.728^2)^{1/2}} \times 0.9807 = 0.7928 \quad (lagging)$$

8-8 LABORATORY EXPERIMENTS

It is possible to develop many experiments for demonstrating the operation and characteristics of thyristor-controlled rectifiers. The following experiments are suggested:

Single-phase half-wave controlled rectifier

Single-phase full-wave controlled rectifier

Three-phase full-wave controlled rectifier

8-8.1 Experiment TC-1

SINGLE-PHASE HALF-WAVE CONTROLLED RECTIFIER

Objective   The objective is to study the operation and characteristics of a single-phase half-wave (thyristor) controlled rectifier under various load conditions.

Textbook   See Ref. 1, Sec. 5-2.

Apparatus   1. One phase-controlled thyristor with ratings of at least 50 A and 400 V, mounted on a heat sink
   2. One diode with ratings of at least 50 A and 400 V, mounted on a heat sink
   3. A firing pulse generator with isolating signals for gating thyristors
   4. An *RL* load
   5. One dual-beam oscilloscope with floating or isolating probes
   6. Ac and dc voltmeters and ammeters and one noninductive shunt
   7. One isolation transformer

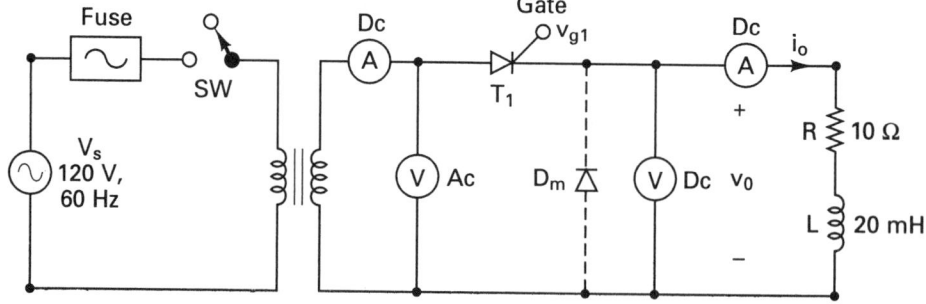

Figure 8-27   Single-phase half-wave controlled rectifier.

Warning Before making any circuit connection, switch the ac power OFF. **Do not** switch the power ON unless the circuit is checked and approved by your lab instructor. **Do not** touch the thyristor heat sinks, which are connected to live terminals.

Part 1: without freewheeling diode

Experimental procedure

1. Set up the circuit as shown in Fig. 8-27. Use the load resistance R only.
2. Connect the measuring instruments as required.
3. Connect the firing pulses to the appropriate thyristors.
4. Set the delay angle to $\alpha = \pi/3$.
5. Observe and record the waveforms of the load voltage v_o and the load current i_o.
6. Measure the average load voltage $V_{o(dc)}$, the rms load voltage $V_{o(rms)}$, the average load current $I_{o(dc)}$, the rms load current $I_{o(rms)}$, the rms input current $I_{s(rms)}$, the rms input voltage $V_{s(rms)}$, and the load power P_L.
7. Measure the conduction angle of the thyristor T_1.
8. Repeat steps 2 to 7 with the load inductance L only.
9. Repeat steps 2 to 7 with both load resistance R and load inductance L.

Part 2: with freewheeling diode

1. Set up the circuit as shown in Fig. 8-27 with a freewheeling diode across the load as shown by the dashed lines.
2. Repeat the steps in Part 1.

Report

1. Present all recorded waveforms and discuss all significant points.
2. Compare the waveforms generated by SPICE with the experimental results, and comment.
3. Compare the experimental results with the results predicted.
4. Calculate and plot the average output voltage $V_{o(dc)}$ and the input power factor PF against the delay angle α.
5. Discuss the advantages and disadvantages of this type of rectifier.
6. Discuss the effects of the freewheeling diode on the performance of the rectifier.

8-8.2 Experiment TC-2

SINGLE-PHASE FULL-WAVE CONTROLLED RECTIFIER

Objective   The objective is to study the operation and characteristics of a single-phase full-wave (thyristor)-controlled rectifier under various load conditions.

Applications   The single-phase full-wave controlled rectifier is used to control power flow in many applications (e.g., power supplies, variable-speed dc motor drives, and input stages of other converters).

Textbook   See Ref. 1, Secs. 5-3 and 5-4.

Apparatus   1. Four phase-controlled thyristors with ratings of at least 50 A and 400 V, mounted on heat sinks
2. One diode with ratings of at least 50 A and 400 V, mounted on a heat sink
3. A firing pulse generator with isolating signals for gating thyristors
4. An *RL* load
5. One dual-beam oscilloscope with floating or isolating probes
6. Ac and dc voltmeters and ammeters and one noninductive shunt

Warning   See Experiment TC-1.

Experimental procedure   Set up the circuit as shown in Fig. 8-28. Repeat the steps in Parts 1 and 2 of Experiment TC-1 on the single-phase half-wave controlled rectifier.

Report   Repeat the steps in Experiment TC-1 for the single-phase half-wave controlled rectifier.

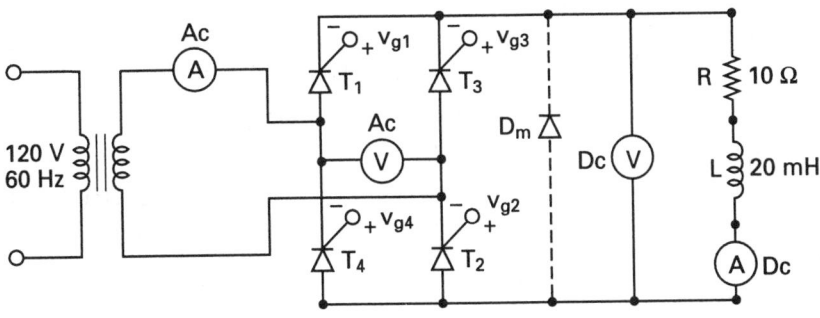

Figure 8-28   Single-phase full-wave controlled rectifier.

8-8.3 Experiment TC-3

THREE-PHASE FULL-WAVE CONTROLLED RECTIFIER

Objective   The objective is to study the operation and characteristics of a three-phase full-wave (thyristor) controlled rectifier under various load conditions.

Applications   The three-phase full-wave controlled rectifier is used to control power flow in many applications (e.g., power supplies, variable-speed dc motor drives, and input stages of other converters).

Textbook   See Ref. 1, Secs. 5-8 and 5-9.

Apparatus   **1.** Six phase-controlled thyristors with ratings of at least 50 A and 400 V, mounted on heat sinks
2. One diode with ratings of at least 50 A and 400 V, mounted on a heat sink
3. A firing pulse generator with isolating signals for gating thyristors
4. An *RL* load
5. One dual-beam oscilloscope with floating or isolating probes
6. Ac and dc voltmeters and ammeters and one noninductive shunt

Warning   See Experiment TC-1.

Experimental procedure   Set up the circuit as shown in Fig. 8-29. Repeat the steps in Parts 1 and 2 for Experiment TC-1 on the single-phase half-wave controlled rectifier.

Report   **1.** Present all recorded waveforms and discuss all significant points.
2. Compare the waveforms generated by SPICE with the experimental results, and comment.
3. Compare the experimental results with the results predicted.

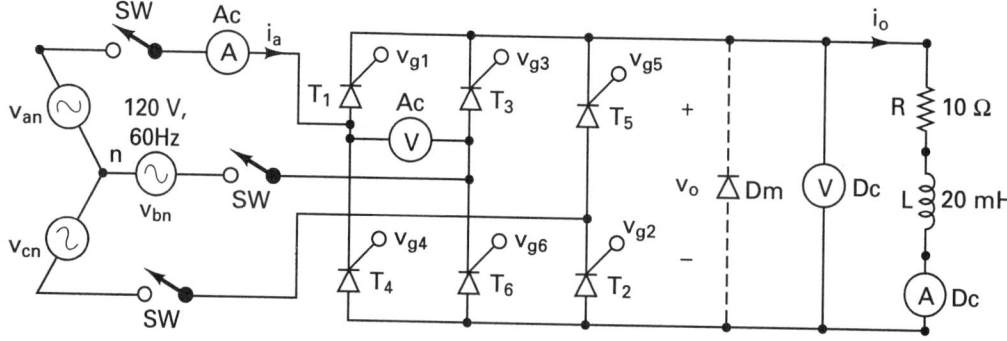

Figure 8-29   Three-phase full-wave controlled rectifier.

4. Discuss the advantages and disadvantages of this type of rectifier.

5. Discuss the effects of the freewheeling diode on the performance of the rectifier.

SUMMARY

The statements for an ac thyristor are as follows:

```
*     Subcircuit call for switched thyristor model:
XT1     NA       NK       +NG          -NG          SCR
        anode   cathode   +control    -control     model
*                         voltage     voltage      name
*     Subcircuit call for PWM control:
XPW     VR       VC       +NG          -NG          PWM
*       ref.    carrier   +control    -control     model
*       input   input     voltage     voltage      name
*     Subcircuit call for sinusoidal PWM control:
XSPW    VR       VS       +NG          -NG          VC              SPWM
*       ref.   sine-wave  +control    -control    rectified        model
*       input   input     voltage     voltage     carrier sine wave  name
```

SUGGESTED READING

1. M. H. Rashid, *Power Electronics: Circuits, Devices, and Applications*, 2nd ed. Englewood Cliffs, N.J.: Prentice Hall, 1993.

2. M. H. Rashid, *SPICE for Circuits and Electronics Using PSpice*. Englewood Cliffs, N.J.: Prentice Hall, 1990, Chap. 4.

3. L. J. Giacoletto, "Simple SCR and TRIAC PSPICE computer models," *IEEE Transactions on Industrial Electronics*, Vol. 36, No. 3, August 1989, pp. 451–455.

4. Goce L. Arsov, "Comments on 'A nonideal macromodel of thyristor for transient analysis in power electronics,'" *IEEE Transactions on Industrial Electronics*, Vol. 39, No. 2, April 1992, pp. 175–176.

5. F. J. Garcia, F. Arizti, and F. J. Aranceta, "A nonideal macromodel of thyristor for transient analysis in power electronics," *IEEE Transactions on Industrial Electronics*, Vol. 37, No. 6, December 1990, pp. 514–520.

6. R. L. Avant and F. C. Y. Lee, "The J3 SCR model applied to resonant converter simulation," *IEEE Transactions on Industrial Electronics*, Vol. 32, No. 1, February 1985, pp. 1–12.

7. P. O. Lauretzen, "Power semiconductor device models for use in circuit simulation," *Conference Proceedings of IEEE-IAS Annual Meeting*, 1990, pp. 1559–1560.

DESIGN PROBLEMS

8-1. Design the single-phase semiconverter of Fig. 8-6(a) with the following specifications:

Ac supply voltage $V_s = 120$ V (rms), 60 Hz

Load resistance $R = 5\ \Omega$

Load inductance $L = 15$ mH

Dc output voltage $V_{o(dc)}$
    = 80% of the maximum value

(a) Determine the ratings of all components and devices under worst-case conditions.

(b) Use SPICE to verify your design.

(c) Provide a cost estimate of the circuit.

8-2. (a) Design an output C filter for the single-phase semiconverter of Problem 8-1. The harmonic content of the load current should be less less than 5% of the value without the filter.

(b) Use SPICE to verify your design in part (a).

8-3. Design the single-phase full converter of Fig. 8-8(a) with the following specifications:

Ac supply voltage V_s = 120 V (rms), 60 Hz

Load resistance R = 5 Ω

Load inductance L = 15 mH

Dc output voltage $V_{o(dc)}$
                = 80% of the maximum value

(a) Determine the ratings of all components and devices under worst-case conditions.

(b) Use SPICE to verify your design.

(c) Provide a cost estimate of the circuit.

8-4. (a) Design an output C filter for the single-phase semiconverter of Problem 8-3. The harmonic content of the load current should be less less than 5% of the value without the filter.

(b) Use SPICE to verify your design in part (a).

8-5. Design the three-phase semiconverter of Fig. 8-12(a) with the following specifications:

Ac supply voltage per phase,
                V_s = 120 V (rms), 60 Hz

Load resistance R = 5 Ω

Load inductance L = 15 mH

Dc output voltage $V_{o(dc)}$
                = 80% of the maximum value

(a) Determine the ratings of all components and devices under worst-case conditions.

(b) Use SPICE to verify your design.

(c) Provide a cost estimate of the circuit.

8-6. (a) Design an output C filter for the three-phase semiconverter of Problem 8-5. The harmonic content of the load current should be less less than 5% of the value without the filter.

(b) Use SPICE to verify your design in part (a).

8-7. Design the three-phase full converter of Fig. 8-14(a) with the following specifications:

Ac supply voltage per phase,
                V_s = 120 V (rms), 60 Hz

Load resistance R = 5 Ω

Load inductance L = 15 mH

Dc output voltage $V_{o(dc)}$
                = 80% of the maximum value

(a) Determine the ratings of all components and devices under worst-case conditions.

(b) Use SPICE to verify your design.

(c) Provide a cost estimate of the circuit.

8-8. (a) Design an output C filter for the converter of Problem 8-7. The harmonic content of the load current should be less less than 5% of the value without the filter.

(b) Use SPICE to verify your design in part (a).

Chapter 9

■■■■■■■■■

AC Voltage Controllers

9-1 INTRODUCTION

The input voltage and the output current of ac voltage controllers pass through zero in every cycle. This simplifies the modeling of a thyristor. A thyristor can be turned on by applying a pulse of short duration, and it is turned off by natural commutation due to the characteristic of the input voltage and the current.

9-2 AC THYRISTOR MODEL

The load current of ac voltage controllers is ac type, and the current of a thyristor always passes through zero. There is no need for the diode D_T in Fig. 8-1(b), and the thyristor model can be simplified to Fig. 9-1. This model can be used as a subcircuit. Switch S_1 is controlled by the controlling voltage V_R connected between nodes 6 and 2. The switch parameters can be adjusted to yield the desired on-state drop of the thyristor. In the following examples, we use the switch parameters RON=0.01, ROFF=10E+5, VON=0.1V, and VOFF=0V. The other parameters are discussed in Section 8-2.

The subcircuit definition for the thyristor model ASCR can be described as follows:

```
*    Subcircuit for ac thyristor model:
.SUBCKT    ASCR     1        2        3        2
*          model    anode    cathode  +control  −control
*          name                       voltage   voltage
S1    1    5    6    2    SMOD        ; Switch
RG    3    4    50
```

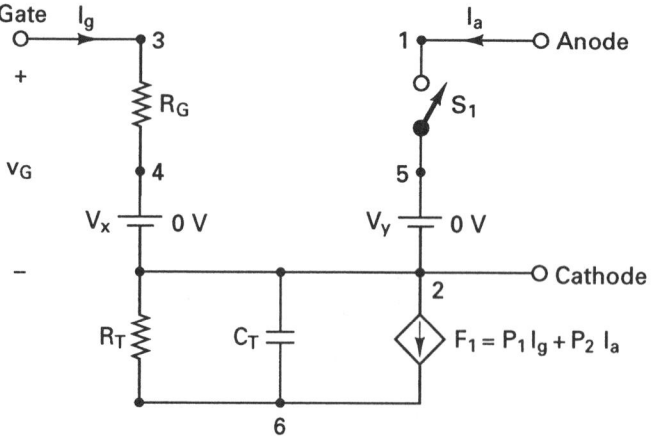

Figure 9-1 Ac thyristor model.

```
VX      4    2    DC    0V
VY      5    2    DC    0V
RT      2    6    1
CT      6    2    10UF
F1      2    6    POLY(2)   VX    VY    0    50    11
.MODEL    SMOD    VSWITCH (RON=0.01 ROFF=10E+5 VON=0.1V VOFF=0V)
.ENDS    ASCR                              ; Ends subcircuit definition
```

9-3 AC VOLTAGE CONTROLLERS

The input to an ac voltage controller is a fixed ac voltage, and its output is a variable ac voltage. When converter switches are turned on, the input voltage is connected to the load. The output voltage, which is varied by varying the conduction time of switches, is discontinuous. The input power factor is low.

9-4 EXAMPLES OF AC VOLTAGE CONTROLLERS

The applications of the ac thyristor model are illustrated by some examples.

Example 9-1

A single-phase full-wave ac voltage controller is shown in Fig. 9-2(a). The input voltage has a peak of 169.7 V, 60 Hz. The load inductance L is 6.5 mH, and the load resistance R is 2.5 Ω. The delay angles are equal: $\alpha_1 = \alpha_2 = 90°$. The gate voltages are shown in Fig. 9-2(b). Use PSpice (a) to plot the instantaneous output voltage v_o and the load current i_o, and (b) to calculate the Fourier coefficients of the input current i_s and the input power factor PF.

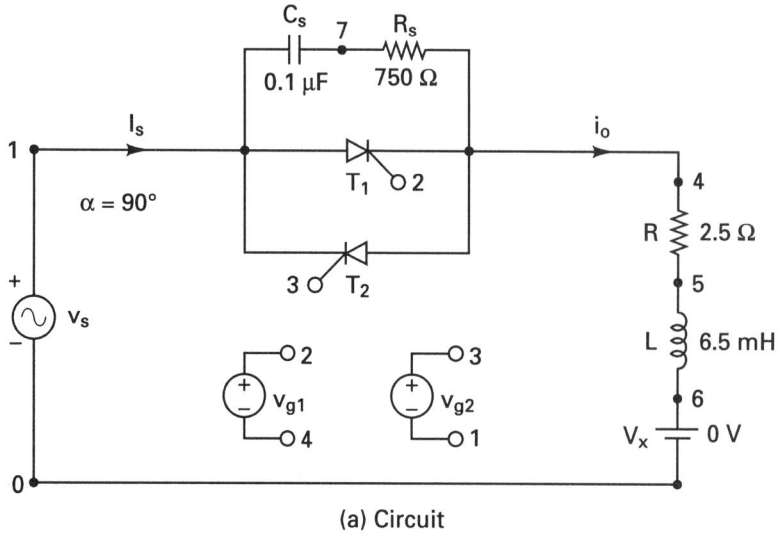

(a) Circuit

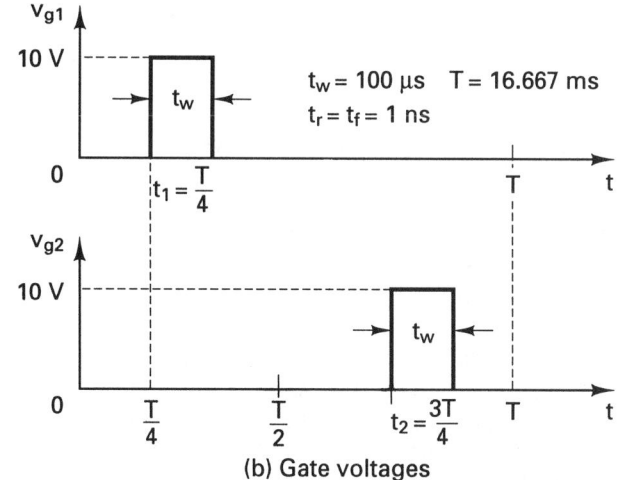

(b) Gate voltages

Figure 9-2 Single-phase full-wave ac voltage-controller for PSpice simulation.

Solution The peak supply voltage, $V_m = 169.7$ V. For $\alpha_1 = \alpha_2 = 90°$, time delay $t_1 = (90/360) \times (1000/60 \text{ Hz}) \times 1000 = 4166.7$ μs. A series snubber with $C_s = 0.1$ μF and $R_s = 750$ Ω is connected across the thyristor to cope with the transient voltage due to the inductive load. The list of the circuit file is as follows:

▪▪▪

Example 9-1   Single-phase ac voltage controller

```
SOURCE    ▪ VS    1    0    SIN (0    169.7V    60HZ)
            Vg1   2    4    PULSE (0V   10V    4166.7US   1NS   1NS   100US   16666.7US)
            Vg2   3    1    PULSE (0V   10V   12500.0US   1NS   1NS   100US   16666.7US)
```

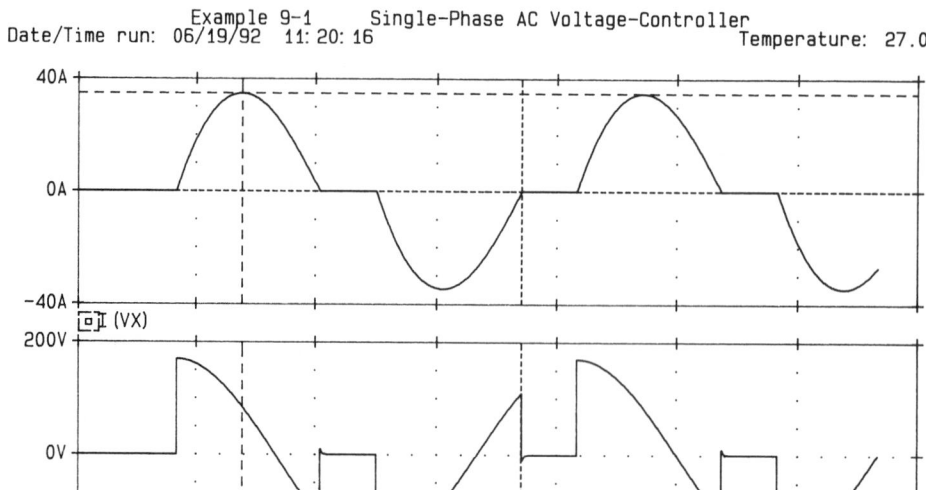

C1 =	6.9277m,	34.719
C2 =	18.529m,	-63.268m
dif=	-11.601m,	34.782

Figure 9-3 Plots for Example 9-1.

```
CIRCUIT  ■ ■ R    4    5    2.5
             L    5    6    6.5MH
             VX   6    0    DC    0V  ; Voltage source to measure the load current
             CS   1    7    0.1UF
             RS   7    4    750
             *   Subcircuit call for thyristor model:
             XT1  1    4    2    4    ASCR      ; Thyristor T1
             XT2  4    1    3    1    ASCR      ; Thyristor T2
             *   Subcircuit ASCR, which is missing, must be inserted.
ANALYSIS ■ ■ ■ .TRAN    20US   33.33MS   0    20US   ; Transient analysis
             .PROBE                               ; Graphics post-processor
             .OPTIONS ABSTOL = 1.00N RELTOL = 1.0M VNTOL = 1.0M ITL5=10000 ;
             *                                              Convergence
             .FOUR    60HZ    I(VX)                ; Fourier analysis
         .END
```

(a) The PSpice plots of instantaneous output voltage V(4) and load current I(VX) are shown in Fig. 9-3. The output voltage and current are discontinuous.

(b) The Fourier series of the input current, which is the same as the current through source VX, is as follows:

FOURIER COMPONENTS OF TRANSIENT RESPONSE I(VX)
DC COMPONENT =   -1.312521E-04

HARMONIC NO	FREQUENCY (HZ)	FOURIER COMPONENT	NORMALIZED COMPONENT	PHASE (DEG)	NORMALIZED PHASE (DEG)
1	6.000E+01	2.883E+01	1.000E+00	-6.244E+01	0.000E+00
2	1.200E+02	1.068E-04	3.704E-06	1.714E+01	7.958E+01
3	1.800E+02	7.966E+00	2.763E-01	-2.494E+00	5.995E+01
4	2.400E+02	9.049E-05	3.139E-06	-1.676E+02	-1.052E+02
5	3.000E+02	2.671E+00	9.265E-02	-1.472E+02	-8.473E+01
6	3.600E+02	3.556E-05	1.233E-06	3.215E+01	9.459E+01
7	4.200E+02	4.161E-01	1.443E-02	3.990E+01	1.023E+02
8	4.800E+02	3.673E-05	1.274E-06	1.672E+02	2.296E+02
9	5.400E+02	5.995E-01	2.079E-02	1.496E+02	2.120E+02

TOTAL HARMONIC DISTORTION =   2.925271E+01 PERCENT

Total harmonic distortion of input current, THD = 29.25% = 0.2925

Displacement angle, $\phi_1 = -62.44°$

Displacement factor, DF = $\cos \phi_1 \doteq \cos(-62.44) = 0.4627$ (lagging)

From Eq. (7-3), the input power factor

$$PF = \frac{1}{(1 + THD^2)^{1/2}} \cos \phi_1 = \frac{1}{(1 + 0.2925^2)^{1/2}} \times 0.4627 = 0.444 \quad \text{(lagging)}$$

Example 9-2

A single-phase half-wave ac voltage controller is supplied from a three-phase wye-connected supply. It is shown in Fig. 9-4(a). The input phase voltage has a peak of 169.7 V, 60 Hz. The load resistance per phase is $R = 2.5 \Omega$. The delay angle is $\alpha = 60°$. The gate voltages are shown in Fig. 9-4(b). Use PSpice (a) to plot the instantaneous output phase voltage v_o, and (b) to calculate the Fourier coefficients of the input phase current i_s and the input power factor PF.

Solution   The peak supply voltage per phase, $V_m = 169.7$ V. For $\alpha = 60°$,

$$\text{time delay } t_1 = \frac{60}{360} \times \frac{1000}{60 \text{ Hz}} \times 1000 = 2777.78 \ \mu s$$

$$\text{time delay } t_2 = \frac{180}{360} \times \frac{1000}{60 \text{ Hz}} \times 1000 = 8333.3 \ \mu s$$

$$\text{time delay } t_3 = \frac{300}{360} \times \frac{1000}{60 \text{ Hz}} \times 1000 = 13,888.9 \ \mu s$$

The list of the circuit file is as follows:

■ ■

Example 9-2   Three-phase half-wave ac voltage controller

SOURCE	■ Van	1	0	SIN(0	169.7V	60HZ)					
	Vbn	2	0	SIN(0	169.7V	60HZ	0	0	-120DEG)		
	Vcn	3	0	SIN(0	169.7V	60HZ	0	0	-240DEG)		
	Vg1	12	4	PULSE (0V	10V	2777.8US	1NS	1NS	100US	16666.7US)
	Vg3	13	7	PULSE (0V	10V	8333.3US	1NS	1NS	100US	16666.7US)
	Vg5	14	9	PULSE (0V	10V	13888.9US	1NS	1NS	100US	16666.7US)

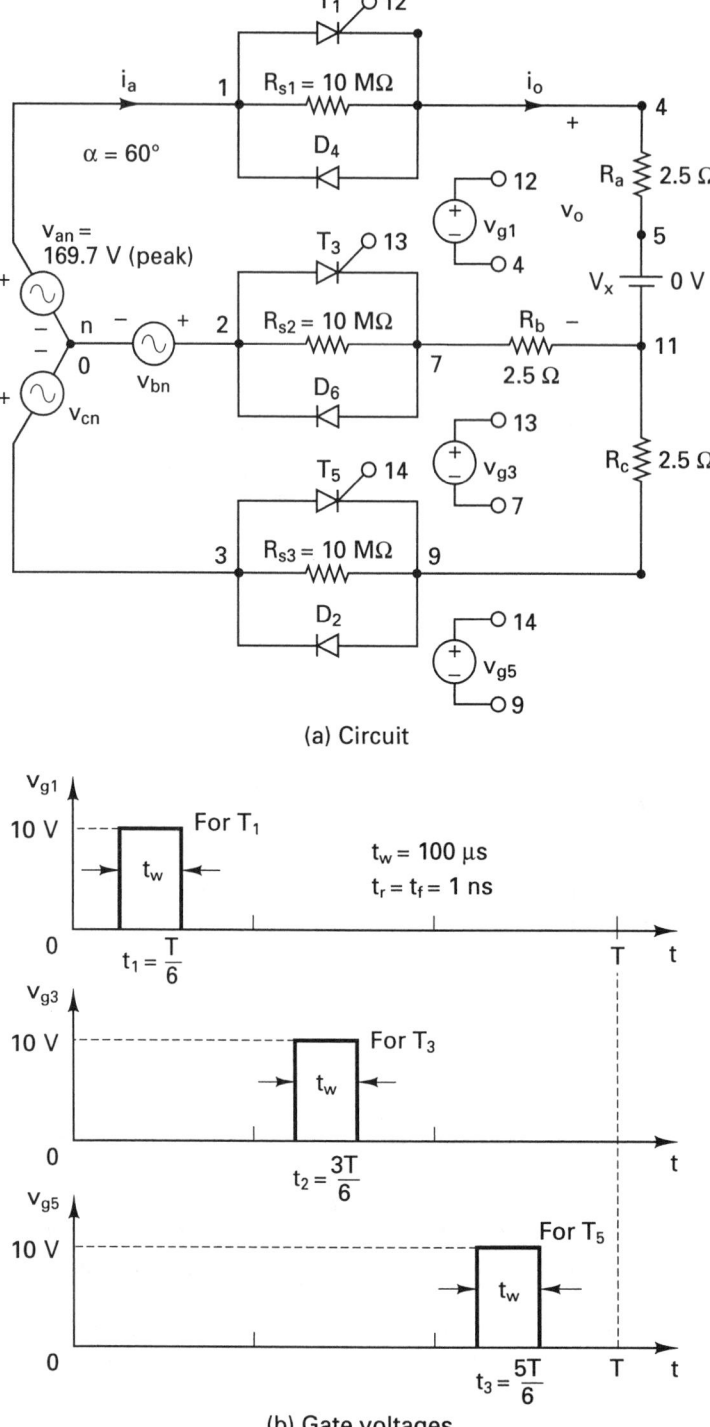

(a) Circuit

(b) Gate voltages

Figure 9-4 Three-phase half-wave ac controller for PSpice simulation.

```
          Rs1   1   4    10MEG
          Rs2   2   7    10MEG
          Rs3   3   9    10MEG
          Ra    4   5    2.5
          VX    5   11   DC    0V  ; To measure load current
          Rb    7   11   2.5
          Rc    9   11   2.5
          *   Subcircuit calls for thyristor model:
          XT1   1   4   12   4   ASCR      ; Thyristor T1
          XT3   2   7   13   7   ASCR      ; Thyristor T3
          XT5   3   9   14   9   ASCR      ; Thyristor T5
          D2    9   3   DMOD             ; Diode
          D4    4   1   DMOD             ; Diode
          D6    7   2   DMOD             ; Diode
          .MODEL   DMOD   D(IS=2.22E-15  BV=1200V  IBV=13E-3  CJO=2PF  TT=1US)
          *   Subcircuit ASCR, which is missing, must be inserted.
```

```
          .TRAN   10US   33.33MS   0   0.1MS     ; Transient analysis
          .PROBE                                 ; Graphics post-processor
          .OPTIONS ABSTOL = 1.00N RELTOL = 0.01 VNTOL = 0.01 ITL5=10000 ;
          *                                                   Convergence
          .FOUR   60HZ   I(VX)
  .END
```

(a) The PSpice plots of the instantaneous output voltage V(4,11) is shown in Fig. 9-5.

(b) The Fourier series of the input current is as follows:

FOURIER COMPONENTS OF TRANSIENT RESPONSE I(VX)

DC COMPONENT = −1.573343E−02

HARMONIC NO	FREQUENCY (HZ)	FOURIER COMPONENT	NORMALIZED COMPONENT	PHASE (DEG)	NORMALIZED PHASE (DEG)
1	6.000E+01	5.881E+01	1.000E+00	−1.195E+01	0.000E+00
2	1.200E+02	1.423E+01	2.419E−01	−1.695E+02	−1.576E+02
3	1.800E+02	3.855E−02	6.555E−04	8.972E+01	1.017E+02
4	2.400E+02	9.580E+00	1.629E−01	1.064E+02	1.184E+02
5	3.000E+02	6.952E+00	1.182E−01	5.972E+01	7.167E+01
6	3.600E+02	3.150E−02	5.356E−04	−9.084E+01	−7.889E+01
7	4.200E+02	3.487E+00	5.930E−02	−6.156E+01	−4.961E+01
8	4.800E+02	3.140E+00	5.339E−02	−1.305E+02	−1.185E+02
9	5.400E+02	3.854E−02	6.554E−04	8.912E+01	1.011E+02

TOTAL HARMONIC DISTORTION = 3.246548E+01 PERCENT

Total harmonic distortion of input current, THD = 32.47% = 0.3247

Displacement angle, $\phi_1 = -11.95°$

Displacement factor, DF = $\cos \phi_1 = \cos(-11.95) = 9.783$ (lagging)

From Eq. (7-3), the input power factor

$$PF = \frac{1}{(1 + THD^2)^{1/2}} \cos \phi_1 = \frac{1}{(1 + 0.3247^2)^{1/2}} \times 0.9783 = 0.93 \quad \text{(lagging)}$$

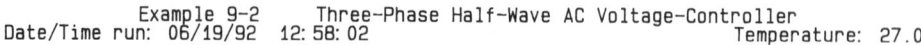

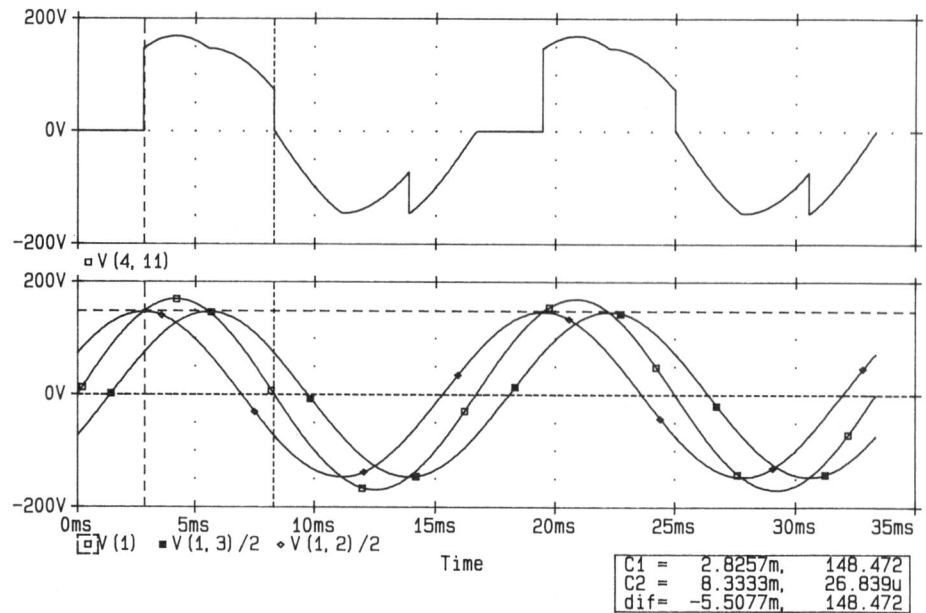

Figure 9-5 Plots for Example 9-2.

Example 9-3

A three-phase full-wave ac voltage controller is supplied from a wye-connected supply. It is shown in Fig. 9-6(a). The input phase voltage has a peak of 169.7 V, 60 Hz. The load resistance per phase is $R = 2.5 \Omega$. The delay angle is $\alpha = 60°$. The gate voltages are shown in Fig. 9-6(b). Use PSpice (a) to plot the instantaneous output phase voltage v_o, and (b) to calculate the Fourier coefficients of the input current i_a and the input power factor PF.

Solution   The peak supply voltage per phase $V_m = 169.7$ V. For $\alpha_1 = 60°$,

$$\text{time delay } t_1 = \frac{60}{360} \times \frac{1000}{60 \text{ Hz}} \times 1000 = 5555.78 \ \mu\text{s}$$

$$\text{time delay } t_3 = \frac{180}{360} \times \frac{1000}{60 \text{ Hz}} \times 1000 = 8333.3 \ \mu\text{s}$$

$$\text{time delay } t_5 = \frac{360}{360} \times \frac{1000}{60 \text{ Hz}} \times 1000 = 13,888.9 \ \mu\text{s}$$

$$\text{time delay } t_2 = \frac{120}{360} \times \frac{1000}{60 \text{ Hz}} \times 1000 = 2777.78 \ \mu\text{s}$$

$$\text{time delay } t_4 = \frac{240}{360} \times \frac{1000}{60 \text{ Hz}} \times 1000 = 11,111.1 \ \mu\text{s}$$

$$\text{time delay } t_6 = \frac{300}{360} \times \frac{1000}{60 \text{ Hz}} \times 1000 = 16,666.7 \ \mu\text{s}$$

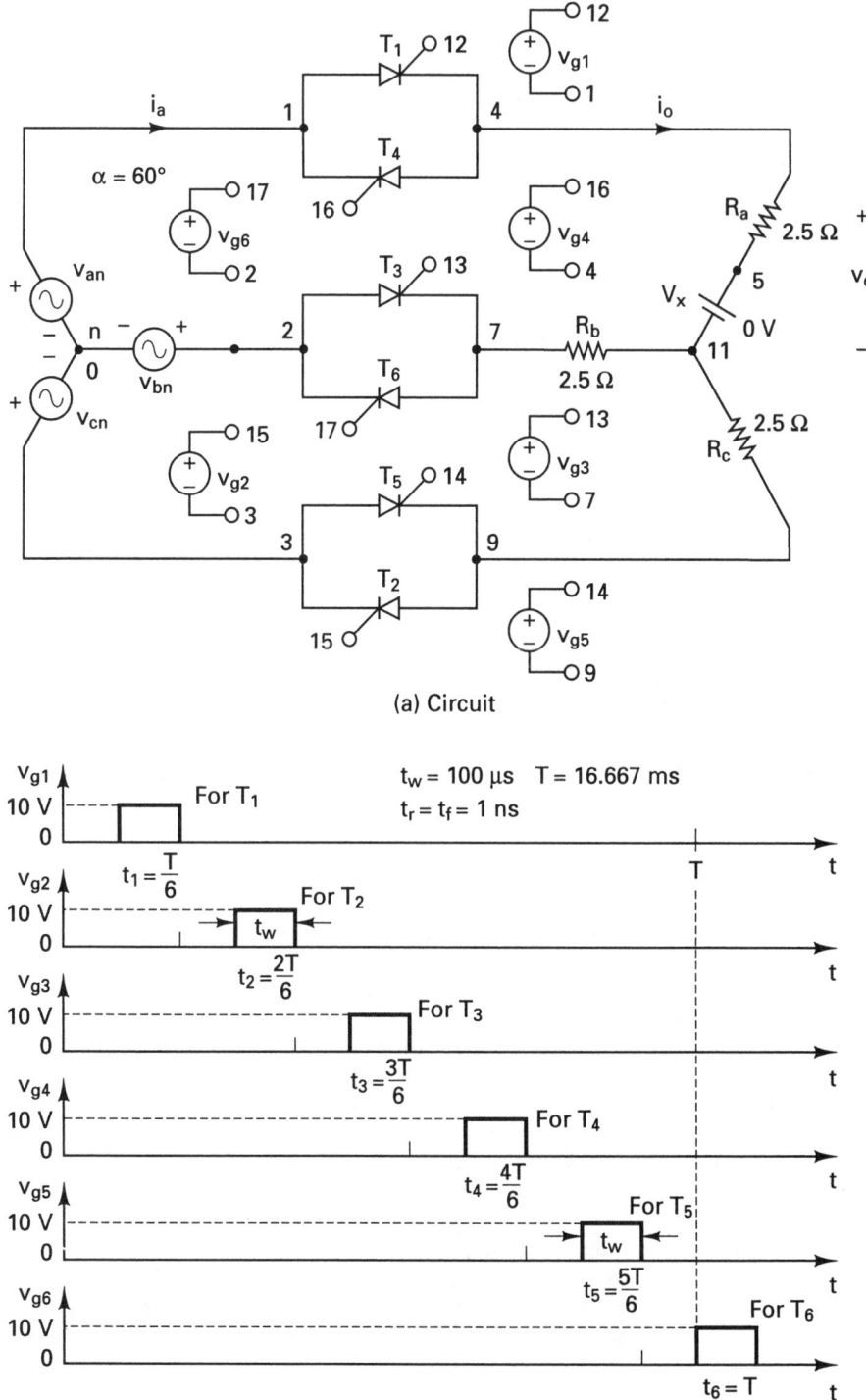

(a) Circuit

(b) Gate voltages

Figure 9-6 Three-phase full-wave ac controller for PSpice simulation.

The list of the circuit file is as follows:

■■■

Example 9-3   Three-phase full-wave ac voltage controller

SOURCE
```
■ Van    1    0    SIN(0    169.7V    60HZ)
  Vbn    2    0    SIN(0    169.7V    60HZ    0    0    -120DEG)
  Vcn    3    0    SIN(0    169.7V    60HZ    0    0    -240DEG)
  Vg1   12    4    PULSE (0V    10V    2777.8US    1NS    1NS    100US    16666.7US)
  Vg3   13    7    PULSE (0V    10V    8333.3US    1NS    1NS    100US    16666.7US)
  Vg5   14    9    PULSE (0V    10V   13888.9US    1NS    1NS    100US    16666.7US)
  Vg2   15    3    PULSE (0V    10V    5555.6US    1NS    1NS    100US    16666.7US)
  Vg4   16    1    PULSE (0V    10V   11111.1US    1NS    1NS    100US    16666.7US)
  Vg6   17    2    PULSE (0V    10V   16666.7US    1NS    1NS    100US    16666.7US)
```
CIRCUIT
```
■■ Ra    4    5    2.5
   VX    5   11    DC    0V   ; To measure load current
   Rb    7   11    2.5
   Rc    9   11    2.5
   *    Subcircuit calls for thyristor model:
   XT1   1    4   12    4    ASCR        ; Thyristor T1
   XT3   2    7   13    7    ASCR        ; Thyristor T3
   XT5   3    9   14    9    ASCR        ; Thyristor T5
   XT2   9    3   15    3    ASCR        ; Thyristor T2
   XT4   4    1   16    1    ASCR        ; Thyristor T4
   XT6   7    2   17    2    ASCR        ; Thyristor T6
   *    Subcircuit ASCR, which is missing, must be inserted.
```
ANALYSIS
```
■■■ .TRAN    0.1MS  33.33MS  0    0.1MS      ; Transient analysis
    .PROBE                                   ; Graphics post-processor
    .OPTIONS ABSTOL = 1.00N RELTOL = 0.01 VNTOL = 0.01 ITL5=10000 ;
    *                                                    Convergence
    .FOUR    60HZ    I(VX)
.END
```
■■

(a) The PSpice plots of instantaneous output phase voltage V(4,11) and input voltages are shown in Fig. 9-7.

(b) The Fourier series of the input phase current is as follows:

```
FOURIER COMPONENTS OF TRANSIENT RESPONSE I(VX)
DC COMPONENT =   3.465652E-05
```

HARMONIC NO	FREQUENCY (HZ)	FOURIER COMPONENT	NORMALIZED COMPONENT	PHASE (DEG)	NORMALIZED PHASE (DEG)
1	6.000E+01	5.354E+01	1.000E+00	-2.701E+01	0.000E+00
2	1.200E+02	6.775E-05	1.266E-06	9.195E+01	1.190E+02
3	1.800E+02	6.944E-02	1.297E-03	8.967E+01	1.167E+02
4	2.400E+02	6.785E-05	1.267E-06	8.835E+01	1.154E+02
5	3.000E+02	1.396E+01	2.608E-01	5.936E+01	8.637E+01
6	3.600E+02	6.873E-05	1.284E-06	8.983E+01	1.168E+02
7	4.200E+02	7.005E+00	1.308E-01	-6.106E+01	-3.404E+01
8	4.800E+02	6.862E-05	1.282E-06	9.200E+01	1.190E+02
9	5.400E+02	6.945E-02	1.297E-03	8.901E+01	1.160E+02

```
TOTAL HARMONIC DISTORTION =   2.917517E+01 PERCENT
```

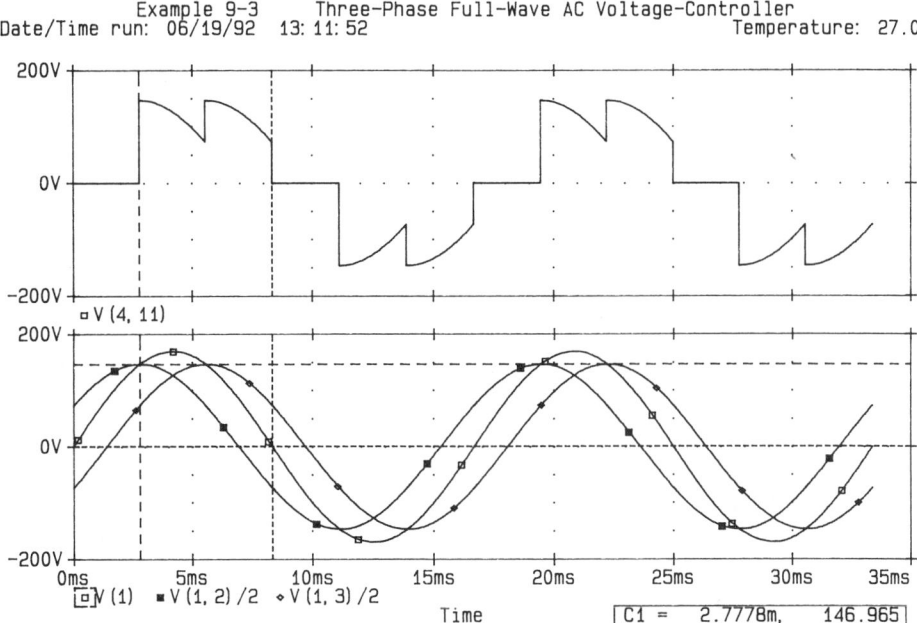

Example 9-3   Three-Phase Full-Wave AC Voltage-Controller
Date/Time run: 06/19/92   13: 11: 52   Temperature: 27.0

Figure 9-7   Plots for Example 9-3.

Total harmonic distortion of input current, THD = 29.18% = 0.2918

Displacement angle, $\phi_1 = -27.01°$

Displacement factor, DF = $\cos \phi_1 = \cos(-27.01) = 0.891$ (lagging)

From Eq. (7-3), the input power factor

$$\text{PF} = \frac{1}{(1 + \text{THD}^2)^{1/2}} \cos \phi_1 = \frac{1}{(1 + 0.2918^2)^{1/2}} \times 0.891 = 0.855 \quad \text{(lagging)}$$

Example 9-4

A three-phase full-wave delta-connected controller is supplied from a wye-connected three-phase supply. It is shown in Fig. 9-8(a). The input phase voltage has a peak of 169.7 V, 60 Hz. The load resistance per phase is $R = 2.5\ \Omega$. The delay angle is $\alpha = 60°$. The gate voltages are shown in Fig. 9-8(b). Use PSpice (a) to plot the instantaneous output phase voltage v_o and the input line current i_a, and (b) to calculate the Fourier coefficients of the output phase current i_o.

Solution   The delay angles are the same as Example 9-3. The list of the circuit file is as follows:

■ ■

Example 9-4   Three-phase full-wave delta-connected ac controller

SOURCE	■ Van	1	0	SIN(0	169.7V	60HZ)		
	Vbn	2	0	SIN(0	169.7V	60HZ	0 0	−120DEG)
	Vcn	3	0	SIN(0	169.7V	60HZ	0 0	−240DEG)

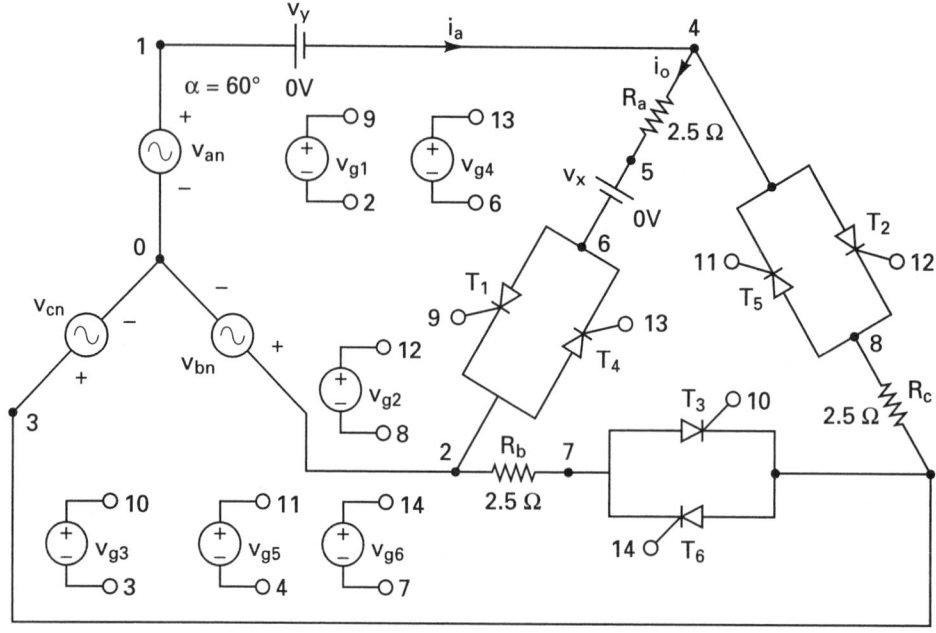

(a) Circuit

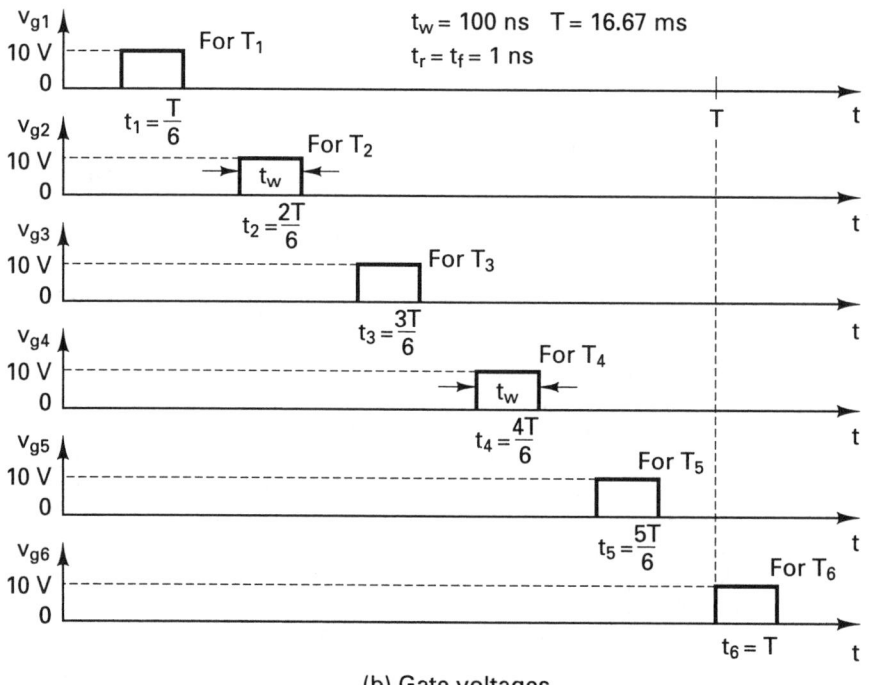

(b) Gate voltages

Figure 9-8 Three-phase full-wave delta-connected controller for PSpice simulation.

```
Vg1    9    2    PULSE (0V    10V     2777.8US    1NS    1NS    100US    16666.7US)
Vg2   12    8    PULSE (0V    10V     5555.6US    1NS    1NS    100US    16666.7US)
Vg3   10    3    PULSE (0V    10V     8333.3US    1NS    1NS    100US    16666.7US)
Vg4   13    6    PULSE (0V    10V    11111.1US    1NS    1NS    100US    16666.7US)
Vg5   11    4    PULSE (0V    10V    13888.9US    1NS    1NS    100US    16666.7US)
Vg6   14    7    PULSE (0V    10V        0.0US    1NS    1NS    100US    16666.7US)
```

CIRCUIT ■ ■
```
Ra    4    5    2.5
VX    5    6    DC     0V  ; Load current ammeter
Rb    2    7    2.5
Rc    3    8    2.5
VY    1    4    DC     0V  ; Line current ammeter
*     Subcircuit calls for thyristor model:
XT1   6    2    9    2    ASCR     ; Thyristor T1
XT3   7    3   10    3    ASCR     ; Thyristor T3
XT5   8    4   11    4    ASCR     ; Thyristor T5
XT2   4    8   12    8    ASCR     ; Thyristor T2
XT4   2    6   13    6    ASCR     ; Thyristor T4
XT6   3    7   14    7    ASCR     ; Thyristor T6
*     Subcircuit ASCR, which is missing, must be inserted.
```

ANALYSIS ■ ■ ■
```
.TRAN    0.1MS  33.33MS  0    0.1MS       ; Transient analysis
.PROBE                                    ; Graphics post-processor
.OPTIONS ABSTOL = 1.00N RELTOL = 0.01 VNTOL = 0.01 ITL5=10000 ;
*                                                    Convergence
.FOUR    60HZ    I(VX)
.END
```

(a) The PSpice plots of the instantaneous output phase voltage V(4,5) and the line current I(VY) are shown in Fig. 9-9. The output voltage and input current are discontinuous.

(b) The Fourier coefficients of the load phase current are:

FOURIER COMPONENTS OF TRANSIENT RESPONSE I(VX)

DC COMPONENT =   −8.217138E−06

HARMONIC NO	FREQUENCY (HZ)	FOURIER COMPONENT	NORMALIZED COMPONENT	PHASE (DEG)	NORMALIZED PHASE (DEG)
1	6.000E+01	6.929E+01	1.000E+00	−2.624E+00	0.000E+00
2	1.200E+02	1.711E−05	2.469E−07	1.321E+01	1.583E+01
3	1.800E+02	3.726E+01	5.377E−01	1.796E+02	1.822E+02
4	2.400E+02	1.747E−05	2.521E−07	1.294E+02	1.321E+02
5	3.000E+02	1.243E+01	1.795E−01	5.899E+01	6.161E+01
6	3.600E+02	1.843E−05	2.660E−07	−1.027E+02	−1.000E+02
7	4.200E+02	1.241E+01	1.791E−01	−6.115E+01	−5.852E+01
8	4.800E+02	2.332E−05	3.366E−07	2.827E+01	3.089E+01
9	5.400E+02	7.466E+00	1.077E−01	1.783E+02	1.809E+02

TOTAL HARMONIC DISTORTION =   6.042025E+01 PERCENT

Example 9-5

A three-phase three-thyristor controller is supplied from a three-phase wye-connected supply. It is shown in Fig. 9-10(a). The input phase voltage has a peak of

---

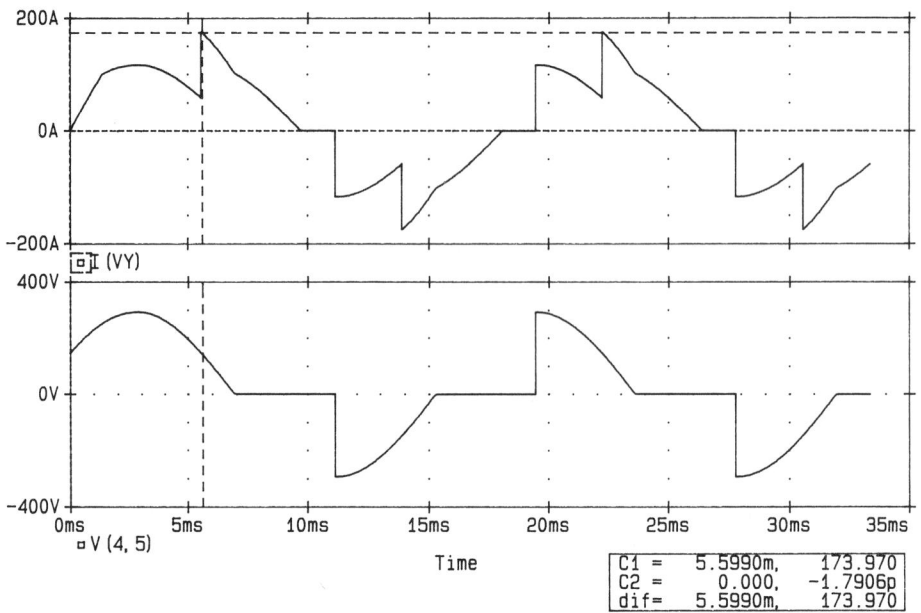

200A

0A

-200A

⊡I (VY)

400V

0V

-400V

0ms   5ms   10ms   15ms   20ms   25ms   30ms   35ms

⊡ V (4, 5)

Time

C1 =	5.5990m,	173.970
C2 =	0.000,	-1.7906p
dif=	5.5990m,	173.970

Figure 9-9 Plots for Example 9-4.

169.7 V, 60 Hz. The load resistance per phase is $R = 2.5\ \Omega$. The delay angle is $\alpha = 30°$. The gate voltages are shown in Fig. 9-10(b). Use PSpice (a) to plot the instantaneous line current i_a and thyristor T_1 current i_{T1}, and (b) to calculate the Fourier coefficients of the line current i_a.

Solution   The peak supply voltage per phase, $V_m = 169.7$ V. For $\alpha = 90°$,

$$\text{time delay } t_1 = \frac{30}{360} \times \frac{1000}{60\ \text{Hz}} \times 1000 = 1388.9\ \mu s$$

$$\text{time delay } t_2 = \frac{150}{360} \times \frac{1000}{60\ \text{Hz}} \times 1000 = 6944.5\ \mu s$$

$$\text{time delay } t_3 = \frac{270}{360} \times \frac{1000}{60\ \text{Hz}} \times 1000 = 12,500\ \mu s$$

The list of the circuit file is as follows:

■ ■

Example 9-5   Three-phase three-thyristor ac controller

```
SOURCE  ■ Van   1   0   SIN(0   169.7V   60HZ)
          Vbn   2   0   SIN(0   169.7V   60HZ   0   0   -120DEG)
          Vcn   3   0   SIN(0   169.7V   60HZ   0   0   -240DEG)
          Vg1   9   7   PULSE (0V   10V   1388.9US   1NS   1NS   100US   16666.7US)
          Vg2   10  8   PULSE (0V   10V   6944.5US   1NS   1NS   100US   16666.7US)
          Vg3   11  5   PULSE (0V   10V   12500.0US  1NS   1NS   100US   16666.7US)
```

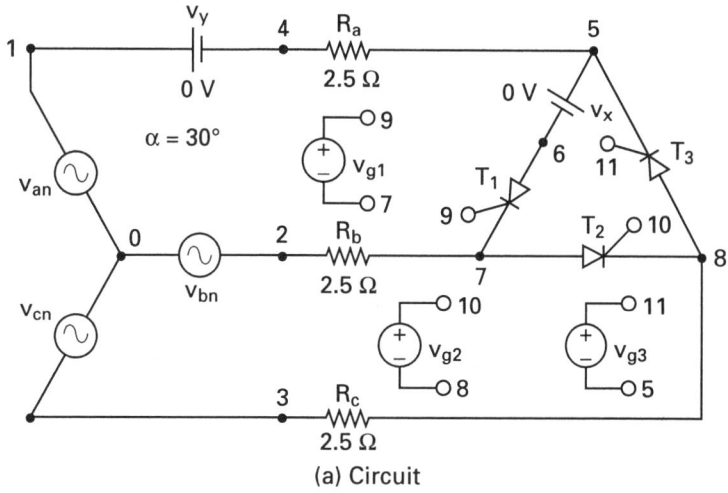

(a) Circuit

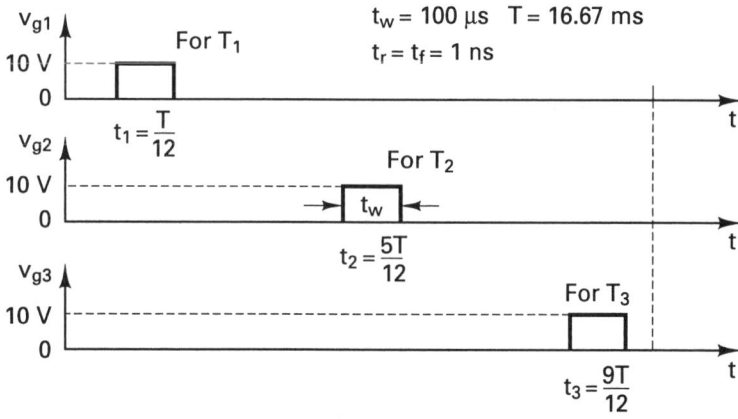

$t_w = 100\ \mu s \quad T = 16.67\ ms$
$t_r = t_f = 1\ ns$

For T$_1$

10 V

0

$t_1 = \dfrac{T}{12}$

For T$_2$

10 V

0

t_w

$t_2 = \dfrac{5T}{12}$

For T$_3$

10 V

0

$t_3 = \dfrac{9T}{12}$

(b) Gate voltages

Figure 9-10   Three-phase three-thyristor controller for PSpice simulation.

CIRCUIT ■ ■					
Ra	4	5	2.5		
VX	5	6	DC	0V	; Load current ammeter
Rb	2	7	2.5		
Rc	3	8	2.5		
VY	1	4	DC	0V	; Line current ammeter
*	Subcircuit calls for thyristor model:				
XT1	6	7	9	7	ASCR ; Thyristor T1
XT2	7	8	10	8	ASCR ; Thyristor T3
XT3	8	5	11	5	ASCR ; Thyristor T5
*	Subcircuit ASCR, which is missing, must be inserted.				

■ ■ ■ .TRAN   10US   33.33MS 0   0.1MS   ; Transient analysis
.PROBE   ; Graphics post-processor
.OPTIONS ABSTOL = 1.00N RELTOL = 0.01 VNTOL = 0.01 ITL5=10000
.FOUR   60HZ   I(VX)

.END
■ ■

(a) The PSpice plots of the instantaneous line current I(VY) and phase voltage V(6,7) are shown in Fig. 9-11.

(b) The Fourier coefficients of the thyristor T_1 current are:

FOURIER COMPONENTS OF TRANSIENT RESPONSE I(VX)

DC COMPONENT =   1.614633E+01

HARMONIC NO	FREQUENCY (HZ)	FOURIER COMPONENT	NORMALIZED COMPONENT	PHASE (DEG)	NORMALIZED PHASE (DEG)
1	6.000E+01	2.614E+01	1.000E+00	8.992E+00	0.000E+00
2	1.200E+02	1.393E+01	5.328E-01	-5.685E+01	-6.584E+01
3	1.800E+02	8.074E+00	3.089E-01	-9.054E+01	-9.953E+01
4	2.400E+02	5.129E+00	1.962E-01	-1.365E+02	-1.455E+02
5	3.000E+02	3.097E+00	1.185E-01	-1.210E+02	-1.300E+02
6	3.600E+02	5.598E+00	2.142E-01	-1.713E+02	-1.803E+02
7	4.200E+02	3.119E+00	1.193E-01	1.192E+02	1.102E+02
8	4.800E+02	2.366E+00	9.053E-02	1.559E+02	1.469E+02
9	5.400E+02	4.030E+00	1.542E-01	8.903E+01	8.004E+01

TOTAL HARMONIC DISTORTION =   7.237835E+01 PERCENT

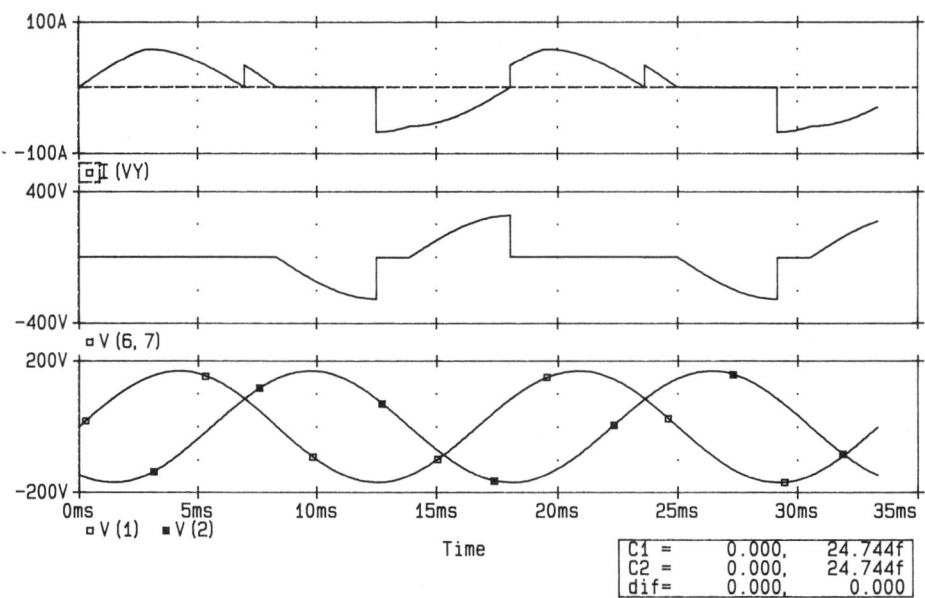

Example 9-5   Three-Phase Three-Thyristor AC Controller
Date/Time run: 06/19/92 13:47:16   Temperature: 27.0

Figure 9-11 Plots for Example 9-5.

Example 9-6

A capacitor of 780 μF is connected across the output of the single-phase full-wave controller of Fig. 9-2(a). This is shown in Fig. 9-12. The input voltage has a peak of 169.7 V, 60 Hz. The load inductance L is 6.5 mH, and the load resistance is $R = 2.5\ \Omega$. The delay angles are equal: $\alpha_1 = \alpha_2 = 60°$. The gate voltages are shown in Fig. 9-12(b). Use PSpice (a) to plot the instantaneous output voltage v_o and the load current i_o, and (b) to calculate the Fourier coefficients of the input current i_a and the input power factor PF.

Solution   The peak supply voltage, $V_m = 169.7$ V. For $\alpha_1 = \alpha_2 = 60°$, time delay $t_1 = (60/360) \times 1000/60$ Hz) $\times 1000 = 2{,}777.7\ \mu$s. A series snubber with $C_s = 0.1\ \mu$F and $R_s = 750\ \Omega$ is connected across the thyristor to cope with the transient voltage due to the inductive load. The list of the circuit file is as follows:

▪ ▪

Example 9-6   Single-phase ac voltage controller with output filter

SOURCE ▪ VS   1   0   SIN (0   169.7V   60HZ)

Vg1   2   4   PULSE (0V   10V   2,777.7US   1NS   1NS   100US   16666.7US)

Vg2   3   1   PULSE (0V   10V   11,111US   1NS   1NS   100US   16666.7US)

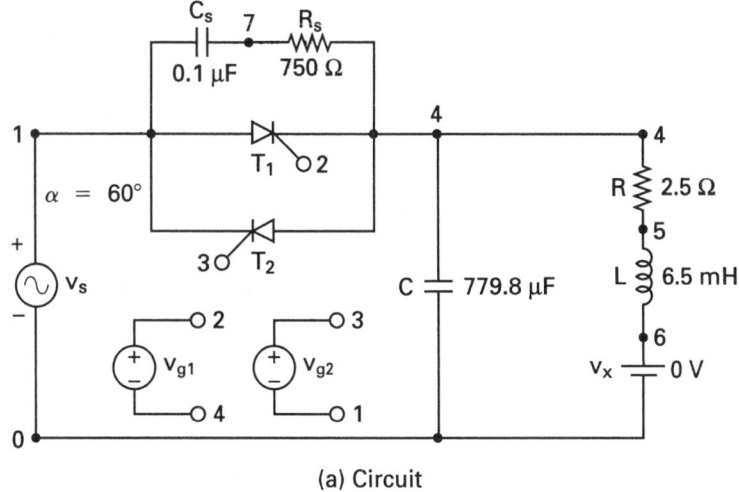

(a) Circuit

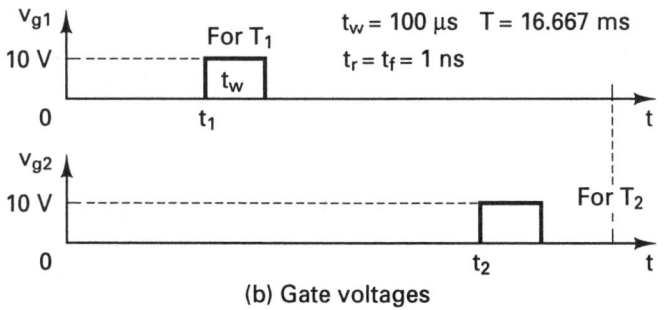

(b) Gate voltages

Figure 9-12   Single-phase full-wave ac controller for PSpice simulation.

```
CIRCUIT  ■ ■  R      4    5    2.5
              L      5    6    6.5MH
              VX     6    0    DC    0V  ; Voltage source to measure the load current
              C      4    0    780UF      ; Output filter capacitance
              CS     1    7    0.1UF
              RS     7    4    750
              *    Subcircuit calls for thyristor model:
              XT1   1    4    2    4    ASCR     ; Thyristor T1
              XT2   4    1    3    1    ASCR     ; Thyristor T2
              *    Subcircuit ASCR, which is missing, must be inserted.
ANALYSIS ■ ■ ■ .TRAN   20US   33.33MS   0   20US   ; Transient analysis
              .PROBE                              ; Graphics post-processor
              .OPTIONS ABSTOL = 1.00N  RELTOL = 1.0M  VNTOL = 1.0M  ITL5=10000
              .FOUR   60HZ   I(VX)                ; Fourier analysis
       .END
```

(a) The PSpice plots of instantaneous output voltage V(4) and load current I(VX) are shown in Fig. 9-13.

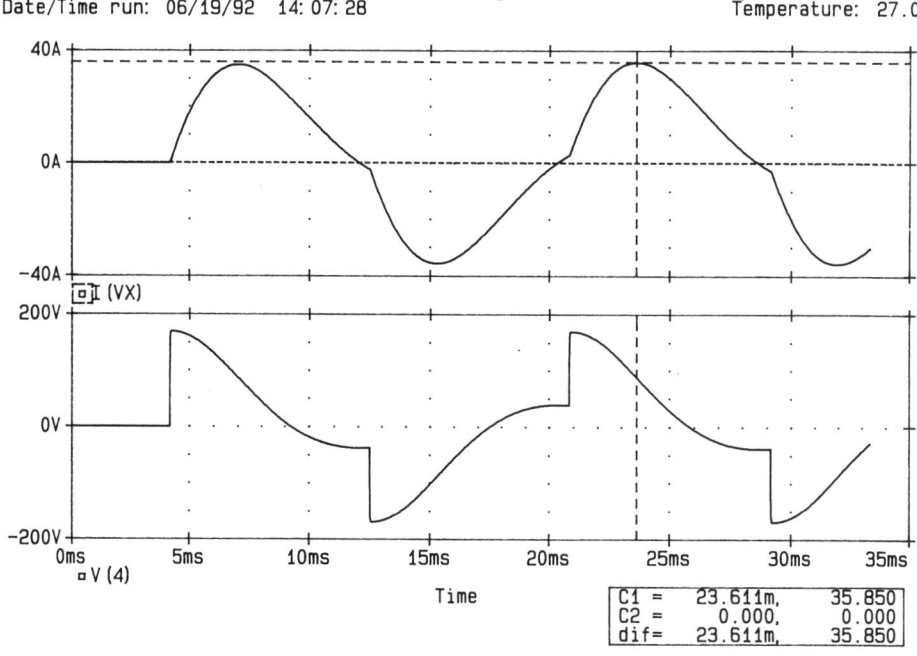

Figure 9-13 Plots for Example 9-6.

(b) The Fourier series of the input current is as follows:

```
FOURIER COMPONENTS OF TRANSIENT RESPONSE I(VX)
DC COMPONENT =   -1.681664E-02
```

HARMONIC NO	FREQUENCY (HZ)	FOURIER COMPONENT	NORMALIZED COMPONENT	PHASE (DEG)	NORMALIZED PHASE (DEG)
1	6.000E+01	3.314E+01	1.000E+00	-6.980E+01	0.000E+00
2	1.200E+02	1.606E-02	4.845E-04	1.053E+02	1.751E+02
3	1.800E+02	4.381E+00	1.322E-01	1.816E+01	8.795E+01
4	2.400E+02	4.887E-03	1.475E-04	4.289E+01	1.127E+02
5	3.000E+02	1.404E+00	4.238E-02	-1.722E+02	-1.024E+02
6	3.600E+02	2.921E-03	8.815E-05	4.669E+01	1.165E+02
7	4.200E+02	7.014E-01	2.116E-02	5.241E+00	7.504E+01
8	4.800E+02	1.879E-03	5.670E-05	2.298E+01	9.278E+01
9	5.400E+02	4.193E-01	1.265E-02	-1.764E+02	-1.066E+02

```
TOTAL HARMONIC DISTORTION =   1.410080E+01 PERCENT
```

Total harmonic distortion of input current, THD = 14.1% = 0.141

Displacement angle, $\phi_1 = -69.88°$

Displacement factor, DF = cos ϕ_1 = cos(-69.88) = 0.344   (lagging)

From Eq. (7-3), the input power factor

$$PF = \frac{1}{(1 + THD^2)^{1/2}} \cos \phi_1 = \frac{1}{(1 + 0.141^2)^{1/2}} \times 0.344 = 0.341 \quad \text{(lagging)}$$

9-5 LABORATORY EXPERIMENTS

The following two experiments are suggested to demonstrate the operation and characteristics of thyristor ac controllers:

Single-phase ac voltage controller

Three-phase ac voltage controller

9-5.1 Experiment AC-1

SINGLE-PHASE AC VOLTAGE CONTROLLER

Objective   The objective is to study the operation and characteristics of a single-phase ac voltage controller under various load conditions.

Applications   The single-phase ac voltage controller is used to control power flow in many applications (e.g., industrial and induction heating, pumps and fans, light dimmers, and food blenders).

Textbook   See Ref. 2, Secs. 6-3 and 6-5.

Apparatus   **1.** Two phase-controlled thyristors with ratings of at least 50 A and 400 V, mounted on heat sinks

2. A firing pulse generator with isolating signals for gating thyristors

3. An RL load

4. One dual-beam oscilloscope with floating or isolating probes

5. Ac voltmeters and ammeters and one noninductive shunt

Warning   Before making any circuit connection, switch the ac power OFF. **Do not** switch the power ON unless the circuit is checked and approved by your lab instructor. **Do not** touch the thyristor heat sinks, which are connected to live terminals.

Experimental   **1.** Set up the circuit as shown in Fig. 9-14. Use a load resistance R only.
procedure   **2.** Connect the measuring instruments as required.

3. Set the delay angle to $\alpha = \pi/3$.

4. Connect the firing pulses to appropriate thyristors.

5. Observe and record the waveforms of the load voltage v_o and the load current i_o.

6. Measure the rms load voltage $V_{o(rms)}$, the rms load current $I_{o(rms)}$, the average thyristor current $I_{T(rms)}$, the rms input voltage $V_{s(rms)}$, and the load power P_L.

7. Measure the conduction angle of thyristor T_1.

8. Repeat steps 2 to 7 with a load inductance L only.

9. Repeat steps 2 to 7 with both load resistance R and load inductance L.

Report   **1.** Present all recorded waveforms and discuss all significant points.

2. Compare the waveforms generated by SPICE with the experimental results, and comment.

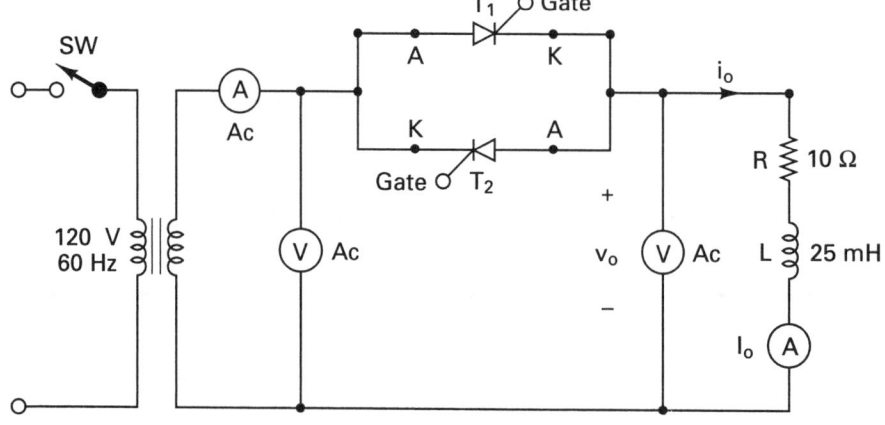

Figure 9-14   Single-phase ac voltage controller.

3. Compare the experimental results with the results predicted.
4. Calculate and plot the rms output voltage V_o against the delay angle α.
5. Discuss the advantages and disadvantages of this type of controller.

9-5.2 Experiment AC-2

THREE-PHASE AC VOLTAGE CONTROLLER

Objective The objective is to study the operation and characteristics of a three-phase ac voltage controller under various load conditions.

Applications The three-phase controller is used to control power flow in many applications (e.g., industrial and induction heating, lighting, speed control of induction motor-driven pumps and fans).

Textbook See Ref. 2, Sec. 6-7.

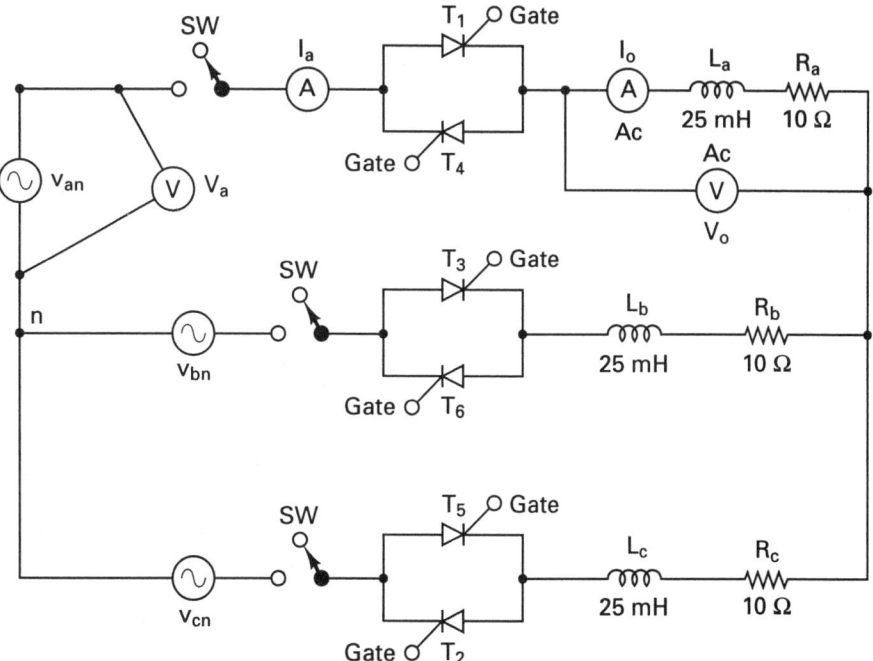

Figure 9-15 Three-phase ac voltage controller.

Apparatus 1. Six phase-controlled thyristors with ratings of at least 50 A and 400 V, mounted on heat sinks

2. A firing pulse generator with isolating signals for gating thyristors

3. *RL* loads

4. One dual-beam oscilloscope with floating or isolating probes

5. Ac voltmeters and ammeters and one noninductive shunt

Warning See Experiment AC-1.

Experimental procedure Repeat the steps in Experiment AC-1.

Report See Experiment AC-1.

SUMMARY

The statements for an ac thyristor are:

```
*    Subcircuit call for switched ac thyristor model:
XT1    NA        NC        +NC         −NC         ASCR
       anode     cathode   +control    −control    model
*                          voltage     voltage     name
```

SUGGESTED READING

1. L. J. Giacoletto, "Simple SCR and TRIAC PSPICE computer models," *IEEE Transactions on Industrial Electronics*, Vol. 36, No. 3, August 1989, pp. 451–455.

2. M. H. Rashid, *Power Electronics: Circuits, Devices, and Applications*, 2nd ed. Englewood Cliffs, N.J.: Prentice Hall, 1993, Chap. 6.

DESIGN PROBLEMS

9-1. It is required to design the single-phase ac voltage controller of Fig. 9-14 with the following specifications:

Ac supply voltage V_s = 120 V (rms), 60 Hz

Load resistance R = 5 Ω

Load inductance L = 15 mH

Rms output voltage $V_{o(rms)}$
   = 75% of the maximum value

(a) Determine the ratings of all devices under worst-case conditions.

(b) Use SPICE to verify your design.

(c) Provide a cost estimate of the circuit.

9-2. (a) Design an output C filter for the single-phase full-wave ac voltage controller of Problem 9-1. The harmonic content of the load current should be less than 10% of the value without the filter.

(b) Use SPICE to verify your design in part (a).

9-3. It is required to design the three-phase ac voltage controller of Fig. 9-15 with the following specifications:

Ac supply voltage per phase
$$V_s = 120 \text{ V (rms)}, 60 \text{ Hz}$$

Load resistance per phase, $R = 5 \text{ } \Omega$

Load inductance per phase, $L = 15 \text{ mH}$

Rms output voltage $V_{o(rms)}$
$$= 75\% \text{ of the maximum value}$$

(a) Determine the ratings of all devices under worst-case conditions.

(b) Use SPICE to verify your design.

(c) Provide a cost estimate of the circuit.

9-4. (a) Design an output C filter for the three-phase ac voltage controller of Problem 9-3. The harmonic content of the load current should be less than 10% of the value without the filter.

(b) Use SPICE to verify your design in part (a).

Chapter 10

■■■■■■■■

DC Thyristor Commutation Circuits

10-1 INTRODUCTION

A thyristor can be turned on by applying a pulse of short duration. Once the thyristor is turned on, the gate pulse has no effect, and it remains on until its current is reduced to zero. It is a latching device. Its behavior can be represented by two transistors—NPN and PNP [1]—and a complex model can be developed to match the thyristor characteristics. To obtain the various waveforms of thyristor commutation circuits with the least circuit complexity, we shall use a very simple model to obtain the various waveforms of thyristor commutation circuits.

10-2 DC THYRISTOR MODEL

Let us assume that the thyristor shown in Fig. 10-1(a) is operated from a dc supply. The switching action of the thyristor can be modeled by a voltage-controlled switch S_T and a diode D_T. This is shown in Fig. 10-1(b). In thyristor commutations, resonant circuits are often used to reverse the voltage on a capacitor and/or to force the current of a thyristor to zero. Diode D_T is used to prevent any negative current flow and to stop the circuit operation when the current falls to zero. Switch S_T is controlled by the gate voltage V_g connected between nodes 3 and 4. To avoid the problem of an open node, the gate voltage V_g is connected through a very high resistance R_g, say 10 MΩ, and it is shown in Fig. 10-1(c). In practice, the gate voltage V_g is connected between the gate and the cathode terminals as shown in Fig. 10-1(d).

To use this model effectively, one *must* know the approximate on-time of the thyristor, because the gate or controlling voltage V_g must be applied for a suffi-

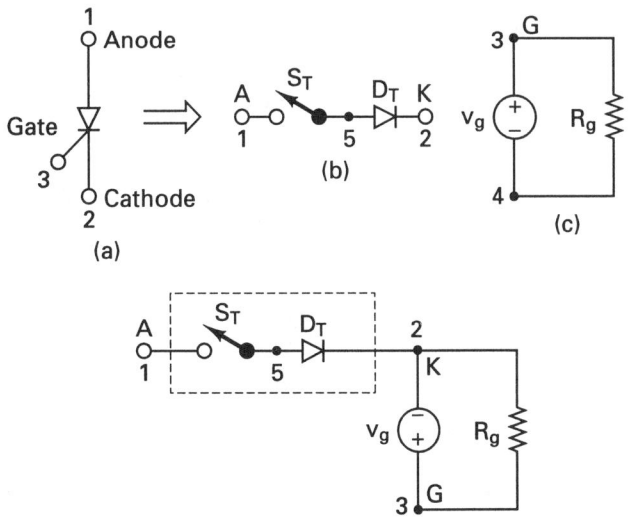

Figure 10-1 Dc thyristor model.

cient period, which is greater than or equal to the actual on-time of the thyristor. Specifically, if the device is expected to be turned off when the current falls to zero, V_g must be applied until the current becomes zero. This thyristor, which is modeled by a diode and a voltage-controlled switch, can be used as a subcircuit. The diode and switch parameters can be adjusted to yield the desired on-state drop of the thyristor. In the following examples, we use the diode parameters IS=2.2E−15, BV=1800V, TT=0, and CJO=0 and the switch parameters RON=0.01, ROFF=10E+6, VON=10V, and VOFF=5V.

Switch S_T is controlled by the controlling voltage V_g connected between nodes 3 and 4. The subcircuit definition for the thyristor model TMOD can be described as follows:

```
*     Subcircuit for switched dc thyristor model:
.SUBCKT    TMOD      1          2          3          4
*          model    anode    cathode    +control    −control
*          name                         voltage     voltage
DT     5    2    DMOD                          ; Switch diode
ST     1    5    3    4    SMOD               ; Switch
.MODEL    DMOD    D(IS=2.2E−15    BV=1200V TT=0 CJO=0) ; Diode model parameters
.MODEL    SMOD    VSWITCH (RON=0.01 ROFF=10E+6 VON=10V VOFF=5V)
.ENDS    TMOD                                ; Ends subcircuit definition
```

We can apply this model to the thyristor circuit in Fig. 10-2(a). Thyristor T_1 is turned on by the voltage V_g connected between the gate and cathode terminals. V_g is shown in Fig. 10-2(b). The list of the PSpice circuit file for determining the transient response is as follows:

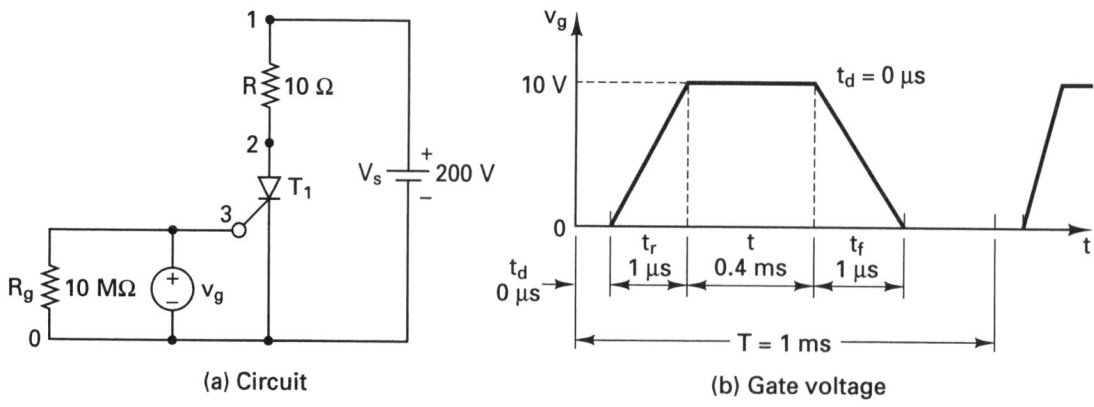

(a) Circuit   (b) Gate voltage

Figure 10-2   Dc thyristor circuit.

```
                ·····································································

              Dc thyristor circuit
SOURCE    ■  VS     1    0    DC      200V
             Vg     3    0    PULSE  (0V  10V  0  1US  1US  0.4MS  1MS)
CIRCUIT   ■ ■ Rg    3    0    10MEG
             R      1    2    10
             *    Subcircuit call for switched dc thyristor model:
             XT1    2    0    3    0    TMOD      ; Thyristor T1
ANALYSIS  ■ ■ ■ .TRAN  1US  5MS               ; Transient analysis from 0 to 5 ms
             .PROBE                            ; Graphics post-processor
             .END
                ·····································································
```

10-3 DC THYRISTOR CHOPPERS

The input to a dc chopper is a fixed dc voltage, and its output is a variable dc voltage. When a main thyristor switch is turned on, the input voltage is connected to the load. The output voltage, which is varied by varying the conduction time of the thyristor, is discontinuous. The turn-off is accomplished by an auxiliary circuit known as a *commutation circuit*.

10-4 EXAMPLES OF DC THYRISTOR CHOPPERS

The applications of a dc thyristor model are illustrated by some examples.

Example 10-1

The dc thyristor circuit of Fig. 10-3(a) has $V_s = 200$ V, $C = 50$ μF, and $L = 10$ μH. The input is a step voltage of $V_s = 200$ V dc. The gate voltage is shown in Fig. 10-3(b). Use PSpice to plot the instantaneous thyristor current i_o and the capacitor voltage v_c.

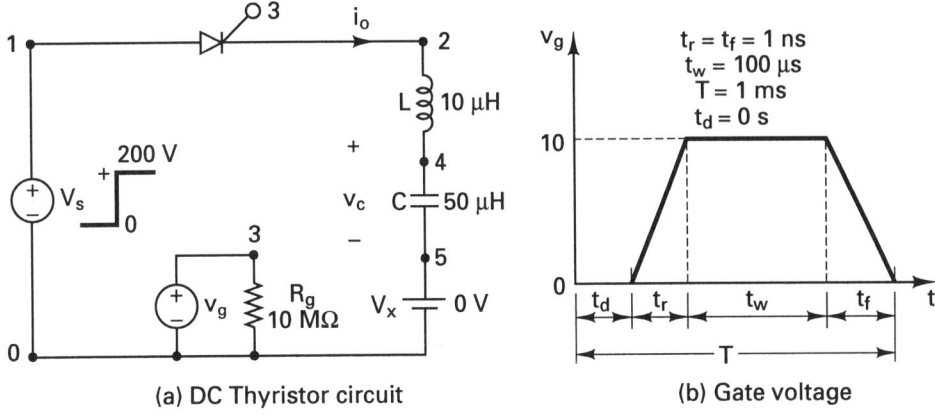

Figure 10-3 Dc thyristor circuit for PSpice simulation.

Solution The list of the circuit file is as follows:

■ ■

Example 10-1 Dc thyristor circuit

SOURCE	■ VS	1	0	PWL (0	0	1NS	200V	1MS	200V)

SOURCE   ■ VS   1   0   PWL (0   0   1NS   200V   1MS   200V)

            Vg   3   0   PULSE (0V 10V   0   1NS   1NS   100US   1MS)

            Rg   3   0   10MEG

CIRCUIT   ■ ■ L   2   4   10UH

            C   4   5   50UF

            VX   5   0   DC   0V ; Voltage source to measure thyristor current

            *   Subcircuit call for switched dc thyristor model:

            XT1 1   2   3   0   TMOD   ; Thyristor T1

            *   Subcircuit TMOD, which is missing, <u>must</u> be inserted.

ANALYSIS ■ ■ ■ .TRAN 0.5US 100US   0 0.5US ; Transient analysis from 0 to 100 μs

         .PROBE   ; Graphics post-processor

.END

■ ■

The PSpice plots of the instantaneous thyristor current I(VX) and capacitor voltage V(4) are shown in Fig. 10-4. The capacitor voltage is $v_c = V_s (1 - \cos \pi t)$, and the peak capacitor current is $I_p = V_s \sqrt{C/L}$.

Example 10-2

The impulse commutated thyristor chopper of Fig. 10-5(a) has $V_s = 200$ V, $R_m = 1\,\Omega$, $L_m = 5$ mH, $C = 20\,\mu$F, $L = 20\,\mu$H, and $L_1 = 25\,\mu$H. The switching frequency is $f_s = 1$ kHz, and the duty cycle is $k = 40\%$. The gate voltages are shown in Fig. 10-5(b). Use PSpice (a) to plot the instantaneous capacitor voltage v_c, the voltage v_{T1} across thyristor T_1, and the output voltage v_o, and (b) to plot the instantaneous load current i_o, the current i_{Dm} through diode D_m, and the current i_{T1} through T_1.

Solution The list of the circuit file is as follows:

■ ■

Example 10-2 Impulse commutated thyristor chopper

SOURCE   ■ VS   1   0   DC   200V

            Vg1 7   0   PULSE (0V 100V   0   1US   1US   0.4MS   1MS)

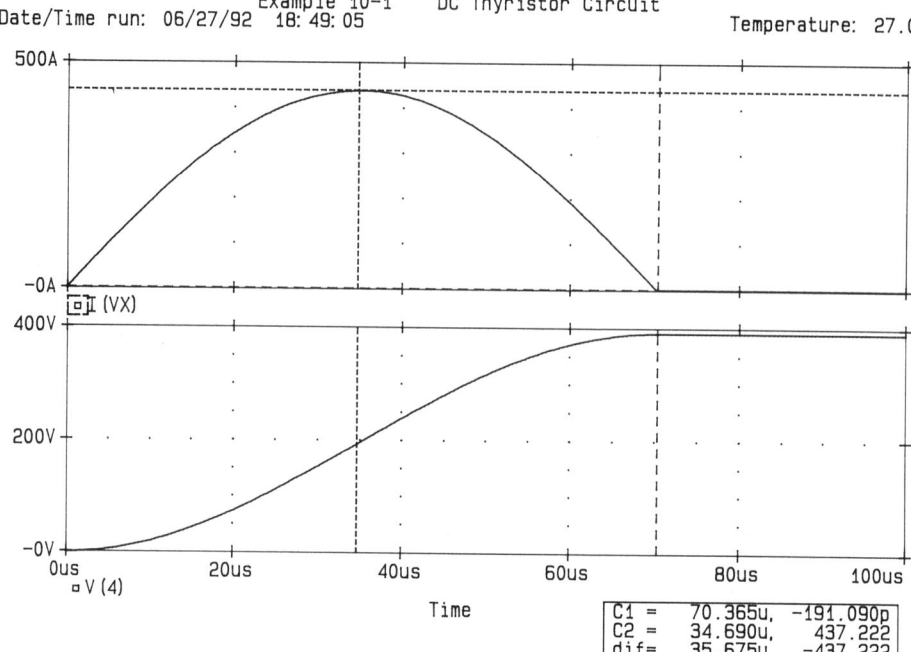

Example 10-1   DC Thyristor Circuit

Temperature: 27.0

Figure 10-4 Plots for Example 10-1.

Vg2	8	0	PULSE	(0V	100V	0.4MS	1US	1US	0.6MS	1MS)
Vg3	9	0	PULSE	(0V	100V	0	1US	1US	0.2MS	1MS)
Rg1	7	0	10MEG							
Rg2	8	0	10MEG							
Rg3	9	0	10MEG							

CIRCUIT ■ ■

```
C       1    2    20UF    IC=200V    ; With initial voltage
L       2    10   20UH
L1      1    3    25UH
D1      4    3    DMOD
DM      0    4    DMOD
.MODEL  DMOD  D(IS=2.22E-15 BV=1200V CJO=1PF TT=0) ; Diode model
RM      4    5    1.0
LM      5    6    5.0MH
VX      6    0    DC    0V          ; Measures load current
VY      11   4    DC    0V          ; Measures current of T1
*    Subcircuit calls for switched thyristor model:
XT1     1    11   7    0    TMOD    ; Thyristor T1
XT2     2    4    8    0    TMOD    ; Thyristor T2
XT3     1    10   9    0    TMOD    ; Thyristor T3
*    Subcircuit TMOD, which is missing, must be inserted.
```

ANALYSIS ■ ■ ■

```
.TRAN   5US    8MS    6MS          ; Transient analysis
.PROBE                            ; Graphics post-processor
```

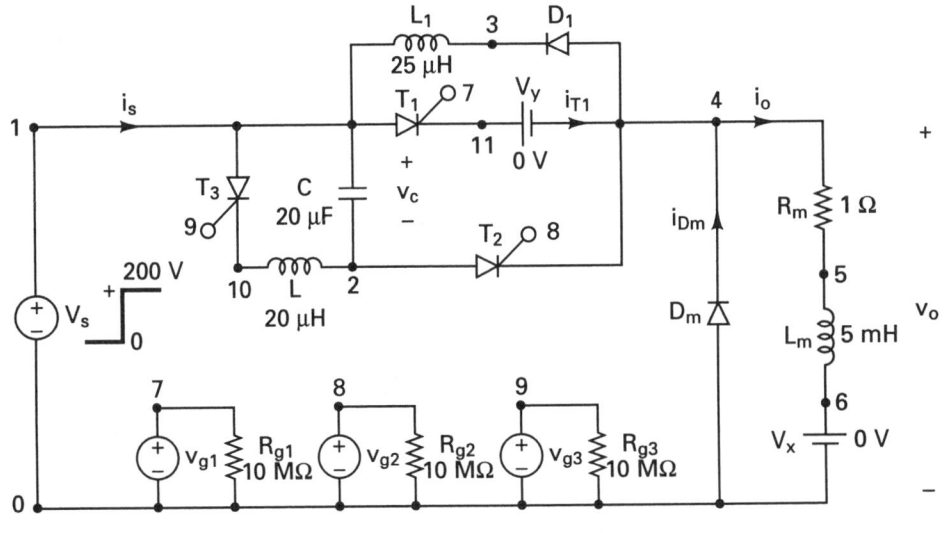

(a) Impulse commutated chopper

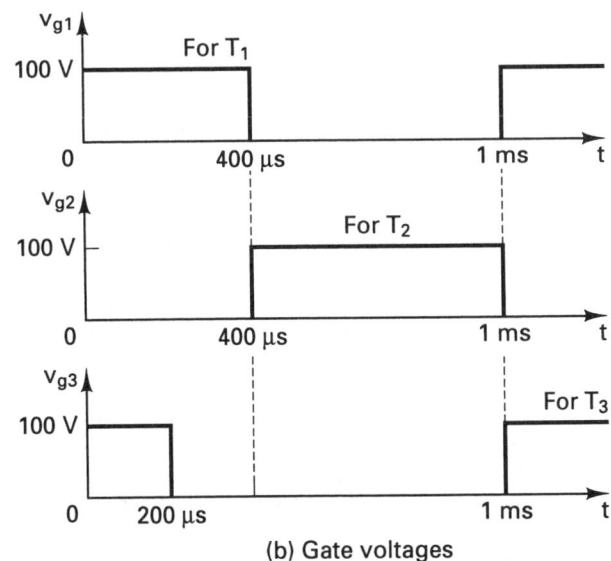

(b) Gate voltages

Figure 10-5 Impulse commutated thyristor chopper for PSpice simulation.

```
.OPTIONS ABSTOL = 1.000U RELTOL = .01 VNTOL = 0.1   ITL5=40000;
*                                                   Convergence
.END
```

(a) The PSpice plots of the instantaneous capacitor voltage V(1,2), the voltage across thyristor T_1 V(1,4), and the output voltage V(4) are shown in Fig. 10-6. The available commutation time can be found from the plot of V(1,4).

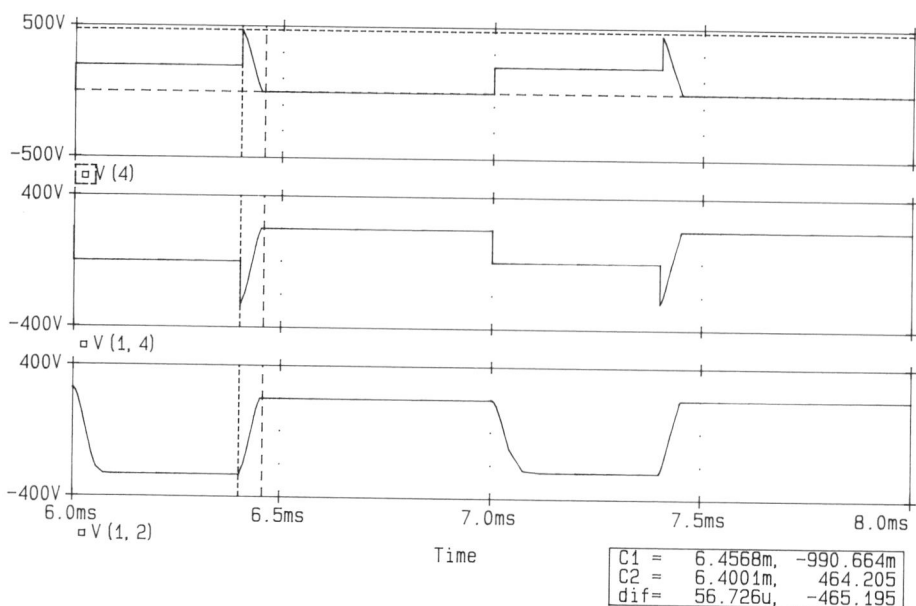

500V

-500V
⊡ V (4)
400V

-400V
□ V (1, 4)
400V

-400V
6.0ms   6.5ms   7.0ms   7.5ms   8.0ms
□ V (1, 2)
                              Time

C1 =	6.4568m,	−990.664m
C2 =	6.4001m,	464.205
dif=	56.726u,	−465.195

Figure 10-6   Plots of voltages for Example 10-2.

     (b) The PSpice plots of the instantaneous load current I(VX), the current through freewheeling diode I(DM), and the current through T_1, I(VY), are shown in Fig. 10-7. Note that the load current has not reached the steady-state condition.

     Note. Diode parasitics will cause switching transients. Students are encouraged to run the circuit files with junction capacitance and transit time (e.g., CJO=2PF and TT=1US).

Example 10-3

     The resonant pulse thyristor chopper of Fig. 10-8(a) has $V_s = 200$ V, $R_m = 1\ \Omega$, $L_m = 5$ mH, $C = 31.2\ \mu$F, and $L = 6.4\ \mu$H. The switching frequency is $f_s = 1$ kHz, and the duty cycle is $k = 40\%$. The gate voltages are shown in Fig. 10-8(b). Use PSpice to plot the instantaneous capacitor voltage v_c, the voltage v_{T1} across thyristor T_1, and the output voltage v_o.

Solution   The list of the circuit file is as follows:

■ ■

Example 10-3   Resonant-pulse thyristor chopper

```
SOURCE   ■ VS    1   0   DC    200V
           Vg1   7   0   PULSE (0V   100V   0       1US   1US   0.4MS   1MS)
           Vg2   8   0   PULSE (0V   100V   0.4MS   1US   1US   0.6MS   1MS)
           Vg3   9   0   PULSE (0V   100V   0       1US   1US   0.2MS   1MS)
           Rg1   7   0   10MEG
           Rg2   8   0   10MEG
           Rg3   9   0   10MEG
```

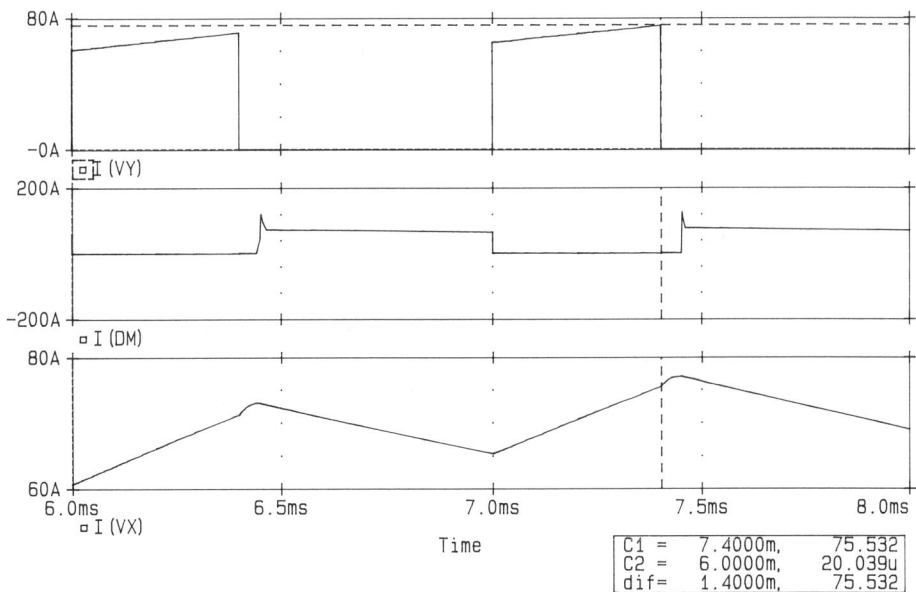

Figure 10-7 Plots of currents for Example 10-2.

```
CIRCUIT   ■ ■ C     1    2     31.2UF   IC=200V  ; With initial voltage
              L     2    3     6.4UH
              D1    4    1     DMOD
              DM    0    4     DMOD
              .MODEL  DMOD  D(IS=2.22E-15 BV=1200V  TT=0)  ; Diode model
              RM    4    5     1.0
              LM    5    6     5.0MH
              VX    6    0     DC    0V              ; Measures load current
              VY    1    10    DC    0V              ; Measures current of T1
              *   Subcircuit calls for switched dc thyristor model:
              XT1  10    4     7    0    TMOD        ; Thyristor T1
              XT2   3    4     8    0    TMOD        ; Thyristor T2
              XT3   1    3     9    0    TMOD        ; Thyristor T3
              *   Subcircuit TMOD, which is missing, must be inserted.
ANALYSIS  ■ ■ ■ .TRAN    1US    6MS    4MS           ; Transient analysis
              .PROBE                                  ; Graphics post-processor
              .OPTIONS ABSTOL = 1.000U RELTOL = .01 VNTOL = 0.1  ITL5=20000 ;
              *                                               Convergence
          .END
```

The PSpice plots of the instantaneous capacitor voltage V(1,2), the voltage V(1,4) across thyristor T_1, and the output voltage V(4) are shown in Fig. 10-9. Note that the capacitor voltage is continuing to rise and has not reached the steady-state condition.

Sec. 10-4   Examples of DC Thyristor Choppers   **225**

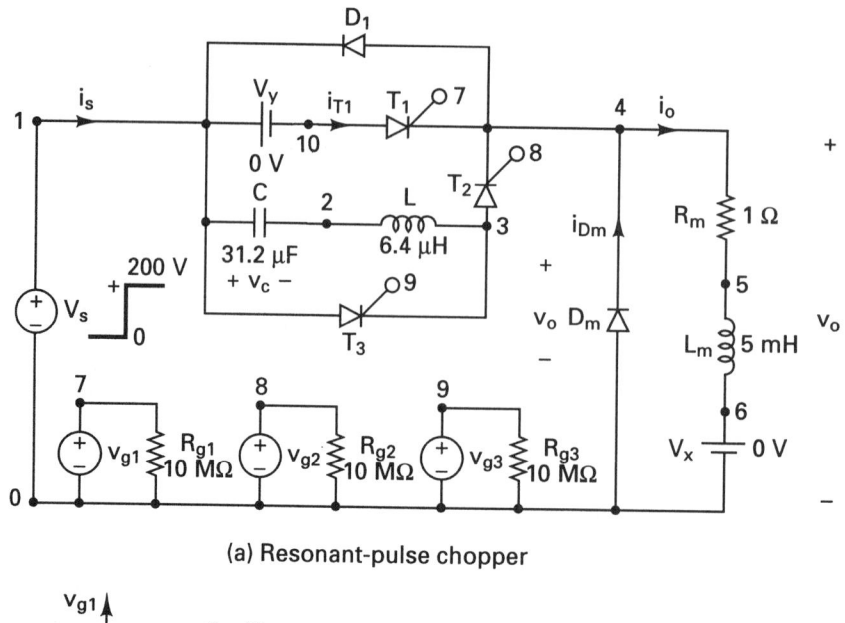

(a) Resonant-pulse chopper

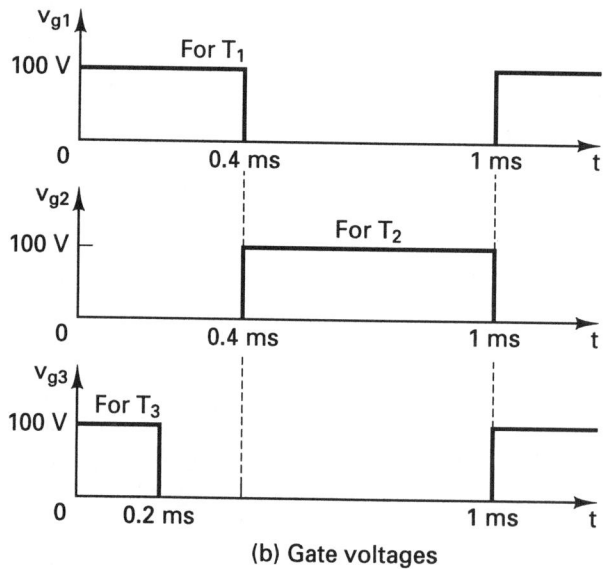

(b) Gate voltages

Figure 10-8 Resonant-pulse chopper for PSpice simulation.

Example 10-4

The complementary commutated thyristor chopper of Fig. 10-10(a) has $V_s = 200$ V, $R_1 = R_2 = 5$ Ω, and $C = 10$ μF. The switching frequency is $f_s = 1$ kHz and the duty cycle is $k = 50\%$. The gate voltages are shown in Fig. 10-10(b). Use PSpice to plot the instantaneous capacitor voltage v_c, the voltage v_{T1} across thyristor T_1, and voltage v_{T2} across thyristor T_2.

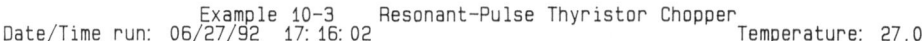

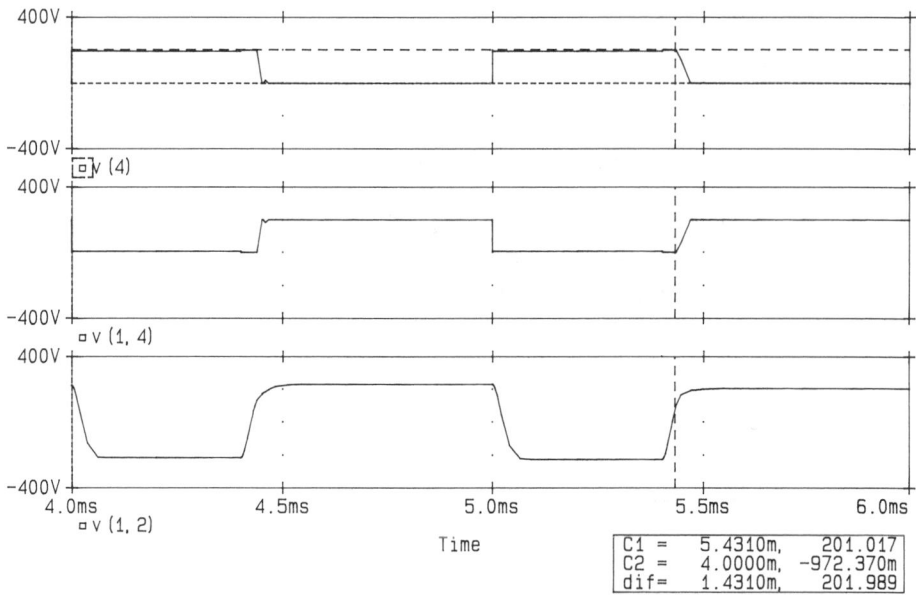

□ v (4)

□ v (1, 4)

□ v (1, 2)

Time

C1 =	5.4310m,	201.017
C2 =	4.0000m,	−972.370m
dif=	1.4310m,	201.989

Figure 10-9   Plots of voltages for Example 10-3.

Solution   The list of the circuit file is as follows:

■■

Example 10-4   Complementary commutation circuit

SOURCE ■ VS   1   0   DC   200V
        Vg1   4   0   PULSE (0V   10V   0   1NS   1NS   0. 5MS   1MS)
        Vg2   5   0   PULSE (0V   10V   0.5MS 1NS   1NS   0. 5MS   1MS)
        Rg1   4   0   10MEG
        Rg2   5   0   10MEG

CIRCUIT ■■ R1   1   2   5
        R2   1   3   5
        C   2   3   10UF
        *   Subcircuit calls for switched thyristor model:
        XT1   2   0   4   0   TMOD   ; Thyristor T1
        XT2   3   0   5   0   TMOD   ; Thyristor T2
        *   Subcircuit TMOD, which is missing, <u>must</u> be inserted.

ANALYSIS ■■■ . TRAN   2US   1.5MS   ; Transient analysis
        . PROBE   ; Graphics post-processor
        . OPTIONS ABSTOL = 1. 000U RELTOL = . 01 VNTOL = 0. 1   ITL5=10000 ;
        *   Convergence
. END ■■■

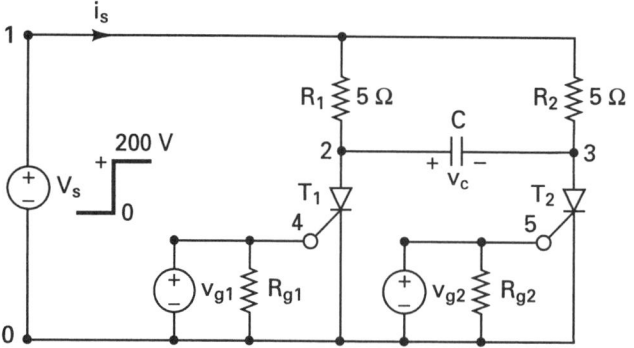

(a) Complementary commutated chopper

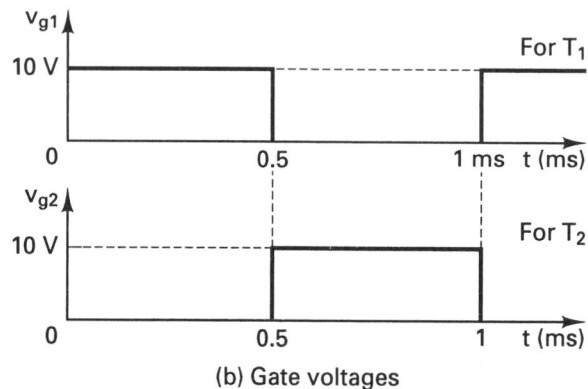

(b) Gate voltages

Figure 10-10 Complementary commutated thyristor chopper for PSpice simulation.

The PSpice plots of the instantaneous capacitor voltage V(2,3), the voltage V(2) across thyristor T_1, and the voltage V(3) across thyristor T_2 are shown in Fig. 10-11.

Example 10-5

The load-side commutated thyristor chopper of Fig. 10-12(a) has $V_s = 200$ V, $R_m = 1$ Ω, $L_m = 5$ mH, $C = 20$ μF, $L = 20$ μH, and $L_1 = 50$ μH. The switching frequency is $f_s = 1$ kHz, and the duty cycle is $k = 40\%$. The gate voltages are shown in Fig. 10-12(b). Use PSpice to plot the instantaneous capacitor voltage v_c, the output voltage v_o, and the load current i_o.

Solution   The list of the circuit file is as follows:

▪ ▪

Example 10-5   Line-side commutated thyristor circuit

```
SOURCE   ▪ VS    1    0    DC     200V
           Vg1   7    4    PULSE (0V  100V   0       1US   1US   0.4MS   1MS)
           Vg2   8    0    PULSE (0V  100V   0.4MS 1US   1US   0.6MS   1MS)
           Vg3   9    10   PULSE (0V  100V   0       1US   1US   0.2MS   1MS)
           Rg1   7    4    10MEG
           Rg2   8    0    10MEG
           Rg3   9    10   10MEG
```

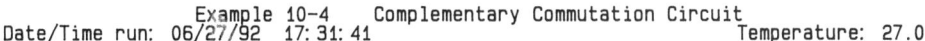

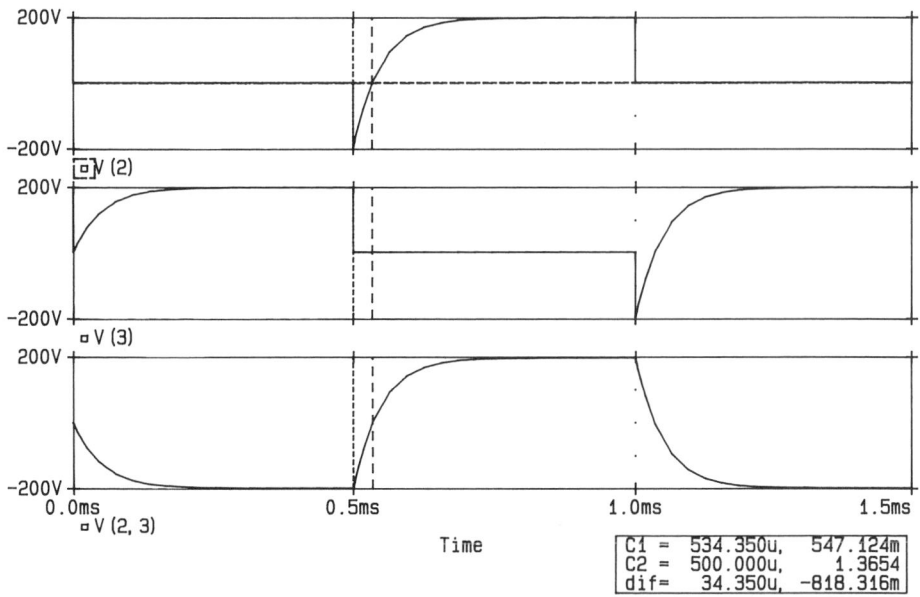

Figure 10-11   Plots of voltages for Example 10-4.

```
CIRCUIT   ■ ■  C      2    3    20UF    IC=200V
               L     10    3    20UH
               L1     1    2    50UH
               DM     0    4    DMOD
               .MODEL    DMOD    D(IS=2.2E-15 BV=1800V TT=0) ; Diode model parameters
               RM     4    5    1.0
               LM     5    6    5.0MH
               VX     6    0    DC    0V           ; Measures load current
               VY     2   11    DC    0V           ; Measures current of transistor T1
               *    Subcircuit calls for switched thyristor model:
               XT1   11    4    7    4    TMOD     ; Thyristor T1
               XT2    3    0    8    0    TMOD     ; Thyristor T2
               XT3    2   10    9   10    TMOD     ; Thyristor T3
               *    Subcircuit TMOD, which is missing, must be inserted.
ANALYSIS  ■ ■ ■ .TRAN   1US   4MS   2MS            ; Transient analysis
               .PROBE                              ; Graphics post-processor
               .OPTIONS ABSTOL = 1.000U RELTOL = .01 VNTOL = 0.1  ITL5=20000
          .END
```

The PSpice plots of the instantaneous capacitor voltage V(2,3), the voltage V(2) across thyristor T_1, and the voltage V(3) across thyristor T_2 are shown in Fig. 10-13. The capacitor voltage is building up continuously and will cause instability.

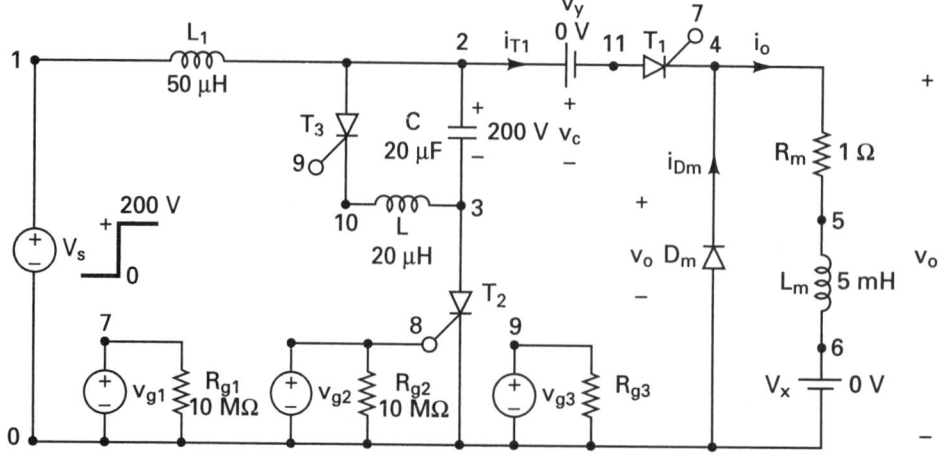

(a) Load-side commutated chopper

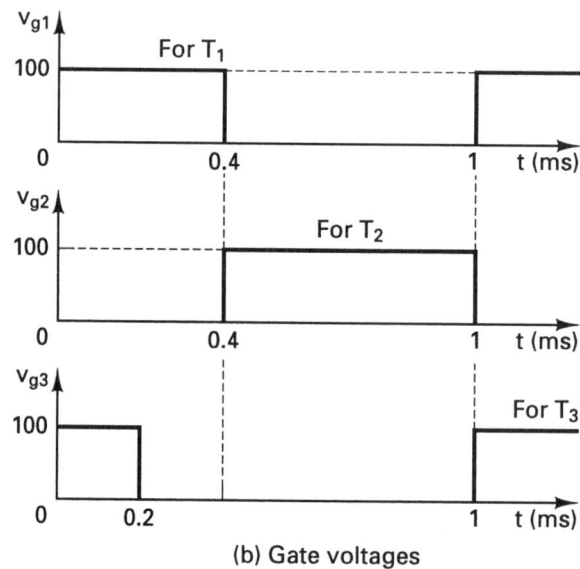

(b) Gate voltages

Figure 10-12 Load-side commutated thyristor chopper for PSpice simulation.

10-5 LABORATORY EXPERIMENTS

It is possible to develop many experiments to demonstrate the techniques of thyristor commutation. The following two experiments should be worthwhile exercises:

Impulse commutated thyristor chopper

Resonant-pulse commutated thyristor chopper

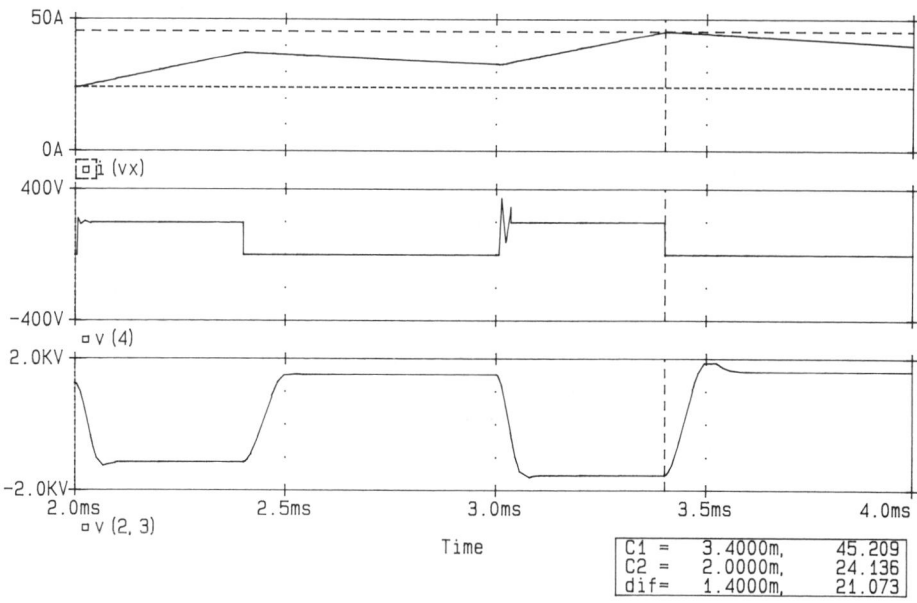

```
50A

 0A
    □ i (vx)
400V

-400V
    □ v (4)
2.0KV

-2.0KV
   2.0ms          2.5ms          3.0ms          3.5ms          4.0ms
    □ v (2, 3)
                            Time         C1 =   3.4000m,    45.209
                                         C2 =   2.0000m,    24.136
                                         dif=   1.4000m,    21.073
```

Figure 10-13   Plots for Example 10-5.

10-5.1 Experiment CP-1

IMPULSE COMMUTATED THYRISTOR CHOPPER

Objective   The objective is to study the technique and characteristics of an impulse commutated thyristor chopper with an *RL* load.

Applications   The impulse commutated thyristor chopper is used for high-powered dc motor control and dc static switches. The impulse commutation techniques are applied to force the current of a circuit to zero in many power electronics circuits.

Textbook   See Ref. 1, Sec. 9-8.1.

Apparatus   **1.** Three inverter-grade thyristors with rating of at least 50 A and 400 V, mounted on heat sinks, with a turn-off time t_q of less than 15 μs
     2. Two fast-recovery diodes with rating of at least 50 A and 400 V, mounted on heat sinks, and a reverse recovery time t_{rr} of less than 2 μs

3. Commutation capacitor $C = 30$ μF, a reversing inductor $L = 10$ μH, and a commutation inductor $L_1 = 20$ μH

4. Firing pulse generator with isolating signals for gating thyristors

5. An RL load

6. One dual-beam oscilloscope with floating or isolating probes

7. Dc voltmeters and ammeters and one noninductive shunt

Warning Before making any circuit connection, switch the dc power OFF. **Do not** switch the power ON unless the circuit is checked and approved by your lab instructor. **Do not** touch the thyristor heat sinks, which are connected to line terminals.

Experimental procedure
1. Set up the circuit as shown in Fig. 10-14. Use a suitable load inductance L_m and a load resistance R_m.

2. Connect the measuring instruments as required.

3. Connect the firing pulses to the appropriate thyristors.

4. Set the duty cycle at 50% and the chopping frequency at $f_c = 1$ kHz.

5. Observe and record the waveforms of the voltage v_c across the commutation capacitor, the voltage v_{T1} across the main thyristor T_1, the output voltage v_o, and the load current i_o.

6. Measure the circuit turn-off time $t_{(off)}$ of the main thyristor T_1 and the reversing time t_r of the capacitor voltage.

7. Measure the average output voltage $v_{o(dc)}$ for duty cycle k from 0 to 1.0 with an increment of 0.1.

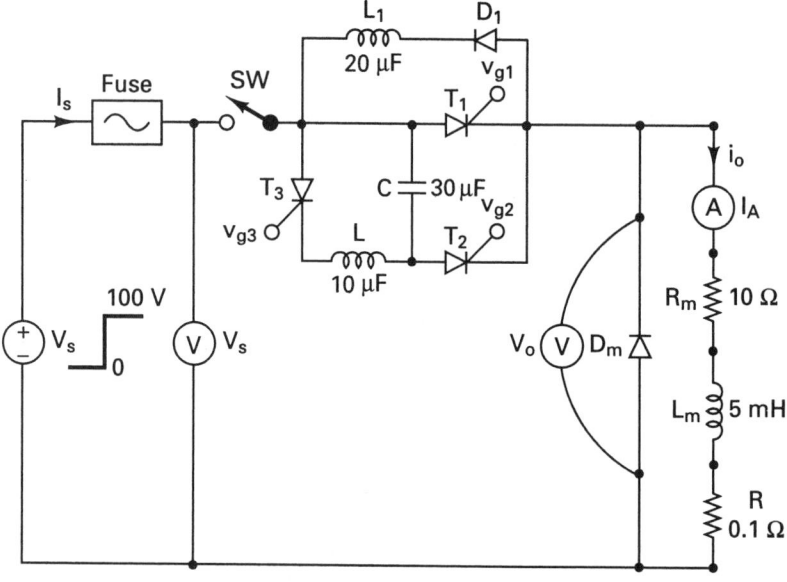

Figure 10-14 Impulse commutated thyristor chopper.

DC Thyristor Commutation Circuits   Chap. 10

1. Present all recorded waveforms and discuss all significant points.

2. Compare and comment on the waveforms generated by SPICE with the experimental results.

3. Plot the average output voltage V_o against the duty cycle k, and compare the experimental results with the results predicted.

4. Discuss the advantages and disadvantages of this type of commutation.

10-5.2 Experiment CP-2

RESONANT-PULSE COMMUTATED THYRISTOR CHOPPER

Objective The objective is to study the technique and characteristics of a resonant-pulse commutated thyristor chopper with an RL load.

Applications The resonant-pulse commutated thyristor chopper is used for high-powered dc motor control and dc static switches. The resonant commutation techniques are used in many power electronics circuits (e.g., high-frequency resonant inverters).

Textbook See Ref. 1, Sec. 7-8.4.

Apparatus 1. Three inverter-grade thyristors with ratings of at least 50 A and 400 V, mounted on heat sinks, with a turn-off time t_q of less than 15 μs

2. Two fast-recovery diodes with ratings of at least 50 A and 400 V, mounted on heat sinks, with a reverse recovery time t_{rr} of less than 2 μs

3. Commutation capacitor $C = 30$ μF, and commutation inductor $L = 10$ μH

4. Firing pulse generator with isolating signals for gating thyristors

5. An RL load

6. One dual-beam oscilloscope with floating or isolating probes

7. Dc voltmeters and ammeters and one noninductive shunt

Warning Before making any circuit connection, switch the dc power OFF. **Do not** switch the power ON unless the circuit is checked and approved by your lab instructor. **Do not** touch the thyristor heat sinks, which are connected to line terminals.

Experimental procedure 1. Set up the circuit as shown in Fig. 10-15. Use a suitable load inductance L_m and a load resistance R_m.

2. Connect the measuring instruments as required.

3. Connect the firing pulses to the appropriate thyristors.

4. Set the duty cycle at 50% and the chopping frequency at $f_c = 1$ kHz.

5. Observe and record the waveforms of the voltage v_c across the commutation capacitor, the voltage v_{T1} across the main thyristor T_1, the output voltage v_o, and the load current i_o.

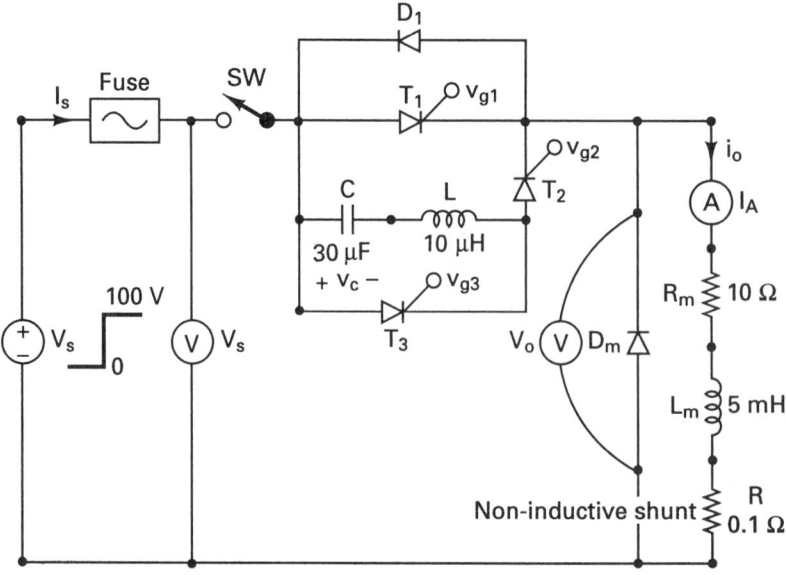

Figure 10-15 Resonant-pulse commutated thyristor chopper.

6. Measure the available turn-off time $t_{(off)}$ of the main thyristor T_1 and the reversing time t_r of the capacitor voltage.

7. Measure the average output voltage $V_{o(dc)}$ for duty cycle k from 0 to 1.0 with an increment of 0.1.

Report 1. Present all recorded waveforms and discuss all significant points.

2. Compare and comment on the waveforms generated by SPICE with the experimental results.

3. Plot the average output voltage $V_{o(dc)}$ against the duty cycle k, and compare the experimental results with the results predicted.

4. Discuss the advantages and disadvantages of this type of commutation.

SUMMARY

The statements for a dc thyristor are:

```
*     Subcircuit call for switched thyristor model:
XT1     NA        NK         +NG        −NG        MODEL
        anode   cathode    +control   −control    model
*                          voltage    voltage     name
*     Subcircuit for switched dc thyristor model:
.SUBCKT   MODEL     NA         NK         +NG        −NG
*         model    anode    cathode    +control   −control
*         name                         voltage    voltage
```

SUGGESTED READING

1. M. H. Rashid, *Power Electronics: Circuits, Devices, and Applications*, 2nd ed. Englewood Cliffs, N.J.: Prentice Hall, 1993, Chap. 9.

2. M. H. Rashid, *SPICE for Circuits and Electronics Using PSpice*. Englewood Cliffs, N.J.: Prentice Hall, 1990, Chap. 4.

DESIGN PROBLEMS

10-1. It is required to design the impulse commutated thyristor chopper of Fig. 10-14 with the following specifications:

Dc supply voltage, $V_s = 100$ V

Load resistance, $R_m = 2\ \Omega$

Load inductance, $L_m = 5$ mH

Minimum output voltage, $V_{o(min)}$, should be less than 5% of V_s

Source stray inductance, $L_s = 4\ \mu H$

Thyristor turn-off time, $t_q = 12\ \mu s$

  (a) Determine the ratings of all components and devices. Plot the output voltage $V_{o(dc)}$ against the duty cycle k.
  (b) Use SPICE to verify your design.
  (c) Provide a cost estimate of the circuit.

10-2. (a) Design an input LC filter for the chopper of Problem 10-1 in order to limit the maximum rms fundamental component of the chopper-generated harmonic current in the supply side to 10% of the average load current. Assume that the load current is ripple free and use the average load current I_A.
  (b) Use SPICE to verify your design in part (a).

10-3. It is required to design the resonant-pulse commutated thyristor chopper of Fig. 10-15 with the following specifications:

Dc supply voltage, $V_s = 100$ V

Load resistance, $R_m = 2\ \Omega$

Load inductance, $L_m = 5$ mH

Minimum average output voltage, $V_{o(min)}$, should be less than 5% of V_s

Source stray inductance, $L_s = 4\ \mu H$

Thyristor turn-off time, $t_q = 15\ \mu s$

  (a) Determine the ratings of all components and devices. Plot the output voltage $V_{o(dc)}$ against the duty cycle k.
  (b) Use SPICE to verify your design.
  (c) Provide a cost estimate of the circuit.

10-4. Design an input LC filter for the chopper of Problem 10-3 in order to limit the maximum rms fundamental component of the chopper-generated harmonic current in the supply side to 10% of the average load current. Assume that the load current is ripple free and use the average load current I_A.

Chapter 11

■■■■■■■■■

DC Choppers

11-1 INTRODUCTION

A dc chopper is a dc-to-dc converter, where both input and output voltages are dc. It uses a power semiconductor device as a switch to turn on and off the dc supply to the load. The switching action can be implemented by a BJT, a MOSFET, or an IGBT.

11-2 DC SWITCH CHOPPER

A chopper switch is shown in Fig. 11-1(a). If switch S_1 is turned on, the supply voltage V_S is connected to the load. If the switch is turned off, the inductive load current i_o is forced to flow through diode D_m. The output voltage and load current are shown in Fig. 11-1(b). The parameters of the switch can be adjusted to model the voltage drop of the chopper. We use the switch parameters RON=0.01, ROFF=10E+6, VON=10V, and VOFF=5V and the diode parameters IS= 2.22E−15, BV=1200V, IBV=13E−3, CJO=1P, and TT=0. The diode parasitics are neglected; however, they will affect the transient behavior.

Example 11-1

A dc chopper switch is shown in Fig. 11-2(a). The dc input voltage is $V_S = 220$ V. The load resistance R is 5 Ω, and the load inductance $L = 7.5$ mH. The chopping frequency is $f_o = 1$ kHz, and the duty cycle of the chopper is $k = 50\%$. The control voltage is shown in Fig. 11-2(b). Use PSpice (a) to plot the instantaneous output voltage v_o, the load current i_o, and the diode current i_{Dm}, (b) to calculate the Fourier coefficients of the load current i_o, and (c) to calculate the Fourier coefficients of the input current i_s.

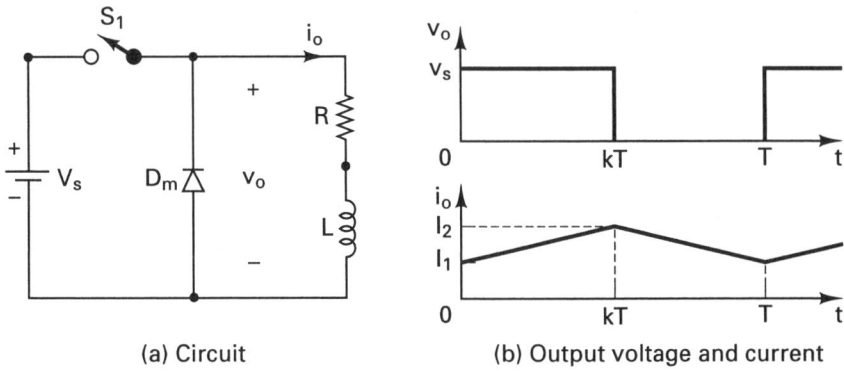

(a) Circuit (b) Output voltage and current

Figure 11-1 Dc switch chopper.

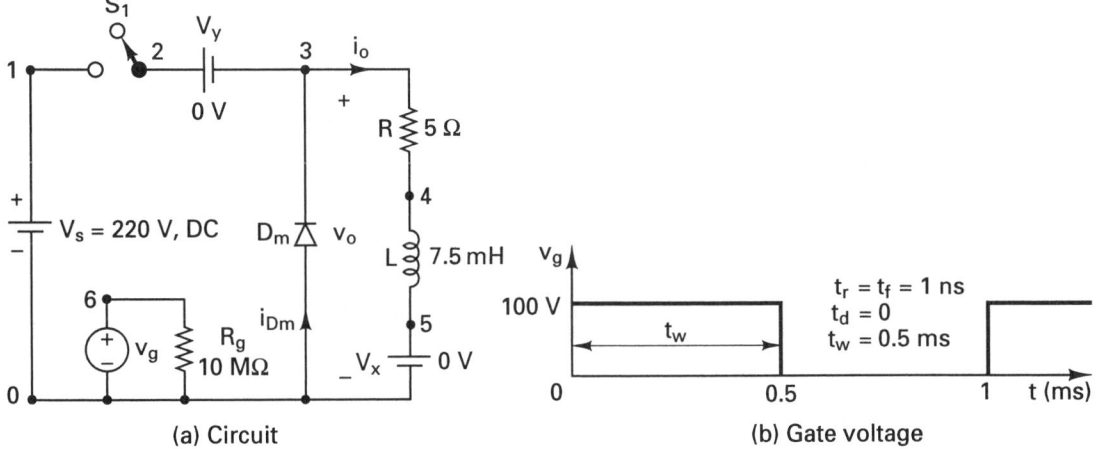

(a) Circuit (b) Gate voltage

Figure 11-2 Dc chopper for PSpice simulation.

Solution The dc supply voltage $V_S = 220$ V. $k = 0.5, f_o = 1$ kHz, $T = 1/f_o = 1$ ms, and $t_{on} = k \times T = 0.5 \times 1$ ms $= 0.5$ ms. The list of the circuit file is as follows:

■ ■

Example 11-1 Chopper circuit

SOURCE	■ VS	1	0	DC	220V			
	Vg	6	0	PULSE (0V	100V	0	1NS	1NS	0.5MS	1MS)
	Rg	6	0	10MEG				
CIRCUIT	■ ■ R	3	4	5				
	L	4	5	7.5MH				
	VX	5	0	DC	0V	; Load battery voltage		
	VY	2	3	DC	0V	; Voltage source to measure chopper current		
	DM	0	3	DMOD		; Freewheeling diode		
	.MODEL	DMOD	D(IS=2.22E-15 BV=1200V CJO=1PF TT=0) ; Diode model					
	S1	1	2	6	0	SMOD	; Switch	
	.MODEL	SMOD	VSWITCH (RON=0.01 ROFF=10E+6 VON=10V VOFF=5V)					

.TRAN   10US   10MS   8MS   ; Transient analysis
            .PROBE   ; Graphics post-processor
            .OPTIONS ABSTOL = 1.00N RELTOL = 0.01 VNTOL = 0.1 ITL5=40000
            .FOUR   1KHZ   I(VX)   I(VY)   ; Fourier analysis
       .END
■ ■

(a) The PSpice plots of the instantaneous input current I(VY), the current through diode I(DM), and the output voltage V(3) are shown in Fig. 11-3. The load current rises when the switch is on and falls when it is off.

(b) The Fourier coefficients of the load current are:

FOURIER COMPONENTS OF TRANSIENT RESPONSE I(VX)

DC COMPONENT =   2.189331E+01

HARMONIC NO	FREQUENCY (HZ)	FOURIER COMPONENT	NORMALIZED COMPONENT	PHASE (DEG)	NORMALIZED PHASE (DEG)
1	1.000E+03	2.955E+00	1.000E+00	−8.439E+01	0.000E+00
2	2.000E+03	2.378E−03	8.046E−04	−1.523E+02	−6.789E+01
3	3.000E+03	3.344E−01	1.132E−01	−8.743E+01	−3.040E+00
4	4.000E+03	5.029E−03	1.702E−03	−1.868E+01	6.570E+01
5	5.000E+03	1.147E−01	3.882E−02	−9.128E+01	−6.896E+00
6	6.000E+03	5.937E−03	2.009E−03	1.115E+02	1.959E+02
7	7.000E+03	6.417E−02	2.171E−02	−9.042E+01	−6.036E+00
8	8.000E+03	2.062E−03	6.979E−04	−1.602E+02	−7.578E+01
9	9.000E+03	3.747E−02	1.268E−02	−9.349E+01	−9.108E+00

TOTAL HARMONIC DISTORTION =   1.222766E+01 PERCENT

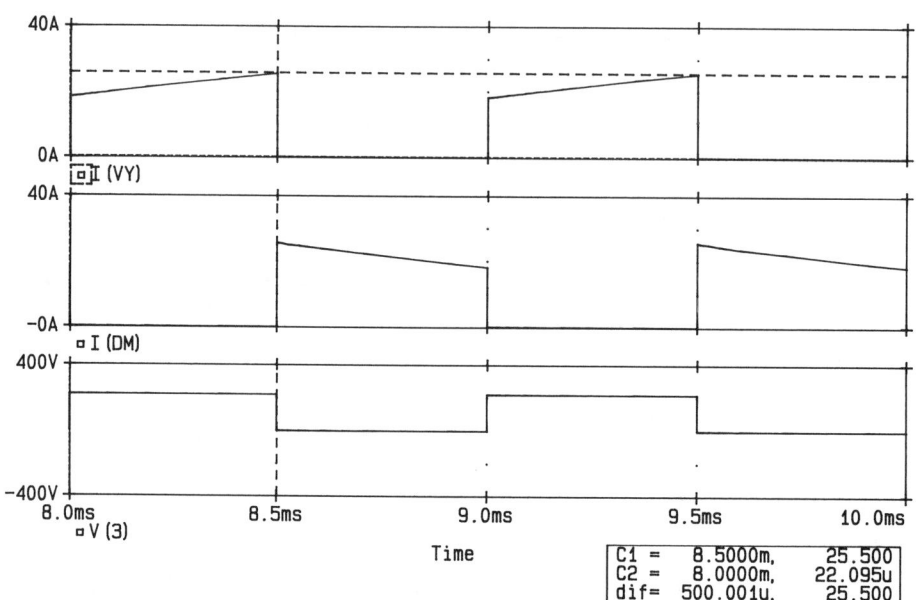

Figure 11-3 Plots for Example 11-1.

(c) The Fourier coefficients of the input current are:

FOURIER COMPONENTS OF TRANSIENT RESPONSE I (VY)

DC COMPONENT =   1.113030E+01

HARMONIC NO	FREQUENCY (HZ)	FOURIER COMPONENT	NORMALIZED COMPONENT	PHASE (DEG)	NORMALIZED PHASE (DEG)
1	1.000E+03	1.414E+01	1.000E+00	−5.237E+00	0.000E+00
2	2.000E+03	1.167E+00	8.254E−02	1.739E+02	1.792E+02
3	3.000E+03	4.652E+00	3.289E−01	1.861E−01	5.423E+00
4	4.000E+03	6.127E−01	4.333E−02	1.647E+02	1.699E+02
5	5.000E+03	2.784E+00	1.969E−01	2.486E+00	7.722E+00
6	6.000E+03	4.410E−01	3.118E−02	1.563E+02	1.615E+02
7	7.000E+03	1.995E+00	1.411E−01	4.250E+00	9.487E+00
8	8.000E+03	3.622E−01	2.561E−02	1.509E+02	1.561E+02
9	9.000E+03	1.552E+00	1.098E−01	5.976E+00	1.121E+01

TOTAL HARMONIC DISTORTION =   4.350004E+01 PERCENT

11-3 BJT SPICE MODEL

SPICE generates a complex model of BJTs. The model equations that are used by SPICE are described in Refs. 1 and 2. If a complex model is not necessary, many model parameters can be ignored by the users, and PSpice assigns default values to the parameters. The PSpice model, which is based on the integral charge-control model of Gummel and Poon [1,2], is shown in Fig. 11-4(a). The static (dc) model that is generated by PSpice is shown in Fig. 11-4(b).

The model statement for NPN transistors has the general form

    .MODEL QNAME NPN (P1=V1 P2=V2 P3=V3 ... PN=VN)

and the general form for PNP transistors is

    .MODEL QNAME PNP (P1=V1 P2=V2 P3=V3 ... PN=VN)

where QNAME is the name of the BJT model. NPN and PNP are the type symbols for NPN and PNP transistors, respectively. QNAME, which is the model name, can begin with any character and its word size is normally limited to eight characters. P1, P2, ... and V1, V2, ... are the parameters and their values, respectively. Table 11-1 shows the model parameters of BJTs. If certain parameters are not specified, PSpice assumes the simple model of Ebers and Moll [3], which is shown in Fig. 11-4(c).

The area factor is used to determine the number of equivalent parallel BJTs of the model specified. The model parameters, which are affected by the area factor, are marked by an asterisk (*) in Table 11-1. A bipolar transistor is modeled as an intrinsic transistor with ohmic resistances in series with the collector (RC/area), the base (RB/area), and the emitter (RE/area). [(area)value] is the relative device area and defaults to 1. For those parameters that have alternative names,

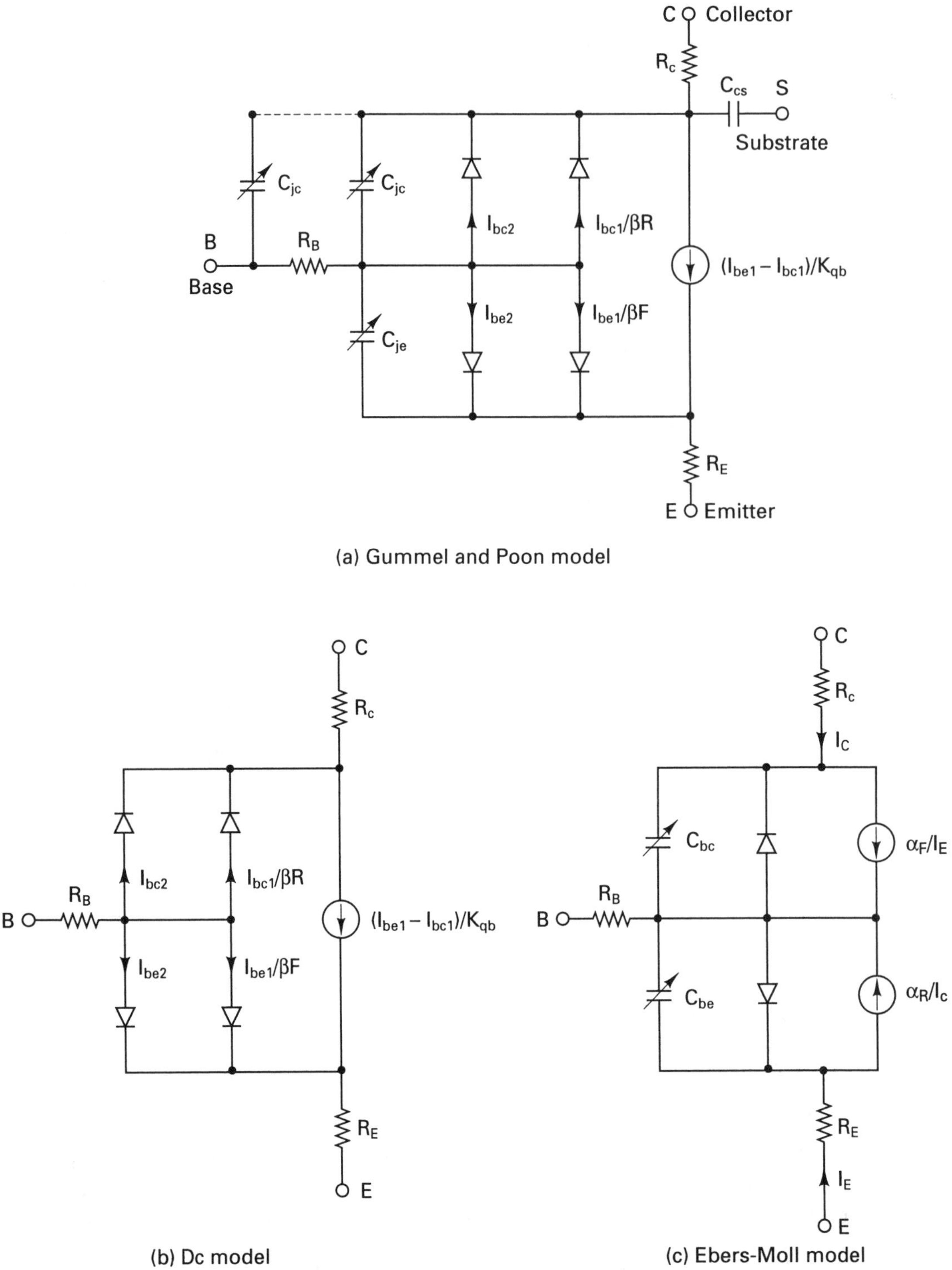

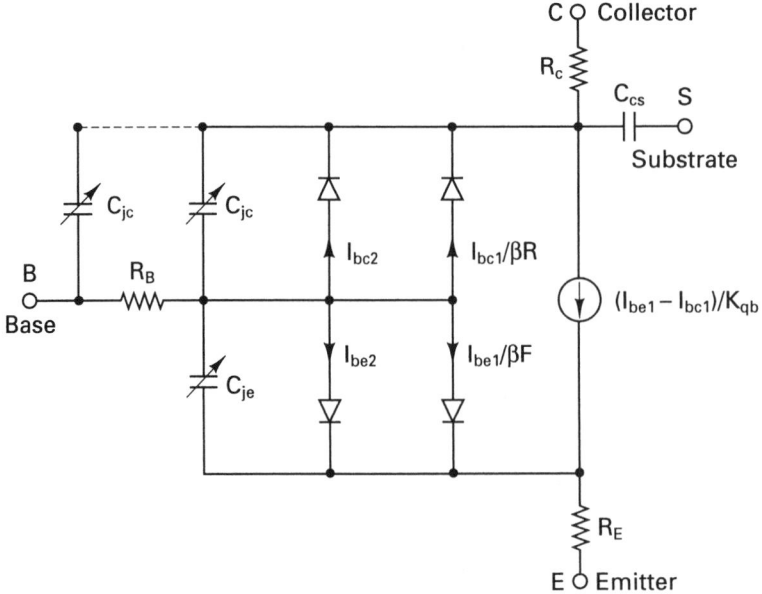

(a) Gummel and Poon model

(b) Dc model

(c) Ebers-Moll model

Figure 11-4 PSpice BJT model.

TABLE 11-1   MODEL PARAMETERS OF BJTS

Name	Area	Model parameter	Unit	Default	Typical
IS	*	*p-n* Saturation current	A	1E−16	1E−16
BF		Ideal maximum forward beta		100	100
NF		Forward current emission co-efficient		1	1
VAF(VA)		Forward Early voltage	V	∞	100
IKF(IK)		Corner for forward beta high-current roll-off	A	∞	10M
ISE(C2)		Base–emitter leakage saturation current	A	0	1000
NE		Base–emitter leakage emission coefficient		1.5	2
BR		Ideal maximum reverse beta		1	0.1
NR		Reverse current emission co-efficient		1	
VAR(VB)		Reverse Early voltage	V	∞	100
IKR	*	Corner for reverse beta high-current roll-off	A	∞	100M
ISC(C4)		Base–collector leakage satura-tion current	A	0	1
NC		Base–collector leakage emission coefficient		2	2
RB	*	Zero-bias (maximum) base resis-tance	Ω	0	100
RBM		Minimum base resistance	Ω	RB	100
IRB		Current at which RB falls half-way to RBM	A	∞	
RE	*	Emitter ohmic resistance	Ω	0	1
RC	*	Collector ohmic resistance	Ω	0	10
CJE	*	Base–emitter zero-bias *p-n* capacitance	F	0	2P
VJE(PE)		Base–emitter built-in potential	V	0.75	0.7
MJE(ME)		Base–emitter *p-n* grading factor	0.33	0.33	
CJC	*	Base–collector zero-bias *p-n* capacitance	F	0	1P
VJC(PC)		Base–collector built-in potential	V	0.75	0.5
MJC(MC)		Base–collector *p-n* grading fac-tor	0.33	0.33	
XCJC		Fraction of C_{bc} connected inter-nal to R_b		1	
CJS(CCS)		Collector–substrate zero-bias *p-n* capacitance	F	0	2PF
VJS(PS)		Collector–substrate built-in potential	V	0.75	
MJS(MS)		Collector–substrate *p-n* grading factor		0	
FC		Forward-bias depletion capacitor coefficient		0.5	
TF		Ideal forward transit time	s	0	0.1NS
XTF		Transit-time bias dependence coefficient		0	

(continued)

TABLE 11-1 (*Continued*)

Name	Area	Model parameter	Unit	Default	Typical
VTF		Transit-time dependency on V_{bc}	V	∞	
ITF		Transit-time dependency on I_c	A	0	
PTF		Excess phase at $1/(2\pi \times TF)$ Hz	degree	0	30°
TR		Ideal reverse transit time	s	0	10NS
EG		Bandgap voltage (barrier height)	eV	1.11	1.11
XTB		Forward and reverse beta temperature coefficient		0	
XTI(PT)		IS temperature effect exponent		3	
KF		Flicker noise coefficient		0	6.6E−16
AF		Flicker noise exponent		1	1

such as VAF and VA (the alternative name is indicated in parentheses), either name may be used.

The parameters ISE (C2) and ISC (C4) may be set to be greater than 1. In this case they are interpreted as multipliers of IS instead of absolute currents: that is, if ISE > 1, it is replaced by ISE*IS, and similarly for ISC. The dc model is defined by (1) parameters BF, C2, IK, and NE, which determine the forward current gain; (2) BR, C4, IKR, and VC, which determine the reverse current gain characteristics; (3) VA and VB, which determine the output conductance for forward and reverse regions; and (4) the reverse saturation current IS.

Base-charge storage is modeled by (1) forward and reverse transit times TF and TR, and nonlinear depletion-layer capacitances, which are determined by CJE, PE, and ME for a base–emitter junction; and (2) CJC, PC, and MC for a base–collector junction. CCS is a constant collector–substrate capacitance. The temperature dependence of the saturation current is determined by the energy gap EG and the saturation current temperature exponent PT.

The parameters that affect the switching behavior of a BJT are the most important ones for power electronics applications: IS, BF, CJE, CJC, and TF. The symbol for a bipolar junction transistor (BJT) is Q. The name of a bipolar transistor must start with Q and it takes the general form

```
Q⟨name⟩ NC NB NE NS QNAME [(area) value]
```

where NC, NB, NE, and NS are the collector, base, emitter, and substrate nodes, respectively. QNAME could be any name of up to eight characters. The substrate node is optional: If not specified, it defaults to ground. Positive current is the current that flows into a terminal. That is, the current flows from the collector node, through the device, to the emitter node for an NPN BJT.

11-4 BJT PARAMETERS

The data sheet for power transistor 2N6546 is shown in Fig. 11-5. SPICE parameters are not available from the data sheet. Some versions of SPICE (e.g., PSpice) support device library files. The software PARTS of PSpice can generate SPICE

Designers Data Sheet

SWITCHMODE SERIES
NPN SILICON POWER TRANSISTORS

The 2N6546 and 2N6547 transistors are designed for high-voltage, high-speed, power switching in inductive circuits where fall time is critical. They are particularly suited for 115 and 220 volt line operated switch-mode applications such as:

- Switching Regulators
- PWM Inverters and Motor Controls
- Solenoid and Relay Drivers
- Deflection Circuits

Specification Features —
High Temperature Performance Specified for:
Reversed Biased SOA with Inductive Loads
Switching Times with Inductive Loads
Saturation Voltages
Leakage Currents

15 AMPERE
NPN SILICON
POWER TRANSISTORS
300 and 400 VOLTS
175 WATTS

Designer's Data for "Worst Case" Conditions

The Designers Data Sheet permits the design of most circuits entirely from the information presented. Limit data — representing device characteristics boundaries — are given to facilitate "worst case" design.

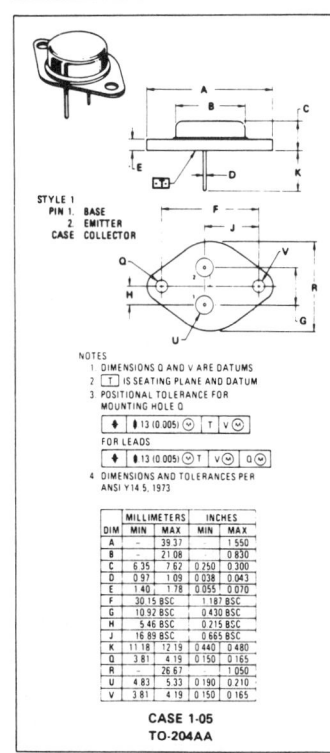

STYLE 1
PIN 1. BASE
2. EMITTER
CASE COLLECTOR

NOTES
1. DIMENSIONS Q AND V ARE DATUMS
2. ⊤ IS SEATING PLANE AND DATUM
3. POSITIONAL TOLERANCE FOR MOUNTING HOLE Q
4. DIMENSIONS AND TOLERANCES PER ANSI Y14 5, 1973

*MAXIMUM RATINGS

Rating	Symbol	2N6546	2N6547	Unit
Collector-Emitter Voltage	$V_{CEO(sus)}$	300	400	Vdc
Collector-Emitter Voltage	$V_{CEX(sus)}$	350	450	Vdc
Collector-Emitter Voltage	V_{CEV}	650	850	Vdc
Emitter Base Voltage	V_{EB}	9.0		Vdc
Collector Current — Continuous — Peak (1)	I_C I_{CM}	15 30		Adc
Base Current — Continuous — Peak (1)	I_B I_{BM}	10 20		Adc
Emitter Current — Continuous — Peak (1)	I_E I_{EM}	25 50		Adc
Total Power Dissipation @ T_C = 25°C @ T_C = 100°C Derate above 25°C	P_D	175 100 1.0		Watts W/°C
Operating and Storage Junction Temperature Range	T_J, T_{stg}	-65 to +200		°C

THERMAL CHARACTERISTICS

Characteristic	Symbol	Max	Unit
Thermal Resistance, Junction to Case	$R_{\theta JC}$	1.0	°C/W
Maximum Lead Temperature for Soldering Purposes: 1/8'' from Case for 5 Seconds	T_L	275	°C

*Indicates JEDEC Registered Data
(1) Pulse Test: Pulse Width = 5.0 ms, Duty Cycle ⩽ 10%.

DIM	MILLIMETERS		INCHES	
	MIN	MAX	MIN	MAX
A	–	39.37	–	1.550
B	–	21.08	–	0.830
C	6.35	7.62	0.250	0.300
D	0.97	1.09	0.038	0.043
E	1.40	1.78	0.055	0.070
F	30.15 BSC		1.187 BSC	
G	10.92 BSC		0.430 BSC	
H	5.46 BSC		0.215 BSC	
J	16.89 BSC		0.665 BSC	
K	11.18	12.19	0.440	0.480
Q	3.81	4.19	0.150	0.165
R	–	26.67	–	1.050
U	4.83	5.33	0.190	0.210
V	3.81	4.19	0.150	0.165

CASE 1-05
TO-204AA

Figure 11-5 Data sheet for transistor 2N6546. (Courtesy of Motorola Inc.)

models from the data sheet parameters of transistors and diodes. Although PSpice allows one to specify many parameters, we shall use only those parameters that affect significantly the output of a power converter [8,9]. From the data sheet we get

$$I_{C(rated)} = 10 \text{ A}$$

$$V_{BE} = 0.8 \text{ V at } I_C = 2 \text{ A}$$

Characteristic	Symbol	Min	Max	Unit
OFF CHARACTERISTICS (1)				
Collector-Emitter Sustaining Voltage (I_C = 100 mA, I_B = 0) 2N6546 2N6547	$V_{CEO(sus)}$	300 400	– –	Vdc

Collector-Emitter Sustaining Voltage
(I_C = 8.0 A, V_{clamp} = Rated V_{CEX}, T_C = 100°C) 2N6546
                                          2N6547
(I_C = 15 A, V_{clamp} = Rated V_{CEO} – 100 V, 2N6546
   T_C = 100°C)   2N6547
	$V_{CEX(sus)}$	350
		450
		200
		300
			–
			–
			–
			–
				Vdc

Collector Cutoff Current (V_{CEV} = Rated Value, $V_{BE(off)}$ = 1.5 Vdc) (V_{CEV} = Rated Value, $V_{BE(off)}$ = 1.5 Vdc, T_C = 100°C)	I_{CEV}	– –	1.0 4.0	mAdc
Collector Cutoff Current (V_{CE} = Rated V_{CEV}, R_{BE} = 50 Ω, T_C = 100°C)	I_{CER}	–	5.0	mAdc
Emitter Cutoff Current (V_{EB} = 9.0 Vdc, I_C = 0)	I_{EBO}	–	1.0	mAdc
SECOND BREAKDOWN				
Second Breakdown Collector Current with base forward biased t = 1.0 s (non-repetitive) (V_{CE} = 100 Vdc)	$I_{S/b}$	0.2	–	Adc
ON CHARACTERISTICS (1)				
DC Current Gain (I_C = 5.0 Adc, V_{CE} = 2.0 Vdc) (I_C = 10. Adc, V_{CE} = 2.0 Vdc)	h_{FE}	12 6.0	60 30	–
Collector-Emitter Saturation Voltage (I_C = 10 Adc, I_B = 2.0 Adc) (I_C = 15 Adc, I_B = 3.0 Adc) (I_C = 10 Adc, I_B = 2.0 Adc, T_C = 100°C)	$V_{CE(sat)}$	– – –	1.5 5.0 2.5	Vdc
Base-Emitter Saturation Voltage (I_C = 10 Adc, I_B = 2.0 Adc) (I_C = 10 Adc, I_B = 2.0 Adc, T_C = 100°C	$V_{BE(sat)}$	– –	1.6 1.6	Vdc
DYNAMIC CHARACTERISTICS				
Current-Gain – Bandwidth Product (I_C = 500 mAdc, V_{CE} = 10 Vdc, f_{test} = 1.0 MHz)	f_T	6.0	28	MHz
Output Capacitance (V_{CB} = 10 Vdc, I_E = 0, f_{test} = 1.0 MHz)	C_{ob}	125	500	pF
SWITCHING CHARACTERISTICS				
Resistive Load				
Delay Time	t_d	–	0.05	µs
Rise Time	t_r	–	1.0	µs
Storage Time	t_s	–	4.0	µs
Fall Time	t_f	–	0.7	µs
Inductive Load, Clamped				
Storage Time	t_s	–	5.0	µs
Fall Time	t_f	–	1.5	µs
			Typical	
Storage Time	t_s		2.0	µs
Fall Time	t_f		0.09	µs

Resistive Load conditions: (V_{CC} = 250 V, I_C = 10 A, I_{B1} = I_{B2} = 2.0 A, t_p = 100 µs, Duty Cycle ≤ 2.0%)

Inductive Load, Clamped conditions: (I_C = 10 A(pk), V_{clamp} = Rated V_{CEX}, I_{B1} = 2.0 A, $V_{BE(off)}$ = 5.0 Vdc, T_C = 100°C)

Typical conditions: (I_C = 10 A(pk), V_{clamp} = Rated V_{CEX}, I_{B1} = 2.0 A, $V_{BE(off)}$ = 5.0 Vdc, T_C = 25°C)

*Indicates JEDEC Registered Data.
(1) Pulse Test: Pulse Width = 300 µs, Duty Cycle = 2%.

Figure 11-5 *(Continued)*

Assuming that $n = 1$ and $V_T = 25.8$ mV, we can apply Eq. (7-1) to find the saturation current I_s:

$$I_C = I_s(e^{V_{BE}/nV_T} - 1)$$
$$2 = I_s(e^{0.8/(25.8 \times 10^{-3})} - 1) \tag{11-1}$$

which gives $I_s = 2.33E-27$ A. Dc current gain at 10 A is $h_{FE} = 6$ to 30. Taking the geometric mean gives BF = BR = $\sqrt{6 \times 30} \approx 13$.

The input capacitance at the base–emitter junction is very small, and its typical value is $0.2 - 1$ pF. Let us assume that C_{je} = CJE = 1 pF. Output

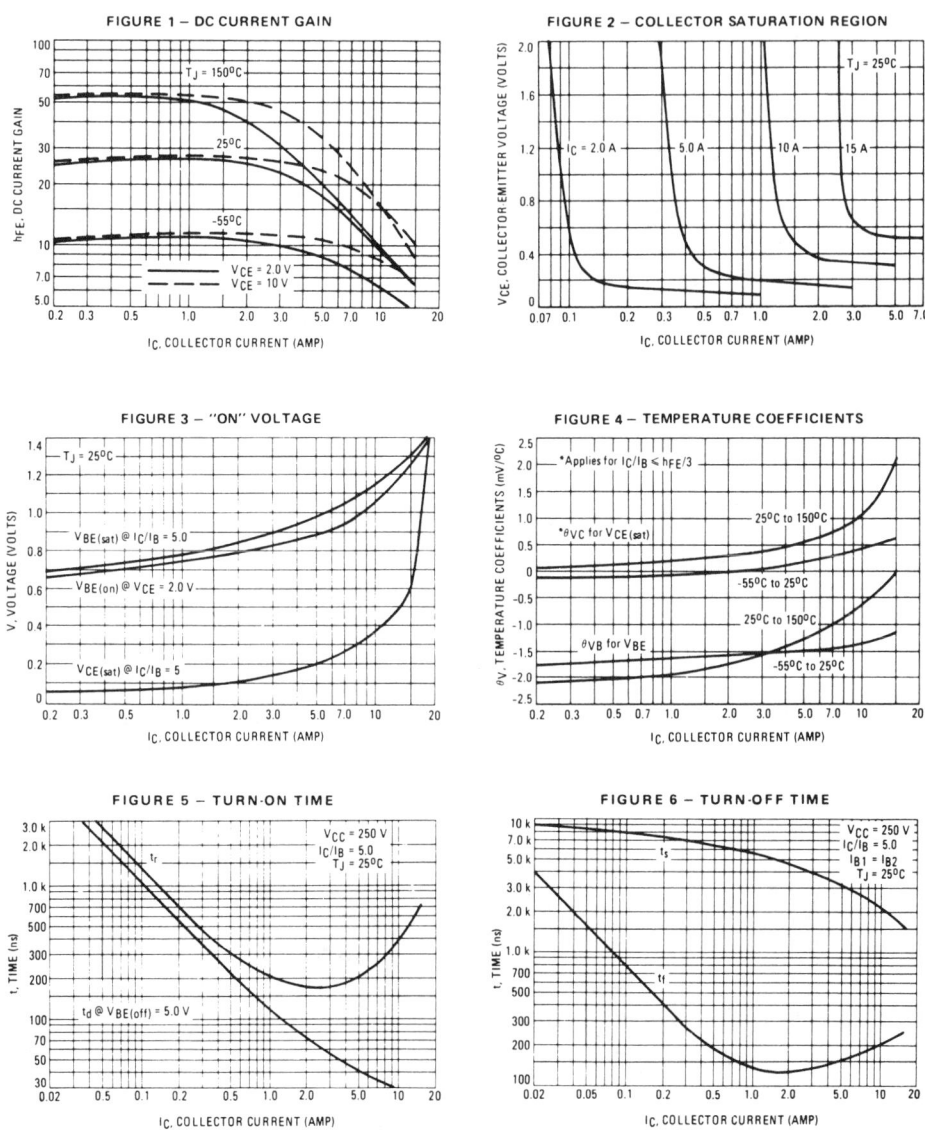

Figure 11-5 *(Continued)*

capacitance, $C_{obo} = 125 - 500$ pF at $V_{CB} = 10$ V, $I_E = 0$ (reverse biased). Taking the geometric mean gives $C_{obo} = C_\mu = \sqrt{125 \times 500} \approx 250$ pF. $C_{\mu o}$ can be found from

$$C_\mu = \frac{C_{\mu o}}{(1 + V_{CB}/V_0)^m} \qquad (11\text{-}2)$$

FIGURE 7 – FORWARD BIAS SAFE OPERATING AREA

FIGURE 8 – REVERSE BIAS SAFE OPERATING AREA

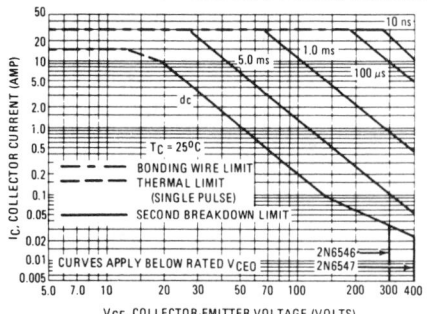

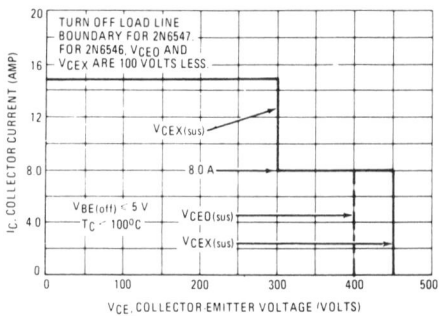

FIGURE 9 – POWER DERATING

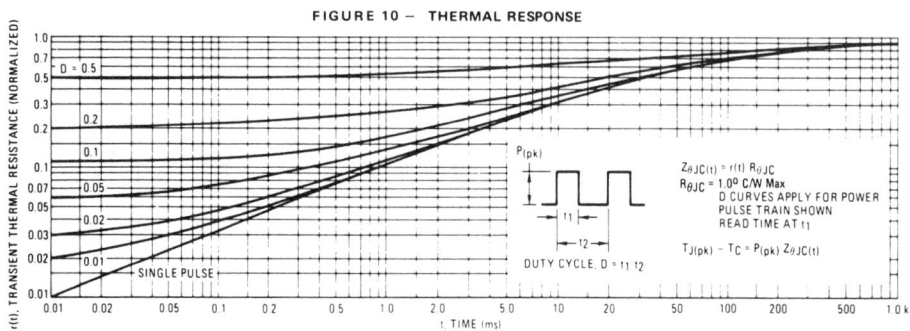

There are two limitations on the power handling ability of a transistor: average junction temperature and second breakdown. Safe operating area curves indicate I_C-V_{CE} limits of the transistor that must be observed for reliable operation; i.e., the transistor must not be subjected to greater dissipation than the curves indicate.

The data of Figure 7 is based on $T_C = 25^oC$; $T_{J(pk)}$ is variable depending on power level. Second breakdown pulse limits are valid for duty cycles to 10% but must be derated when $T_C \geqslant 25^oC$. Second breakdown limitations do not derate the same as thermal limitations. Allowable current at the voltages shown on Figure 7 may be found at any case temperature by using the appropriate curve on Figure 9.

$T_{J(pk)}$ may be calculated from the data in Figure 10. At high case temperatures, thermal limitations will reduce the power that can be handled to values less than the limitations imposed by second breakdown.

FIGURE 10 – THERMAL RESPONSE

Figure 11-5 *(Continued)*

where $m = \frac{1}{3}$ and $V_0 = 0.75$ V. From Eq. (11-2), $C_{\mu o} = CJC = 607.3$ pF at $V_{CB} = 10$ V.

The transition frequency $f_{T(min)} = 6$ MHz at $V_{CE} = 10$ V, $I_C = 500$ mA. The transition period is $\tau_T = 1/2\pi f_T = 1/(2\pi \times 6$ MHz$) = 26,525.8$ ps. Thus $V_{CB} \approx V_{CE} - V_{BE} = 10 - 0.7 = 9.3$ V, and Eq. (11-2) gives $C_\mu = 255.7$ pF. The transconductance g_m is

$$g_m = \frac{I_C}{V_T}$$

$$= \frac{500 \text{ mA}}{25.8 \text{ mV}} = 19.38 \text{ A/V}$$

(11-3)

The transition period τ_T is related to forward transit time τ_F by

or

$$\tau_T = \tau_F + \frac{C_{je}}{g_m} + \frac{C_\mu}{g_m}$$

(11-4)

$$26,525.8 \text{ ps} = \tau_F + \frac{1 \text{ pF}}{19.38} + \frac{255.7 \text{ pF}}{19.38}$$

which gives $\tau_F = 26,512.6$ ps. Thus the PSpice model statement for transistor 2N6546 is

```
.MODEL  2N6546  NPN (IS=6.83E-14 BF=13 CJE=1PF CJC=607.3PF TF=26.5NS)
```

This model can be used to plot the characteristic of the MOSFET. It may be necessary to modify the parameter values to conform to the actual characteristics.

Note. It is often necessary to adjust the base resistance RB and/or base (control) voltage Vg so that the transistor is driven into saturation.

11-5 EXAMPLES OF BJT CHOPPERS

The applications of the SPICE BJT model are illustrated by some examples.

Example 11-2

A BJT buck chopper is shown in Fig. 11-6(a). The dc input voltage is $V_S = 12$ V. The load resistance R is 5 Ω. The filter inductance is $L = 145.84$ μH, and the filter capacitance is $C = 200$ μF. The chopping frequency is $f_c = 25$ kHz, and the duty cycle of the chopper is $k = 42\%$. The control voltage is shown in Fig. 11-6(b). Use PSpice (a) to plot the instantaneous load current i_o, the input current i_s, the diode voltage v_D, and the output voltage v_C, and (b) to calculate the Fourier coefficients of the input current i_s.

Solution The dc supply voltage $V_S = 12$ V. $k = 0.42$, $f_c = 25$ kHz, $T = 1/f_c = 40$ μs, and $t_{on} = k \times T = 16.7$ μs. An initial value for C is assigned to reach steady state faster. The list of the circuit file is as follows:

■ ■

Example 11-2 BJT buck chopper

```
SOURCE  ■ VS   1   0   DC    12V
          VY   1   2   DC    0V   ; Voltage source to measure input current
          Vg   7   3   PULSE (0V  30V   0   0.1NS  0.1NS  16.7US  40US)
```

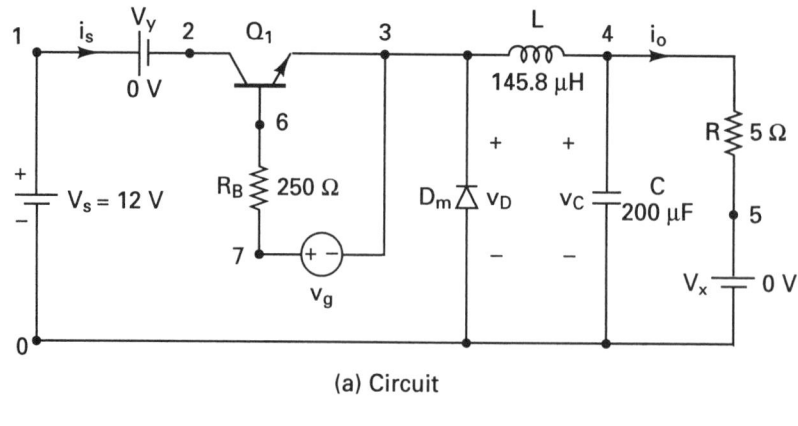

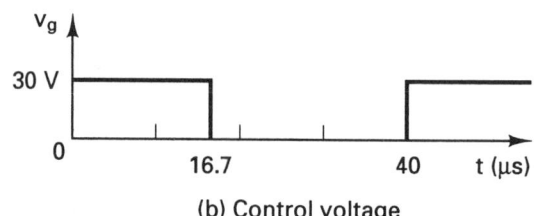

(b) Control voltage

Figure 11-6 BJT buck chopper for PSpice simulation.

```
CIRCUIT   ■■■ RB    7    6    250              ; Transistor base resistance
              R     4    5    5
              L     3    4    145.8UH
              C     4    0    200UF   IC=3V ; Initial voltage
              VX    5    0    DC    0V       ; To measure load current
              DM    0    3    DMOD                 ; Freewheeling diode
              .MODEL DMOD  D(IS=2.22E-15 CJO=1200V CJO=0 TT=0)
              Q1    2    6    3    3    2N6546      ; BJT switch
              .MODEL 2N6546 NPN(IS=6.83E-14 BF=13 CJE=1PF CJC=607.3PF TF=26.5NS)
ANALYSIS  ■■■ .TRAN  2US   2.1MS   2MS   UIC      ; Transient analysis
              .PROBE                               ; Graphics post-processor
              .OPTIONS ABSTOL = 1.00N  RELTOL = 0.01  VNTOL = 0.1  ITL5=40000
              .FOUR   25KHZ  I(VY)                 ; Fourier analysis
          .END
```

(a) The PSpice plots of the instantaneous load current I(VX), the input current I(VY), the diode voltage V(3), and the output voltage V(5) are shown in Fig. 11-7.
(b) The Fourier coefficients of the input current are:

FOURIER COMPONENTS OF TRANSIENT RESPONSE I(VY)

DC COMPONENT = 3.938953E-01

HARMONIC NO	FREQUENCY (HZ)	FOURIER COMPONENT	NORMALIZED COMPONENT	PHASE (DEG)	NORMALIZED PHASE (DEG)
1	2.500E+04	5.919E-01	1.000E+00	1.792E+02	0.000E+00
2	5.000E+04	2.021E-01	3.415E-01	-1.106E+02	-2.898E+02

HARMONIC NO	FREQUENCY (HZ)	FOURIER COMPONENT	NORMALIZED COMPONENT	PHASE (DEG)	NORMALIZED PHASE (DEG)
3	7.500E+04	1.545E-01	2.611E-01	-1.260E+02	-3.053E+02
4	1.000E+05	1.299E-01	2.195E-01	-6.211E+01	-2.413E+02
5	1.250E+05	6.035E-02	1.020E-01	-6.698E+01	-2.462E+02
6	1.500E+05	9.647E-02	1.630E-01	-2.055E+01	-1.998E+02
7	1.750E+05	4.065E-02	6.869E-02	1.750E+01	-1.617E+02
8	2.000E+05	6.771E-02	1.144E-01	1.779E+01	-1.614E+02
9	2.250E+05	4.600E-02	7.772E-02	7.940E+01	-9.984E+01

TOTAL HARMONIC DISTORTION = 5.420088E+01 PERCENT

Example 11-3

A BJT buck–boost chopper is shown in Fig. 11-8(a). The dc input voltage is $V_S =$ 12 V. The load resistance R is 5 Ω. The inductance is $L = 150$ μH, and the filter capacitance is $C = 220$ μF. The chopping frequency is $f_c = 25$ kHz, and the duty cycle of the chopper is $k = 25\%$. The control voltage is shown in Fig. 11-8(b). Use PSpice to plot the instantaneous the output voltage v_C, the capacitor current i_C, the inductor current i_L, and the inductor voltage v_L.

Solution The dc supply voltage $V_S = 12$ V. $k = 0.25$, $f_c = 25$ kHz, $T = 1/f_c = 40$ μs, and $t_{on} = k \times T = 0.25 \times 40 = 10$ μs. The list of the circuit file is as follows:

■ ■

Example 11-3   BJT buck–boost chopper

```
SOURCE  ■ VS    1    0    DC    12V
          VY    1    2    DC    0V     ; Voltage source to measure input current
          Vg    7    3    PULSE (0V    40V    0    1NS    1NS    10US    40US)
```

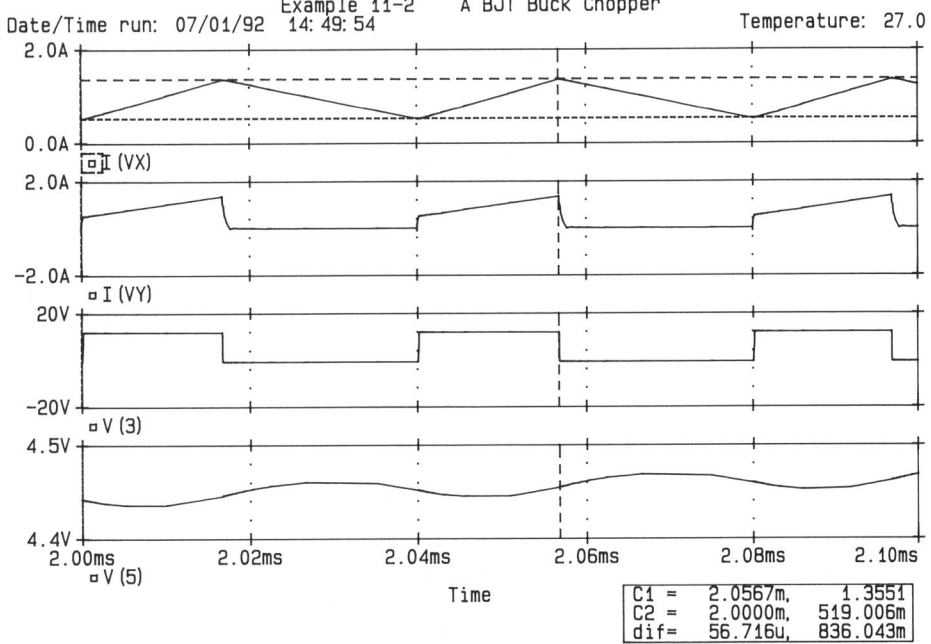

Example 11-2   A BJT Buck Chopper
Date/Time run: 07/01/92 14:49:54   Temperature: 27.0

C1 =	2.0567m,	1.3551
C2 =	2.0000m,	519.006m
dif=	56.716u,	836.043m

Figure 11-7 Plots for Example 11-2.

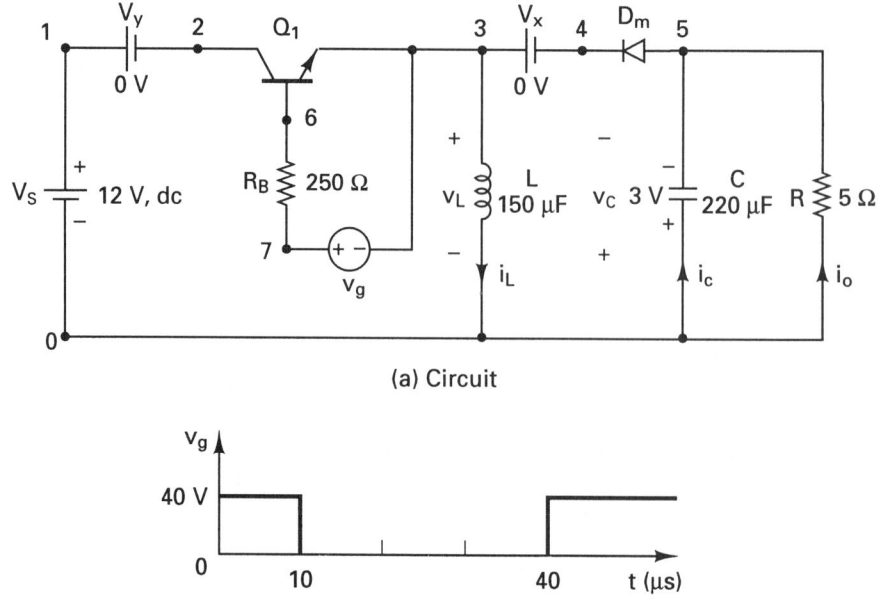

(a) Circuit

(b) Control voltage

Figure 11-8 BJT buck–boost chopper for PSpice simulation.

```
CIRCUIT  ■ ■  RB      7    6    250          ; Transistor base resistance
             R       5    0    5
             L       3    0    150UH
             C       5    0    220UF  IC=-3V  ; Initial voltage on capacitor
             VX      3    4    DC    0V   ; Voltage source to measure diode current
             DM      5    4    DMOD              ; Freewheeling diode
             .MODEL  DMOD  D(IS=2.22E-15 BV=1200V IBV=13E-3 CJO=0 TT=0) ; Diode model
             Q1      2    6    3    3    2N6546       ; BJT switch
             .MODEL  2N6546  NPN (IS=6.83E-14 BF=13 CJE=1PF CJC=607.3PF TF=26.5NS)
ANALYSIS ■ ■ ■ .TRAN   1US  1MS   750US   UIC         ; Transient analysis
             .PROBE                              ; Graphics post-processor
             .OPTIONS ABSTOL = 1.00N  RELTOL = 0.01  VNTOL = 0.1  ITL5=40000
             .FOUR    25KHZ   I(VY)               ; Fourier analysis
     .END
■ ■ ■ ■ ■ ■ ■ ■ ■ ■ ■ ■ ■ ■ ■ ■ ■ ■ ■ ■ ■ ■ ■ ■ ■ ■ ■ ■ ■ ■ ■ ■ ■ ■ ■ ■ ■ ■ ■ ■ ■ ■ ■ ■
```

The PSpice plots of the instantaneous output voltage V(5), the capacitor current I(C), the inductor current I(L), and the inductor voltage V(3) are shown in Fig. 11-9.

Example 11-4

A BJT cuk chopper is shown in Fig. 11-10(a). The dc input voltage is $V_S = 12$ V. The load resistance R is 5 Ω. The inductances are $L_1 = 200$ μH and $L_2 = 150$ μH. The capacitance are $C_1 = 200$ μF and $C_2 = 220$ μF. The chopping frequency is $f_c = 25$ kHz, and the duty cycle of the chopper is $k = 25\%$. The control voltage is shown

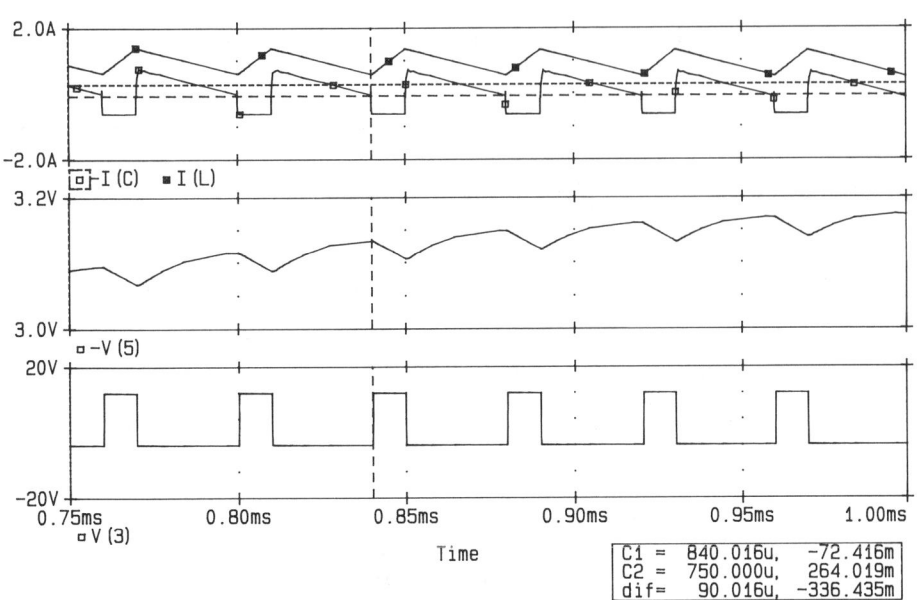

Figure 11-9 Plots for Example 11-3.

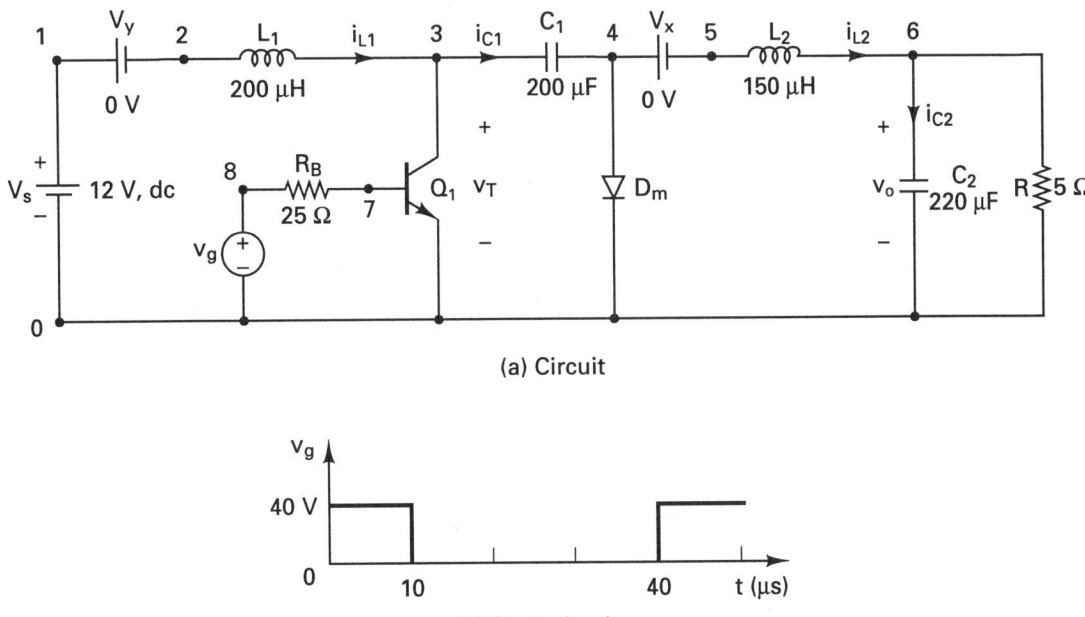

(a) Circuit

(b) Control voltage

Figure 11-10 BJT cuk chopper for PSpice simulation.

in Fig. 11-10(b). Use PSpice to plot the instantaneous capacitor current i_{C1}, the capacitor current i_{C2}, the inductor current i_{L1}, the inductor current i_{L2}, and the transistor voltage v_T.

Solution The dc supply voltage $V_S = 12$ V. An initial capacitor voltage $V_C = 3$ V. $k = 0.25, f_c = 25$ kHz, $T = 1/f_c = 40$ μs, and $t_{on} = k \times T = 0.25 \times 40 = 10$ μs. The list of the circuit file is as follows:

■■

Example 11-4   BJT cuk chopper

SOURCE ■ VS   1   0   DC   12V
         VY   1   2   DC   0V   ; Voltage source to measure input current
         Vg   8   0   PULSE (0V   40V   0   1NS   1NS   10US   40US)
CIRCUIT ■■ RB   8   7   25   ; Transistor base resistance
         R   6   0   5
         L1   2   3   200UH
         C1   3   4   200UF
         L2   5   6   150UF
         C2   6   0   220UF   IC=-3V   ; Initial conditions
         VX   4   5   DC   0V   ; Voltage source to measure current of L2
         DM   4   0   DMOD   ; Freewheeling diode
         .MODEL DMOD D(IS=2.22E-15 BV=1200V IBV=13E-3 CJO=0 TT=0) ; Diode model
         Q1   3   7   0   0   2N6546   ; BJT switch
         .MODEL 2N6546 NPN (IS=6.83E-14 BF=13 CJE=1PF CJC=607.3PF TF=26.5NS)

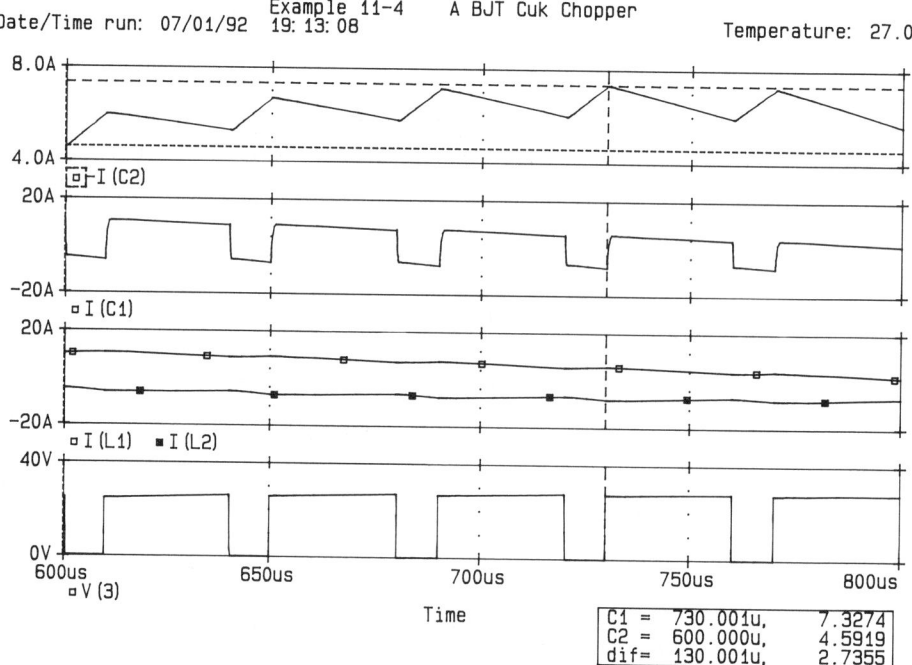

Figure 11-11 Plots for Example 11-4.

■ ■ ■ .TRAN   2US   800US   600US   UIC   ; Transient analysis
              .PROBE   ; Graphics post-processor
              .OPTIONS ABSTOL = 1.00N RELTOL = 0.01 VNTOL = 0.1 ITL5=40000
              .FOUR   25KHZ   I(VY)   ; Fourier analysis
          .END

The PSpice plots of the instantaneous capacitor current I(C1), the capacitor current I(C2), the inductor current I(L1), the inductor current I(L2), and the transistor voltage V(3) are shown in Fig. 11-11. The output voltage has not reached steady state.

11-6 MOSFET CHOPPERS

The PSpice model of an n-channel MOSFET [4–6] is shown in Fig. 11-12(a). The static (dc) model that is generated by PSpice is shown in Fig. 11-12(b). The model parameters for a MOSFET device and the default values assigned by PSpice are

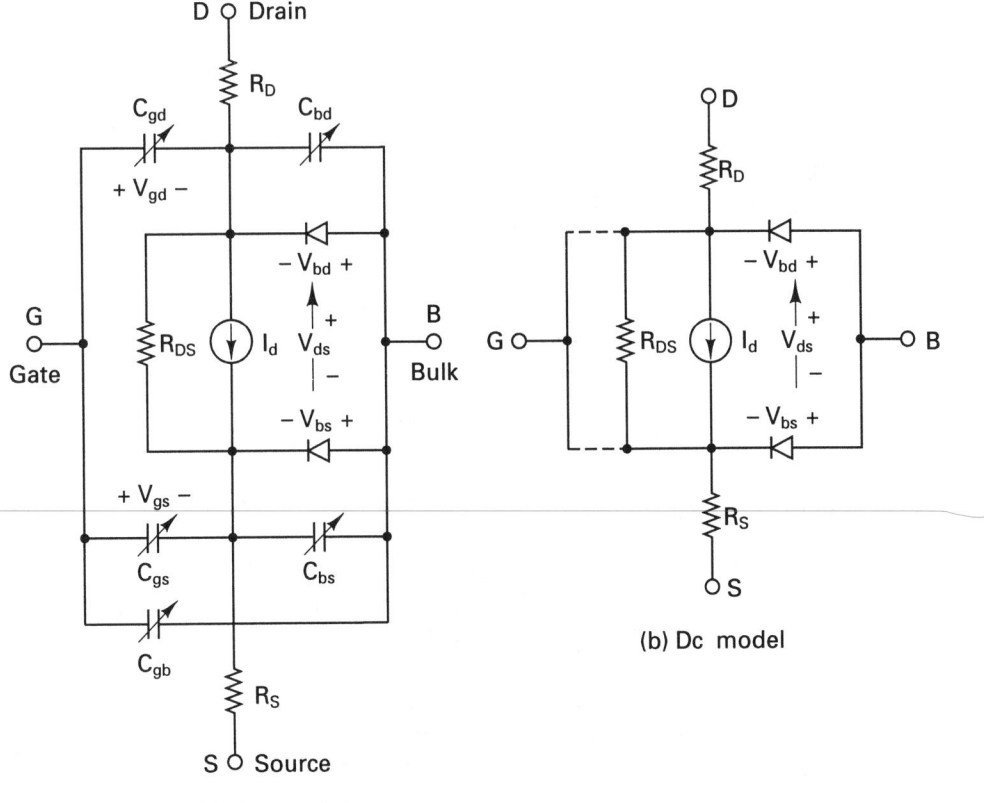

(a) PSpice model

(b) Dc model

Figure 11-12   PSpice n-channel MOSFET model.

given in Table 11-2. The model equations of MOSFETs that are used by PSpice are described in Refs. 4 and 6.

The model statement of *n*-channel MOSFETs has the general form

`.MODEL   MNAME   NMOS  (P1=V1   P2=V2  P3=V3  ...  PN=VN)`

and the statement for *p*-channel MOSFETs has the form

`.MODEL   MNAME   PMOS  (P1=V1  P2=V2  P3=V3  ...  PN=VN)`

where MNAME is the model name. It can begin with any charac.er and its word size is normally limited to eight characters. NMOS and PMOS are the type symbols of *n*-channel and *p*-channel MOSFETs, respectively. P1, P2, . . . and V1, V2, . . . are the parameters and their values, respectively.

TABLE 11-2   MODEL PARAMETERS OF MOSFETS

Name	Model parameter	Unit	Default	Typical
LEVEL	Model type (1, 2, or 3)		1	
L	Channel length	m	DEFL	
W	Channel width	m	DEFW	
LD	Lateral diffusion (length)	m	0	
WD	Lateral diffusion (width)	m	0	
VTO	Zero-bias threshold voltage	V	0	0.1
KP	Transconductance	A/V^2	$2E-5$	$2.5E-5$
GAMMA	Bulk threshold parameter	$V^{1/2}$	0	0.35
PHI	Surface potential	V	0.6	0.65
LAMBDA	Channel-length modulation (LEVEL = 1 or 2)	V^{-1}	0	0.02
RD	Drain ohmic resistance	Ω	0	10
RS	Source ohmic resistance	Ω	0	10
RG	Gate ohmic resistance	Ω	0	1
RB	Bulk ohmic resistance	Ω	0	1
RDS	Drain–source shunt resistance	Ω	∞	
RSH	Drain–source diffusion sheet resistance	Ω/square	0	20
IS	Bulk *p-n* saturation current	A	$1E-14$	$1E-15$
JS	Bulk *p-n* saturation current/area	A/m^2	0	$1E-8$
PB	Bulk *p-n* potential	V	0.8	0.75
CBD	Bulk–drain zero-bias *p-n* capacitance	F	0	5PF
CBS	Bulk–source zero-bias *p-n* capacitance	F	0	2PF
CJ	Bulk *p-n* zero-bias bottom capacitance/length	F/m^2	0	
CJSW	Bulk *p-n* zero-bias perimeter capacitance/length	F/m	0	
MJ	Bulk *p-n* bottom grading coefficient		0.5	
MJSW	Bulk *p-n* sidewall grading coefficient		0.33	
FC	Bulk *p-n* forward-bias capacitance coefficient		0.5	
CGSO	Gate–source overlap capacitance/ channel width	F/m	0	

(continued)

TABLE 11-2 (*Continued*)

Name	Model parameter	Unit	Default	Typical
CGDO	Gate–drain overlap capacitance/ channel width	F/m	0	
CGBO	Gate–bulk overlap capacitance/ channel length	F/m	0	
NSUB	Substrate doping density	$1/cm^3$	0	
NSS	Surface state density	$1/cm^2$	0	
NFS	Fast surface state density	$1/cm^2$	0	
TOX	Oxide thickness	m	∞	
TPG	Gate material type: $+1$ = opposite of substrate -1 = same as substrate 0 = aluminum		$+1$	
XJ	Metallurgical junction depth	m	0	
UO	Surface mobility	$cm^2/V \cdot s$	600	
UCRIT	Mobility degradation critical field (LEVEL = 2)	V/cm	1E4	
UEXP	Mobility degradation exponent (LEVEL = 2)		0	
UTRA	(*not used*) Mobility degradation transverse field coefficient			
VMAX	Maximum drift velocity	m/s	0	
NEFF	Channel charge coefficient (LEVEL = 2)		1	
XQC	Fraction of channel charge attributed to drain		1	
DELTA	Width effect on threshold		0	
THETA	Mobility modulation (LEVEL = 3)	V^{-1}	0	
ETA	Static feedback (LEVEL = 3)		0	
KAPPA	Saturation field factor (LEVEL = 3)		0.2	
KF	Flicker noise coefficient		0	$1E-26$
AF	Flicker noise exponent		1	1.2

L and W are the channel length and width, respectively. AD and AS are the drain and source diffusion areas. L is decreased by twice LD to get the effective channel length. W is decreased by twice WD to get the effective channel width. L and W can be specified on the device, the model, or on the .OPTION statement. The value on the device supersedes the value on the model, which supersedes the value on the .OPTION statement.

AD and AS are the drain and source diffusion areas. PD and PS are the drain and source diffusion perimeters. The drain–bulk and source–bulk saturation currents can be specified either by JS, which is multiplied by AD and AS, or by IS, which is an absolute value. The zero-bias depletion capacitance can be specified by CJ, which is multiplied by AD and AS, and by CJSW, which is multiplied by PD and PS. Alternatively, these capacitances can be set by CBD and CBS, which are absolute values.

A MOSFET is modeled as an intrinsic MOSFET with ohmic resistances in series with the drain, source, gate, and bulk (substrate). There is also a shunt resistance (RDS) in parallel with the drain–source channel. NRD, NRS, NRG, and NRB are the relative resistivities of the drain, source, gate, and substrate in squares. These parasitic (ohmic) resistances can be specified by RSH, which is multiplied by NRD, NRS, NRG, and NRB, respectively; or, alternatively, the absolute values of RD, RS, RG, and RB can be specified directly.

PD and PS default to 0. NRD and NRS default to 1. NRG and NRB default to 0. Defaults for L, W, AD, and AS may be set in the .OPTIONS statement. If AD or AS defaults are not set, they also default to 0. If L or W defaults are not set, they default to 100 μm.

The dc characteristics are defined by parameters VTO, KP, LAMBDA, PHI, and GAMMA, which are computed by PSpice by using the fabrication-process parameters NSUB, TOX, NSS, NFS, TPG, and so on. The values of VTO, KP, LAMDA, PHI, and GAMMA, which are specified on the model statement, superseded the values calculated by PSpice based on fabrication-process parameters. *VTO is positive for enhancement-type n-channel MOSFET and for depletion-type p-channel MOSFET. VTO is negative for enhancement-type p-channel MOSFET and for depletion-type n-channel MOSFET.*

PSpice incorporates three MOSFET device models. The LEVEL parameter selects between different models for the intrinsic MOSFET. If LEVEL = 1, the Shichman–Hodges model [5] is used. If LEVEL = 2, an advanced version of the Shichman–Hodges model, which is a geometry-based analytical model and incorporates extensive second-order effects [6], is used. If LEVEL = 3, a modified version of the Shichman–Hodges model, which is a semiempirical short-channel model [7], is used.

The LEVEL 1 model, which employs fewer fitting parameters, gives approximates results. However, it is useful for a quick, rough estimate of circuit performance and is normally adequate for the analysis of power electronic circuits. The LEVEL 2 model, which can take into consideration various parameters, requires considerable CPU time for the calculations and could cause convergence problems. The LEVEL 3 model introduces a smaller error than that of the LEVEL 2 model, and the CPU time is also approximately 25% less. The LEVEL 3 model is designed for MOSFETs with a short channel.

The parameters that affect the switching behavior of a BJT in power electronics applications are

```
L     W     VTO     KP     CGSO     CGDO
```

The symbol for a metal-oxide silicon field-effect transistor (MOSFET) is M. The name of a MOSFET must start with M and takes the general form

```
M⟨name⟩  ND   NG   NS   NB   MNAME
+        [L=⟨value] [W=⟨value⟩]
+        [AD=⟨value⟩]  [AS=⟨value⟩]
+        [PD=⟨value⟩]  [PS=⟨value⟩]
+        [NRD=⟨value⟩]  [NRS=⟨value⟩]
+        [NRG=⟨value⟩]  [NRB=⟨value⟩]
```

where ND, NG, NS, and NB are the drain, gate, source, and bulk (or substrate) nodes, respectively. MNAME is the model name and can begin with any character; its word size is normally limited to eight characters. Positive current is the current that flows into a terminal. That is, the current flows from the drain node, through the device, to the source node for an *n*-channel MOSFET.

11-7 MOSFET PARAMETERS

The data sheet for an *n*-channel MOSFET type IRF150 is shown in Fig. 11-13. The library file of the student version of PSpice supports a model of this MOSFET as follows:

```
.MODEL  IRF150  NMOS (TOX=100N PHI=.6 KP=20.53U W=.3
+    L=2U VTO=2.831 RD=1.031M RDS=444.4K CBD=3.229N PB=.8 MJ=.5
+    CGSO=9.027N CGDO=1.679N RG=13.89 IS=194E-18 N=1 TT=288N)
```

However, we shall generate approximate values of some parameters [8, 9]. From the data sheet we get $I_{DSS} = 250$ μA at $V_{CS} = 0$ V, $V_{DS} = 100$ V. $V_{Th} = 2$ to 4 V. Geometric mean, $V_{Th} = $ VTO $ = \sqrt{2 \times 4} = 2.83$ V. The constant K_p can be found from

$$I_D = K_p(V_{GS} - V_{Th})^2 \qquad (11\text{-}5)$$

For $I_D = I_{DSS} = 250$ μA, and $V_{Th} = 2.83$ V, Eq. (11-5) gives $K_p = 250$ μA/2.83^2 = 31.2 μA/V^2. K_p is related to channel length L and channel width W by

$$K_p = \frac{\mu_a C_o}{2} \left(\frac{W}{L}\right) \qquad (11\text{-}6)$$

where C_o is the capacitance per unit area of the oxide layer, a typical value for a power MOSFET being 3.5×10^{-8} F/cm^2 at a thickness of 0.1 μm, and μ_a is the surface mobility of electrons, 600 cm^2/(V $\cdot$ s).

The ratio W/L can be found from

$$\frac{W}{L} = \frac{2K_p}{\mu_a C_o} = \frac{2 \times 31.2 \times 10^{-6}}{600 \times 3.5 \times 10^{-8}} = 3$$

Let $L = 1$ μm and $W = 3000$ μm $= 3$ mm. $C_{rss} = 350 - 500$ pF at $V_{GS} = 0$, $V_{DS} = 25$ V. Geometric mean, $C_{rss} = C_{gd} = \sqrt{350 \times 500} = 418.3$ pF at $V_{DG} = 25$ V.

For a MOSFET, the values of C_{gs} and C_{gd} remain relatively constant with changing V_{GS} or V_{DS}. They are determined mainly by the thickness and type of the insulating oxide. Although, the curves of the capacitances versus drain-source voltage show some variations, we will assume constant capacitances. Thus, $C_{gdo} = 418.3$ pF. $C_{iss} = 2000$ to 3000 pF. Geometric mean $C_{iss} = \sqrt{2000 \times 3000} = 2450$ pF. Since C_{iss} is measured at $V_{GS} = 0$ V, $C_{gs} = C_{gso}$. That is,

$$C_{iss} = C_{gso} + C_{gd}$$

HEXFET® TRANSISTORS IRF150

N-Channel

IRF151
IRF152
IRF153

100 Volt, 0.055 Ohm HEXFET

The HEXFET® technology is the key to International Rectifier's advanced line of power MOSFET transistors. The efficient geometry and unique processing of the HEXFET design achieve very low on-state resistance combined with high transconductance and great device ruggedness.

The HEXFET transistors also feature all of the well established advantages of MOSFETs such as voltage control, freedom from second breakdown, very fast switching, ease of paralleling, and temperature stability of the electrical parameters.

They are well suited for applications such as switching power supplies, motor controls, inverters, choppers, audio amplifiers, and high energy pulse circuits.

Features:

- Fast Switching
- Low Drive Current
- Ease of Paralleling
- No Second Breakdown
- Excellent Temperature Stability

Product Summary

Part Number	V_{DS}	$R_{DS(on)}$	I_D
IRF150	100V	0.055Ω	40A
IRF151	60V	0.055Ω	40A
IRF152	100V	0.08Ω	33A
IRF153	60V	0.08Ω	33A

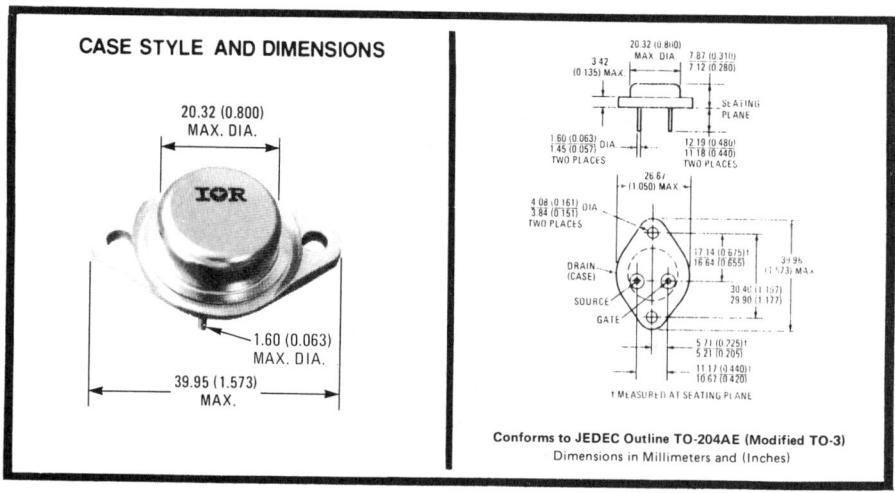

CASE STYLE AND DIMENSIONS

Conforms to JEDEC Outline TO-204AE (Modified TO-3)
Dimensions in Millimeters and (Inches)

Figure 11-13 Data sheet for MOSFET type IRF150. (Courtesy of International Rectifier.)

which gives $C_{gso} = C_{iss} - C_{sd} = 2450 - 418.3 = 2032$ pF $= 2.032$ nF. Thus the PSpice model statement for MOSFET IRF150 is

```
.MODEL IRF150 NMOS (VTO=2.83 KP=31.2U L=1U W=3U CGDO=0.418N CGSO=2.032N)
```

The model can be used to plot the characteristic of the MOSFET. It may be

IRF150, IRF151, IRF152, IRF153 Devices

Absolute Maximum Ratings

	Parameter	IRF150	IRF151	IRF152	IRF153	Units
V_{DS}	Drain - Source Voltage ①	100	60	100	60	V
V_{DGR}	Drain - Gate Voltage (R_{GS} = 20 kΩ) ①	100	60	100	60	V
I_D @ T_C = 25°C	Continuous Drain Current	40	40	33	33	A
I_D @ T_C = 100°C	Continuous Drain Current	25	25	20	20	A
I_{DM}	Pulsed Drain Current ③	160	160	132	132	A
V_{GS}	Gate - Source Voltage	± 20				V
P_D @ T_C = 25°C	Max. Power Dissipation	150	(See Fig. 14)			W
	Linear Derating Factor	1.2	(See Fig. 14)			W/K
I_{LM}	Inductive Current, Clamped	(See Fig. 15 and 16) L = 100μH				A
		160	160	132	132	
T_J T_{stg}	Operating Junction and Storage Temperature Range	−55 to 150				°C
	Lead Temperature	300 (0.063 in. (1.6mm) from case for 10s)				°C

Electrical Characteristics @ T_C = 25°C (Unless Otherwise Specified)

	Parameter	Type	Min.	Typ.	Max.	Units	Test Conditions
BV_{DSS}	Drain - Source Breakdown Voltage	IRF150 IRF152	100	--	--	V	V_{GS} = 0V
		IRF151 IRF153	60	--	--	V	I_D = 250μA
$V_{GS(th)}$	Gate Threshold Voltage	ALL	2.0	--	4.0	V	V_{DS} = V_{GS}; I_D = 250μA
I_{GSS}	Gate-Source Leakage Forward	ALL	--	--	100	nA	V_{GS} = 20V
I_{GSS}	Gate-Source Leakage Reverse	ALL	--	--	-100	nA	V_{GS} = -20V
I_{DSS}	Zero Gate Voltage Drain Current	ALL	--	--	250	μA	V_{DS} = Max. Rating, V_{GS} = 0V
			--	--	1000	μA	V_{DS} = Max. Rating x 0.8, V_{GS} = 0V, T_C = 125°C
$I_{D(on)}$	On-State Drain Current ②	IRF150 IRF151	40	--	--	A	V_{DS} › $I_{D(on)}$ x $R_{DS(on)}$ max.; V_{GS} = 10V
		IRF152 IRF153	33	--	--	A	
$R_{DS(on)}$	Static Drain-Source On-State Resistance ②	IRF150 IRF151	--	0.045	0.055	Ω	V_{GS} = 10V, I_D = 20A
		IRF152 IRF153	--	0.06	0.08	Ω	
g_{fs}	Forward Transconductance ②	ALL	9.0	11	--	S (℧)	V_{DS} › $I_{D(on)}$ x $R_{DS(on)}$ max.; I_D = 20A
C_{iss}	Input Capacitance	ALL	--	2000	3000	pF	V_{GS} = 0V, V_{DS} = 25V, f = 1.0 MHz
C_{oss}	Output Capacitance	ALL	--	1000	1500	pF	See Fig. 10
C_{rss}	Reverse Transfer Capacitance	ALL	--	350	500	pF	
$t_{d(on)}$	Turn-On Delay Time	ALL	--	--	35	ns	V_{DD} = 24V, I_D = 20A, Z_o = 4.7Ω
t_r	Rise Time	ALL	--	--	100	ns	See Figure 17.
$t_{d(off)}$	Turn-Off Delay Time	ALL	--	--	125	ns	(MOSFET switching times are essentially independent of operating temperature.)
t_f	Fall Time	ALL	--	--	100	ns	
Q_g	Total Gate Charge (Gate-Source Plus Gate-Drain)	ALL	--	63	120	nC	V_{GS} = 10V, I_D = 50A, V_{DS} = 0.8 Max. Rating. See Fig. 18 for test circuit. (Gate charge is essentially independent of operating temperature.)
Q_{gs}	Gate-Source Charge	ALL	--	27	--	nC	
Q_{gd}	Gate-Drain ("Miller") Charge	ALL	--	36	--	nC	
L_D	Internal Drain Inductance	ALL	--	5.0	--	nH	Measured between the contact screw on header that is closer to source and gate pins and center of die. / Modified MOSFET symbol showing the internal device inductances.
L_S	Internal Source Inductance	ALL	--	12.5	--	nH	Measured from the source pin, 6 mm (0.25 in.) from header and source bonding pad.

Thermal Resistance

R_{thJC}	Junction-to-Case	ALL	--	--	0.83	K/W	
R_{thCS}	Case-to-Sink	ALL	--	0.1	--	K/W	Mounting surface flat, smooth, and greased.
R_{thJA}	Junction-to-Ambient	ALL	--	--	30	K/W	Free Air Operation

Figure 11-13 *(Continued)*

Source-Drain Diode Ratings and Characteristics

I_S	Continuous Source Current (Body Diode)	IRF150 IRF151	—	—	40	A	Modified MOSFET symbol showing the integral reverse P-N junction rectifier.
		IRF152 IRF153	—	—	33	A	
I_{SM}	Pulse Source Current (Body Diode) ③	IRF150 IRF151	—	—	160	A	
		IRF152 IRF153	—	—	132	A	
V_{SD}	Diode Forward Voltage ②	IRF150 IRF151	—	—	2.5	V	$T_C = 25°C$, $I_S = 40A$, $V_{GS} = 0V$
		IRF152 IRF153	—	—	2.3	V	$T_C = 25°C$, $I_S = 33A$, $V_{GS} = 0V$
t_{rr}	Reverse Recovery Time	ALL	—	600	—	ns	$T_J = 150°C$, $I_F = 40A$, $dI_F/dt = 100A/\mu s$
Q_{RR}	Reverse Recovered Charge	ALL	—	3.3	—	μC	$T_J = 150°C$, $I_F = 40A$, $dI_F/dt = 100A/\mu s$
t_{on}	Forward Turn-on Time	ALL					Intrinsic turn-on time is negligible. Turn-on speed is substantially controlled by $L_S + L_D$.

① $T_J = 25°C$ to $150°C$.   ② Pulse Test: Pulse width ≤ 300μs, Duty Cycle ≤ 2%.   ③ Repetitive Rating: Pulse width limited by max. junction temperature. See Transient Thermal Impedance Curve (Fig. 5).

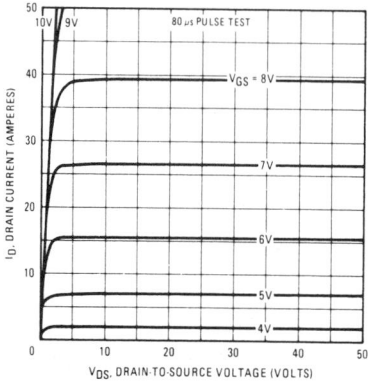

Fig. 1 — Typical Output Characteristics

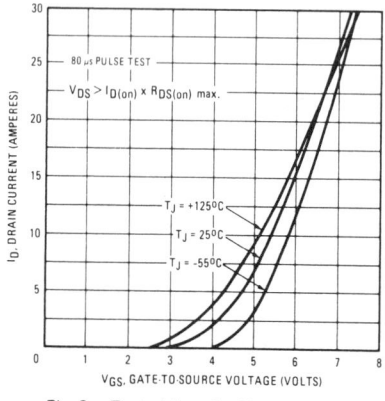

Fig. 2 — Typical Transfer Characteristics

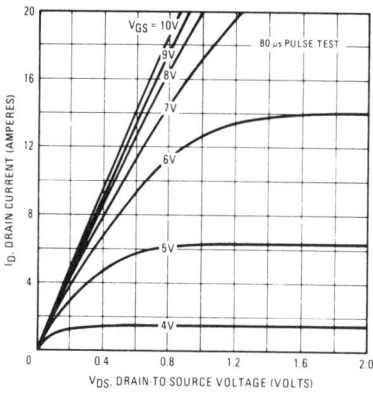

Fig. 3 — Typical Saturation Characteristics

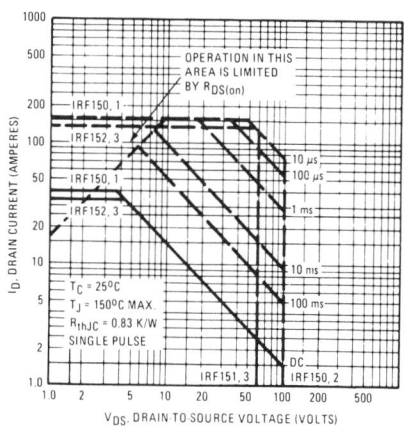

Fig. 4 — Maximum Safe Operating Area

Figure 11-13 (*Continued*)

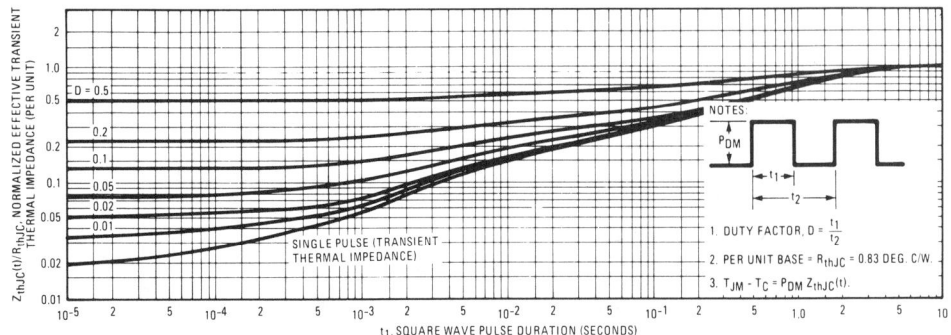

Fig. 5 — Maximum Effective Transient Thermal Impedance, Junction-to-Case Vs. Pulse Duration

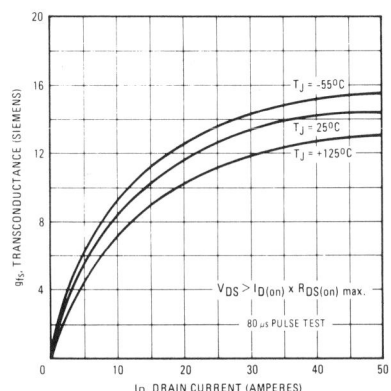

Fig. 6 — Typical Transconductance Vs. Drain Current

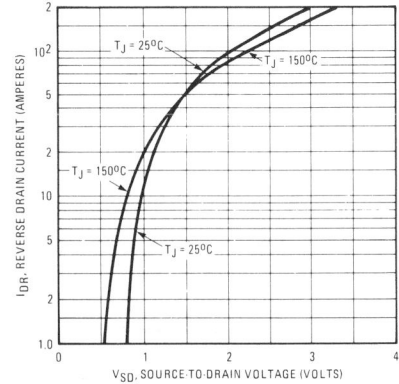

Fig. 7 — Typical Source-Drain Diode Forward Voltage

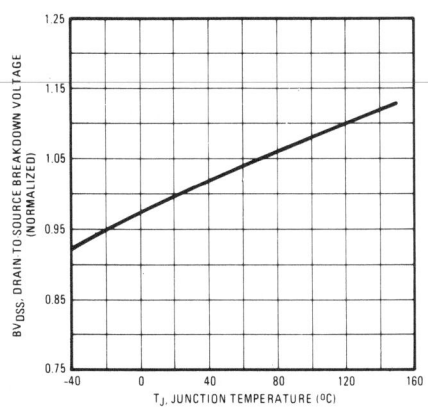

Fig. 8 — Breakdown Voltage Vs. Temperature

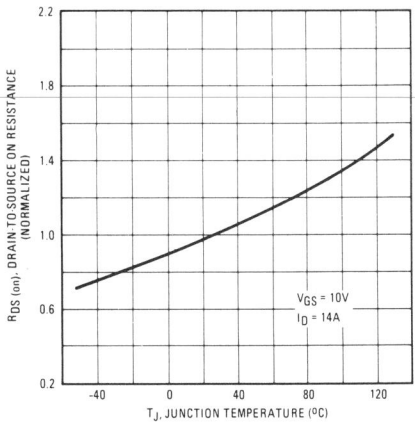

Fig. 9 — Normalized On-Resistance Vs. Temperature

Figure 11-13 *(Continued)*

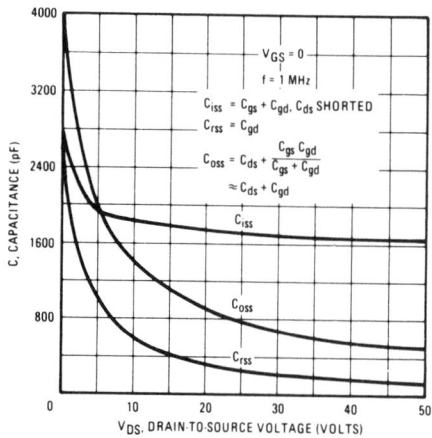

Fig. 10 — Typical Capacitance Vs. Drain-to-Source Voltage

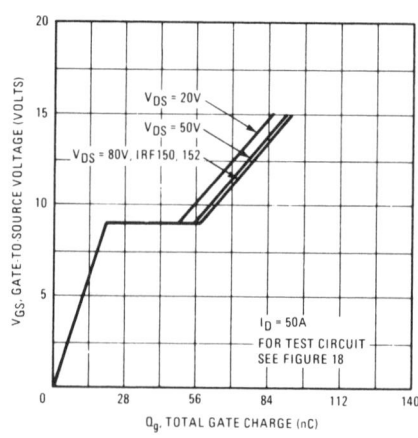

Fig. 11 — Typical Gate Charge Vs. Gate-to-Source Voltage

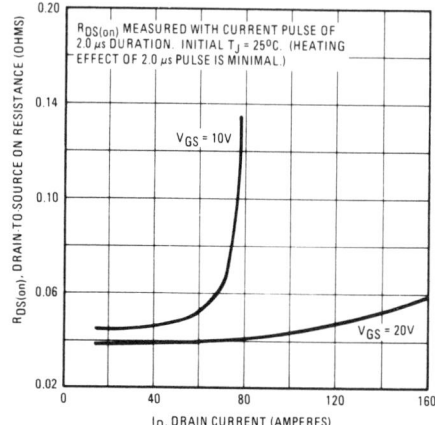

Fig. 12 — Typical On-Resistance Vs. Drain Current

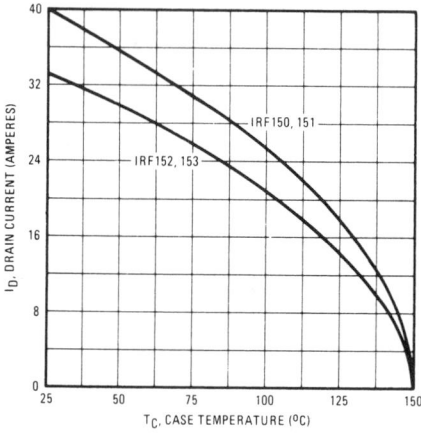

Fig. 13 — Maximum Drain Current Vs. Case Temperature

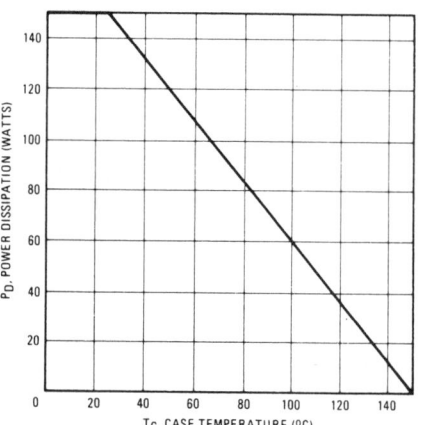

Fig. 14 — Power Vs. Temperature Derating Curve

Figure 11-13 (*Continued*)

IRF150, IRF151, IRF152, IRF153 Devices

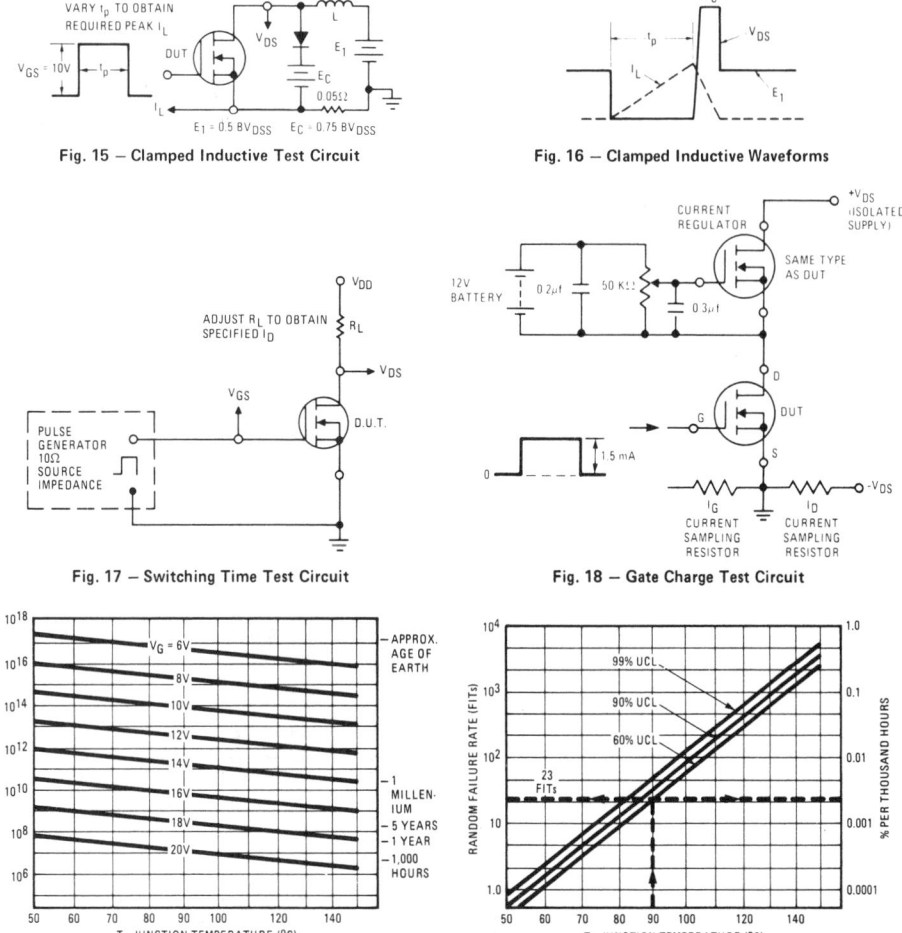

Fig. 15 – Clamped Inductive Test Circuit

Fig. 16 – Clamped Inductive Waveforms

Fig. 17 – Switching Time Test Circuit

Fig. 18 – Gate Charge Test Circuit

*Fig. 19 – Typical Time to Accumulated 1% Failure

*Fig. 20 – Typical High Temperature Reverse Bias (HTRB) Failure Rate

*The data shown is correct as of April 15, 1984. This information is updated on a quarterly basis; for the latest reliability data, please contact your local IR field office.

Figure 11-13 (*Continued*)

necessary to modify the parameter values to conform with the actual characteristics. It should be noted that the parameters differ from those given in the PSpice library, because their values are dependent on the constants used in derivations. Students are encouraged to run the following circuit file with the PSpice library model and compare the results.

Note. The gate (control) voltage Vg should be adjusted to drive the MOSFET into saturation.

Example 11-5

A MOSFET boost chopper is shown in Fig. 11-14(a). The dc input voltage is $V_S = 5$ V. The load resistance R is 100 Ω. The inductance is $L = 150$ μH, and the filter capacitance is $C = 220$ μF. The chopping frequency is $f_c = 25$ kHz, and the duty cycle of the chopper is $k = 66.7\%$. The control voltage is shown in Fig. 11-14(b). Use PSpice to plot the instantaneous output voltage v_C, the input current i_s, and the MOSFET voltage v_T.

Solution   The dc supply voltage $V_S = 5$ V. An initial capacitor voltage $V_C = 12$ V. $k = 0.667$, $f_c = 25$ kHz, $T = 1/f_c = 40$ μs, and $t_{on} = k \times T = 0.667 \times 40 = 26.7$ μs. The list of the circuit file is as follows:

■ ■

Example 11-5   MOSFET boost chopper

SOURCE ■ VS	1	0	DC	5V	
VY	1	2	DC	0V	; Voltage source to measure input current
Vg	7	0	PULSE (0V	20V 0 1NS 1NS 26.7US 40US)	
Rg	7	0	10MEG		
CIRCUIT ■ ■ ■ RB	7	6	250	; Transistor base resistance	
R	5	0	100		
C	5	0	220UF	IC=12V ; With initial condition	
L	2	3	150UH		
VX	4	5	DC	0V	; Voltage source to inductor current

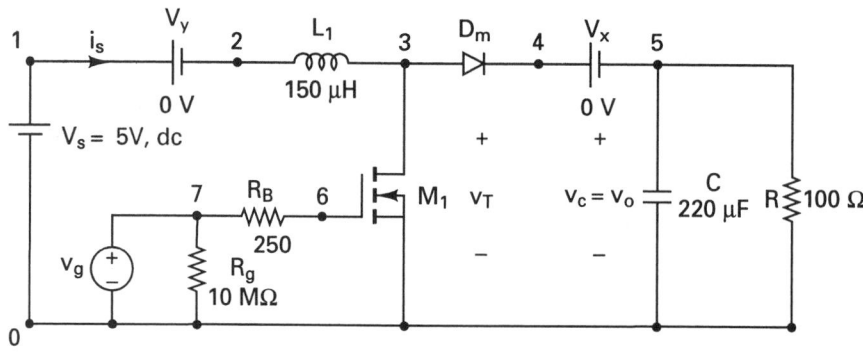

(a) Circuit

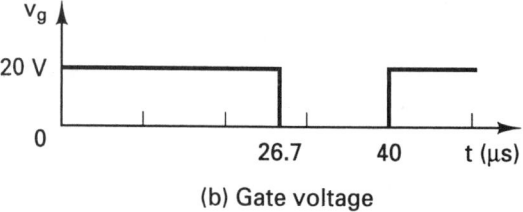

(b) Gate voltage

Figure 11-14   MOSFET boost chopper for PSpice simulation.

```
DM      3    4    DMOD                      ; Freewheeling diode
.MODEL DMOD  D(IS=2.22E-15 BV=1200V IBV=13E-3 CJO=0 TT=0) ; Diode model
M1      3    6    0    0    IRF150            ; MOSFET switch
.MODEL IRF150  NMOS (VTO=2.83 KP=31.2U L=1U W=3.0U
+       CGDO=0.418N CGSO=2.032N)             ; MOSFET parameters
```

ANALYSIS ■ ■ ■ `.TRAN  2US 3.6MS 3.4MS  UIC ; Transient analysis with initial conditions`

```
.PROBE                                      ; Graphics post-processor
.OPTIONS ABSTOL = 1.00N  RELTOL = 0.01  VNTOL = 0.1  ITL5=40000
.FOUR    25KHZ    I(VY)                      ; Fourier analysis
.END
```
■ ■

The PSpice plots of the instantaneous MOSFET voltage V(3), the input current I(VY), and the output voltage V(5) are shown in Fig. 11-15. The output voltage has not reached steady state.

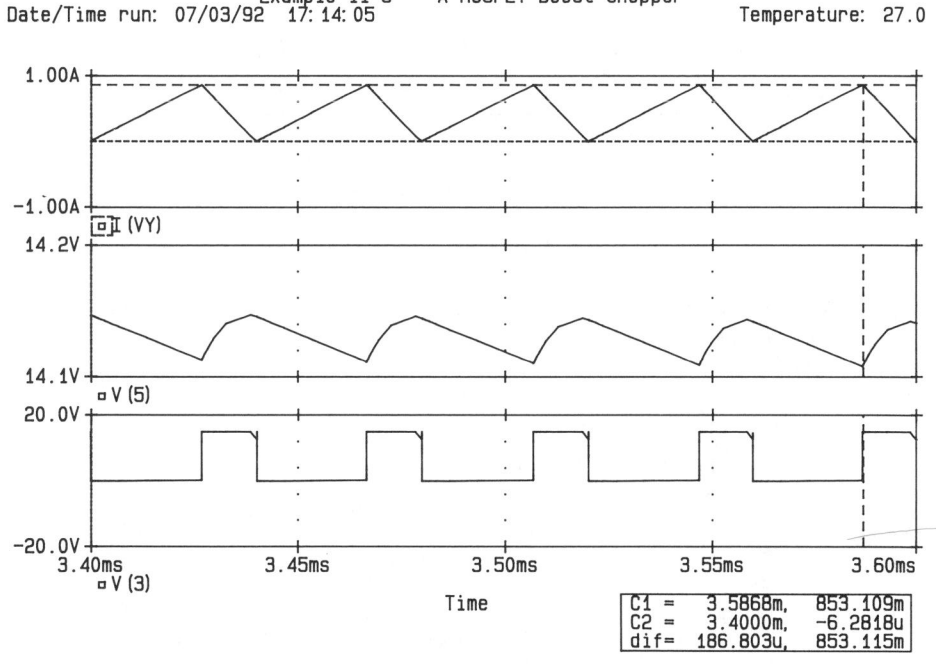

Figure 11-15 Plots for Example 11-5.

11-9 IGBT MODEL

The IGBT shown in Fig. 11-16(a) behaves as a MOSFET from the input side and as a BJT from the output side. It is shown in Fig. 11-16(a). The modeling of an IGBT is very complex [8]. Its characteristics can be simulated approxi-

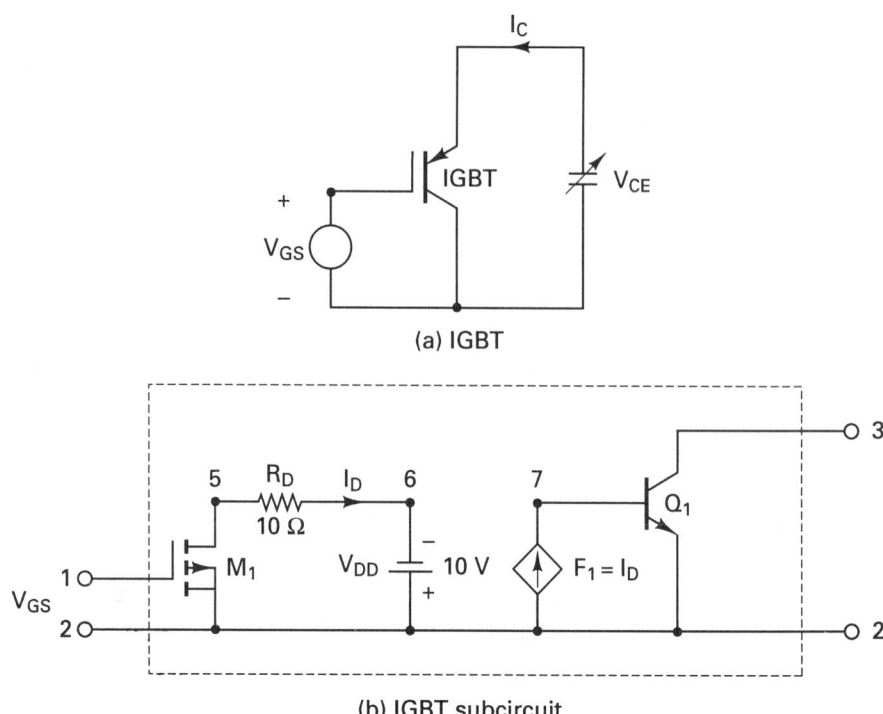

(a) IGBT

(b) IGBT subcircuit

Figure 11-16 IGBT circuit model.

mately by controlling a BJT from a MOSFET as shown in Fig. 11-16(b). It can be represented as a subcircuit IGBT, whose definition is described as follows:

```
*      Subcircuit for IGBT model:
.SUBCKT   IGBT      1         2          3          2
*          model   -gate     +gate      collector  emitter
*          name    voltage   voltage    terminal   terminal
M1    5    1    2    2    MMOD          ; input MOSFET
.MODEL   MMOD    PMOS (TOX=100N PHI=.6 KP=20.53U W=.3
+     L=2U VTO=-2.83 RD=1.031M RDS=444.4K CBD=3.229N PB=.8 MJ=.5
+     CGSO=9.027N CGDO=1.679N RG=13.89 IS=194E-18 N=1 TT=288N)
F1    7    2    VDD    0.2           ; Base current generator
Q1    3    7    2    QMOD            ; Output BJT
.MODEL   QMOD    NPN  (BF=50 CJE=1PF)  ; BJT parameters
VDD   2    6    DC    10V           ; MOSFET dc bias voltage
RD    5    6    10                  ; MOSFET drain resistance
.ENDS   IGBT                        ; Ends subcircuit definition
```

Example 11-6

Use PSpice to plot the output characteristics (V_{CE} versus I_C) of the IGBT for $V_{CE} = 0$ to 10 V and $V_{GS} = -4$ V to -2 V.

Solution The list of the circuit file is as follows:

■ ■

Example 11-6   IGBT characteristics

CIRCUIT
```
■ VCE    2    0    DC    20V
  VGS    1    0    DC    -20V
  XIGBT  1    0    2    0    IGBT
  *   Subcircuit IGBT, which is missing, must be inserted.
```
ANALYSIS
```
■ ■ .DC  VCE 0 10V  0.1V VGS LIST -4V -3.5V -3.4V -3.25V -3V -2V
   .PROBE                                   ; Graphics post-processor
   .OPTIONS ABSTOL = 1.00N  RELTOL = 0.01  VNTOL = 0.1  ITL5=40000
.END
```
■ ■

The output characteristics are shown in Fig. 11-17 for various gate voltages. The model parameters of the BJT and the MOSFET can be adjusted to give the approximate characteristics of the IGBT.

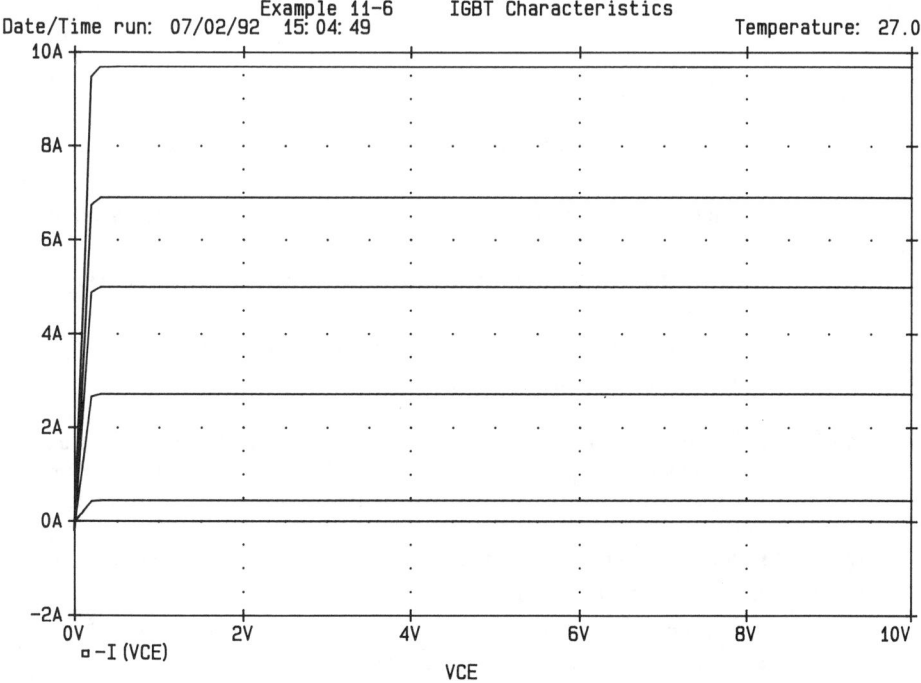

Figure 11-17   IGBT characteristics for Example 11-6.

The following two experiments are suggested to demonstrate the operation and characteristics of dc choppers:

> Dc buck chopper
>
> Dc boost chopper

11-10.1 Experiment TP-1

DC BUCK CHOPPER

Objective The objective is to study the operation and characteristics of a dc buck chopper under various load conditions.

Applications The dc buck (step-down) chopper is used to control power flow in many applications (e.g., power supplies, dc motor control, and input stages to inverters).

Textbook See Ref. 12, Secs. 9-3 and 9-7.

Apparatus
1. One BJT/MOSFET with ratings of at least 50 A and 500 V, mounted on a heat sink
2. One fast-recovery diode with ratings of at least 50 A and 500 V, mounted on a heat sink
3. A firing pulse generator with isolating signals for gating the BJT
4. An *RL* load
5. One dual-beam oscilloscope with floating or isolating probes
6. Dc voltmeters and ammeters and one noninductive shunt

Warning Before making any circuit connection, switch the dc power OFF. **Do not** switch the power ON unless the circuit is checked and approved by your lab instructor. **Do not** touch the transistor heat sinks, which are connected to live terminals.

Experimental procedure
1. Set up the circuit as shown in Fig. 11-18. Use a load resistance R only.
2. Connect the measuring instruments as required.
3. Set the chopping frequency to $f_c = 1$ kHz and the duty cycle to $k = 50\%$.
4. Connect the firing pulse to the BJT/MOSFET.
5. Observe and record the waveforms of the load voltage v_o, the load current i_o, and the input current i_s.
6. Measure the average load voltage $V_{o(dc)}$, the average load current $I_{o(dc)}$, the rms transistor current $I_{T(rms)}$, the average input current $I_{s(dc)}$, and the load power P_L.

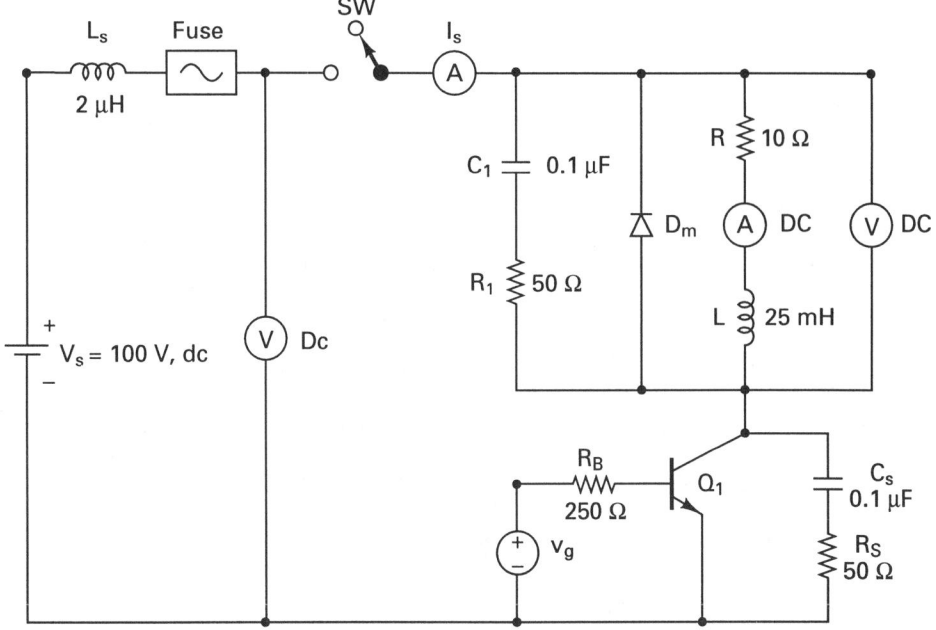

Figure 11-18 BJT step-down chopper.

7. Repeat steps 2 to 6 with a load inductance L only.

8. Repeat steps 2 to 6 with both load resistance R and load inductance L.

Report **1.** Present all recorded waveforms and discuss all significant points.

2. Compare the waveforms generated by SPICE with the experimental results, and comment.

3. Compare the experimental results with the results predicted.

4. Calculate and plot the average output voltage $V_{o(dc)}$ against the duty cycle.

5. Discuss the advantages and disadvantages of this type of chopper.

11-10.2 Experiment TP-2

DC BOOST CHOPPER

Objective The objective is to study the operation and characteristics of a dc boost chopper under various load conditions.

Applications The dc boost (step-up) chopper is used to control power flow in many applications (e.g., power supplies, dc motor control, and input stages to inverters).

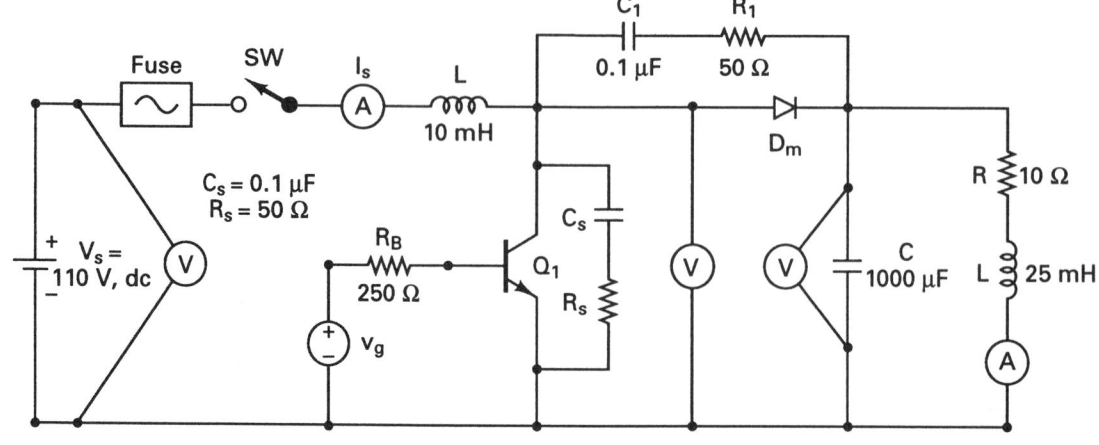

Figure 11-19 BJT step-up chopper.

Textbook   See Ref. 12, Secs. 9-4 and 9-7.

Apparatus   See Experiment TP-1.

Warning   See Experiment TP-1.

Experimental Set up the circuit as shown in Fig. 11-19. Repeat the steps for the Experiment TP-1.
procedure

Report   See Experiment TP-1.

SUMMARY

The statements for BJTS are:

```
Q⟨name⟩ NC  NB  NE  NS  QNAME   [ (area) value]
.MODEL  QNAME  NPN  (P1=V1 P2=V2 P3=V3 ... PN=VN)
.MODEL  QNAME  NPN  (P1=V1 P2=V2 P3=V3 ... PN=VN)
```

The statements for MOSFETs are

```
M⟨name⟩ ND  NG  NS  NB  MNAME
+       [L=⟨value] [W=⟨value⟩]
+       [AD=⟨value⟩] [AS=⟨value⟩]
+       [PD=⟨value⟩] [PS=⟨value⟩]
+       [NRD=⟨value⟩] [NRS=⟨value⟩]
+       [NRG=⟨value⟩] [NRB=⟨value⟩]
.MODEL  MNAME  NMOS  (P1=V1 P2=V2 P3=V3 ... PN=VN)
.MODEL  MNAME  PMOS  (P1=V1 P2=V2 P3=V3 ... PN=VN)
```

SUGGESTED READING

1. Ian Getreu, *Modeling the Bipolar Transistor*, Part # 062-2841-00. Beaverton, Ore.: Tektronix Inc., 1979.

2. H. K. Gummel and H. C. Poon, "An integral charge control model for bipolar transistors, *Bell Systems Technical Journal*, Vol. 49, January 1970, pp. 115–120.

3. J. J. Ebers and J. J. Moll, "Large signal behavior of junction transistors," *Proceedings of the IRE*, Vol. 42, December 1954, pp. 1161–1172.

4. H. Schichman and D. A. Hodges, "Modeling and simulation of insulated gate field effect transistor switching circuits," *IEEE Journal of Solid-State Circuits*, Vol. SC-3, September 1968, pp. 285–289.

5. J. F. Meyer, "MOS models and circuit simulation," *RCA Review*, Vol. 32, March 1971, pp. 42–63.

6. A. Vladimirescu and S. Liu, *The Simulation of MOS Integrated Circuits Using SPICE2*, Memorandum M80/7, February 1980, University of California, Berkeley.

7. S. Hangeman, "Behavioral modeling and PSpice simulation SMPS control loops: Parts I and II," *PCIM Magazine*, April–May 1990.

8. P. O. Lauretzen, "Power semiconductor device models for use in circuit simulation," *Conference Proceedings of the IEEE-IAS Annual Meeting*, 1990, pp. 1559–1560.

9. S. Natarajan, "An effective approach to obtain model parameters for BJTs and FETs from data books," *IEEE Transactions on Education*, Vol. 35, No. 2, 1992, pp. 164–169.

10. G. Fay and J. Sutor, "Power FET models are easy and accurate," *Powertechnics Magazine*, August 1987, pp. 16–21.

11. C. F. Wheatley, H. R. Ronan, Jr., and G. M. Dolny, "Spicing up SPICE II software for power MOSFET modeling," *Powertechnics Magazine*, August 1987, pp. 28–32.

12. M. H. Rashid, *Power Electronics: Circuits, Devices, and Applications*, 2nd ed. Englewood Cliffs, N.J.: Prentice Hall, 1993.

DESIGN PROBLEMS

11-1. It is required to design the step-up chopper of Fig. 11-14 with the following specifications:

Dc supply voltage, $V_s = 110$ V

Load resistance, $R = 5$ Ω

Load inductance, $L = 15$ mH

Dc output voltage $V_{o(dc)}$ is 80% of the maximum value

Peak-to-peak load ripple current should be less than 10% of the load average value, $I_{o(dc)}$

   (a) Determine the ratings of all devices and components under worst-case conditions.

   (b) Use SPICE to verify your design.

   (c) Provide a cost estimate of the circuit.

11-2. **(a)** Design an input LC filter for the chopper of Problem 11-1. The harmonic content of the input current should be less than 10% of the value without the filter.

   (b) Use SPICE to verify your design in part (a).

11-3. It is required to design a step-up chopper of Fig. 11-14 with the following specifications:

Dc supply voltage, $V_s = 24$ V

Load resistance, $R = 5$ Ω

Load inductance, $L = 15$ mH

Dc output voltage $V_{o(dc)}$ is 150% of the maximum value

Average load current, $I_{o(dc)} = 2.5$ A

Peak-to-peak load ripple current should be less than 10% of the load average value, $I_{o(dc)}$

(a) Determine the ratings of all devices under worst-case conditions.
(b) Use SPICE to verify your design.
(c) Provide a cost estimate of the circuit.

11-4. (a) Design an output C filter for the chopper of Problem 11-3. The harmonic content of the output voltage should be less than 10% of the value without the filter.
(b) Use SPICE to verify your design in part (a).

11-5. It is required to design a buck regulator with the following specifications:

Dc input voltage, $V_s = 15$ V

Average output voltage, $V_{o(dc)} = 10$V

Average load current, $I_{o(dc)} = 2.5$ A

Peak-to-peak output ripple voltage should be less than 50 mV

Switching frequency, $f_c = 20$ kHz

Peak-to-peak inductor ripple current should be less than 0.5 A

(a) Determine the ratings of all devices and components under worst-case conditions.
(b) Use SPICE to verify your design.
(c) Provide a cost estimate of the circuit.

11-6. It is required to design a boost regulator with the following specifications:

Dc input voltage, $V_s = 12$ V

Average output voltage, $V_{o(dc)} = 24$ V

Average load current, $I_{o(dc)} = 2.5$ A

Switching frequency, $f_c = 20$ kHz

Peak-to-peak output ripple voltage should be less than 50 mV

Peak-to-peak inductor ripple current should be less than 0.5 A

(a) Determine the ratings of all devices and components under worst-case conditions.
(b) Use SPICE to verify your design.
(c) Provide a cost estimate of the circuit.

11-7. It is required to design a buck–boost regulator with the following specifications:

Dc input voltage, $V_s = 15$ V

Average output voltage, $V_{o(dc)} = 10$ V

Average load current, $I_{o(dc)} = 2.5$ A

Switching frequency, $f_c = 20$ kHz

Peak-to-peak output ripple voltage should be less than 50 mV

Peak-to-peak ripple current of inductor should be less than 0.2 A

(a) Determine the ratings of all devices and components under worst-case conditions.
(b) Use SPICE to verify your design.
(c) Provide a cost estimate of the circuit.

11-8. It is required to design a Cúk regulator with the following specifications:

Dc input voltage, $V_s = 15$ V

Average output voltage, $V_{o(dc)} = 10$ V

Average load current, $I_{o(dc)} = 2.5$ A

Switching frequency, $f_c = 20$ kHz

Peak-to-peak ripple voltages of capacitors should be less than 50 mV

Peak-to-peak ripple currents of inductors should be less than 0.2 A

(a) Determine the ratings of all devices and components under worst-case conditions.
(b) Use SPICE to verify your design.
(c) Provide a cost estimate of the circuit.

Chapter 12

■■■■■■■■■

Pulse-Width-Modulated Inverters

12-1 INTRODUCTION

The input to an inverter is dc and the output is ac. The power semiconductor devices perform the switching action and the desired output is obtained by varying their turn-on and turn-off times. They must have controllable turn-on and turn-off characteristics. The commonly used devices are BJTs, MOSFETs, IGBTs, GTOs, MCTs, and forced commutated thyristors. We use PSpice switches, BJTs, and MOSFETs to simulate the characteristics of the following inverters:

> Voltage-source inverters
> Current-source inverters

12-2 VOLTAGE-SOURCE INVERTERS

Two types of control are commonly used for varying the output of an inverter: pulse-width modulation (PWM) and sinusoidal pulse-width modulation (SPWM). The number of pulses per half-cycle usually ranges from 1 to 15.

In PWM control, the conduction angles of power semiconductor devices are generated by comparing a reference signal v_r with a carrier signal v_c as shown in Fig. 12-1. The pulse width δ can be varied by varying the carrier voltage v_c. This technique can be implemented by an op-amp circuit as shown in Fig. 8-17(b). The input voltages to the comparator are v_r and v_c, and its output is the conduction angle δ for which the switch remains on. The modulation index M is defined by

$$M = \frac{A_c}{A_r} \tag{12-1}$$

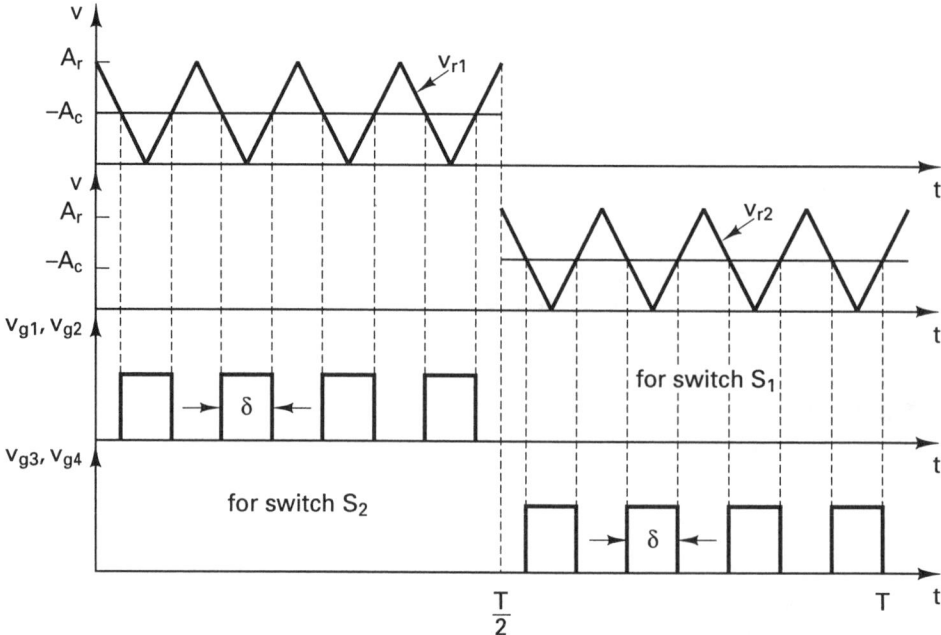

Figure 12-1 Pulse-width modulation (PWM) control.

where A_r is the peak value of reference signal v_r and A_c is the peak value of carrier signal v_c.

The PWM modulator can be used as a subcircuit to generate control signals for a triangular reference voltage of one or more pulses per half-cycle and a dc carrier signal. The subcircuit definition for the modulator model PWM can be described as follows:

```
*    Subcircuit for PWM control:
.SUBCKT   PWM      1       2       3       4
*         model    ref.    carrier +control -control
*         name     input   input   voltage  voltage
R1     1     5     1K
R2     2     5     1K
RIN    5     0     2MEG
RF     5     6     100K
RO     6     3     75
CO     3     4     10PF
E1     6     4     0     5     2E+5       ; Voltage-controlled voltage source
.ENDS     PWM                            ; Ends subcircuit definition
```

In SPWM control, the carrier signal is a rectified sine wave as shown in Fig. 12-2. PSpice generates only a sine wave. Thus we can use a precision recti-

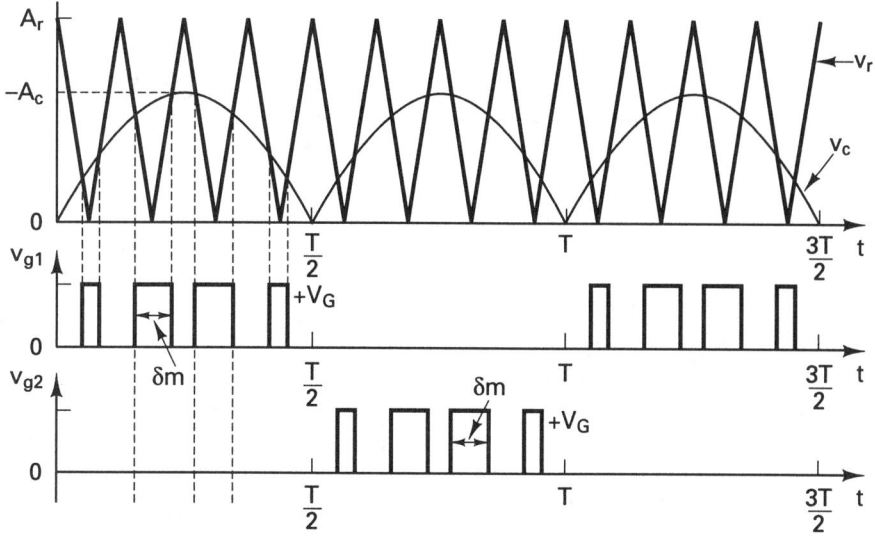

Figure 12-2 Gate voltages with a sinusoidal PWM control.

fier to convert a sine-wave signal to rectified sine-wave pulses, and use a comparator to generate a PWM waveform. This can be implemented by the circuit shown in Fig. 12-3.

We can use subcircuit SPWM for the generation of control signals. The subcircuit definition for the modulator model SPWM can be described as follows:

```
*    Subcircuit for sinusoidal PWM control:
.SUBCKT    SPWM       1        2        3        4        8
*          model    ref.    carrier  +control  -control  rectified
*          name     input    input   voltage   voltage   carrier sine wave
R1     1     9    1K
R2     8     9    1K
RF     9     3    100K
R3     2     10   50K
R4     6     7    50K
R5     10    6    50K
R6     2     7    100K
R7     7     8    100K
CO     3     4    10PF            ; Capacitor to aid convergence
D1     6     5    DMD
D2     5     10   DMD
.MODEL     DMD       D            ; Default model parameters
X1     10    0     5     0     OPAMP    ; Call subcircuit for op-amp A1
X2     7     0     8     0     OPAMP    ; Call subcircuit for op-amp A2
X3     9     0     3     4     OPAMP    ; Call subcircuit for op-amp A3
.SUBCKT    OPAMP      1        5        2        3
*          name     -vi      +vi      +vo      -vo
```

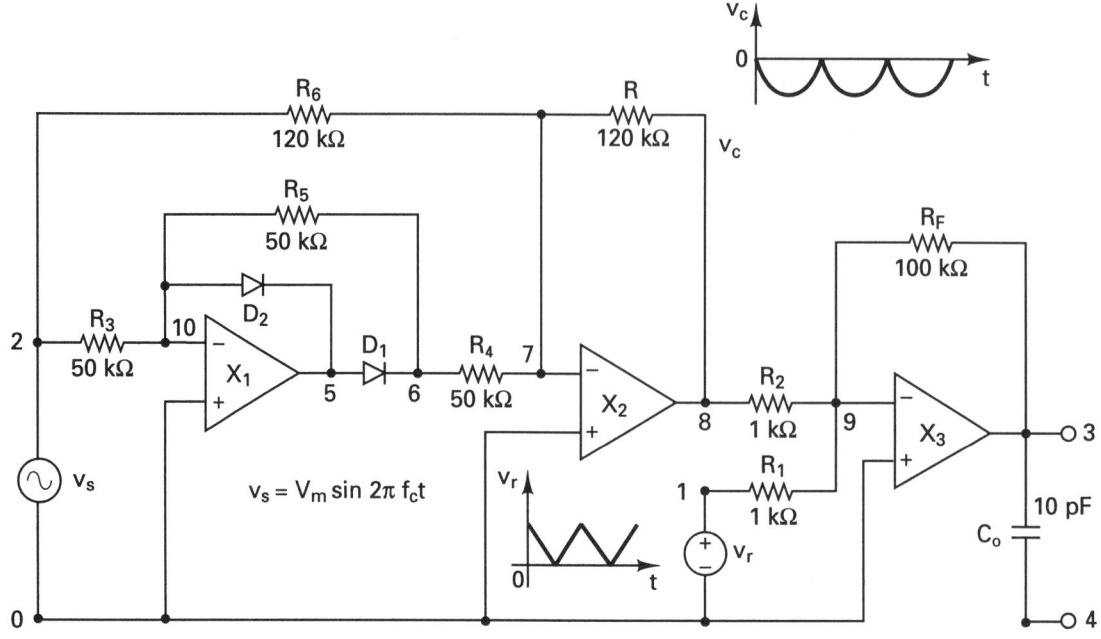

(a) Precision rectifier and comparator

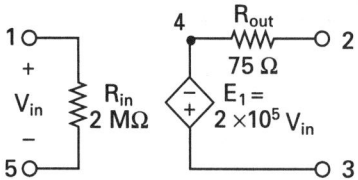

(b) Op-amp model

Figure 12-3 Sinusoidal PWM generator.

```
RIN    1    5    2MEG
RO     4    2    75
E1     3    4    1    5    2E+5    ; Voltage-controlled voltage source
.ENDS    OPAMP                    ; Ends subcircuit OPAMP definition
.ENDS    SPWM                     ; Ends subcircuit PWM definition
```

Example 12-1

A single-phase bridge inverter is shown in Fig. 12-4. The dc input voltage is 100 V. It is operated at an output frequency of $f_o = 60$ Hz with PWM control and four pulses per half-cycle. The modulation index $M = 0.6$. The load is purely resistive with $R = 2.5\ \Omega$. Use PSpice (a) to plot the instantaneous output voltage v_o, the instantaneous

---

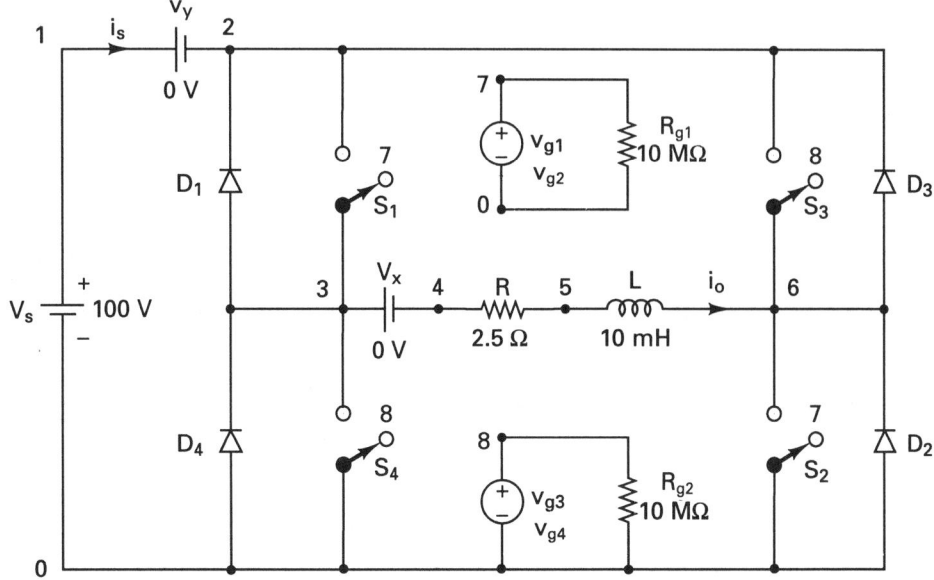

Figure 12-4 Single-phase bridge inverter.

carrier and reference voltages, and (b) to calculate the Fourier coefficients of output voltage v_o. Use voltage-controlled switches to perform the switching action.

Solution $P = 4$, $M = 0.6$. Assuming that $A_r = 10$ V, $A_c = 10 \times 0.6 = 6$ V. We shall model a power device as a voltage-controlled switch as shown in Fig. 12-5. The subcircuit definition for the switched transistor model STMOD can be described as follows:

```
*      Subcircuit for switched transistor model:
.SUBCKT   STMOD    1        2        3         4
*         model    anode    cathode  +control  -control
*         name                       voltage   voltage
DT     5    2    DMOT                 ; Switch diode
ST     1    5    3    4    SMOD       ; Switch
.MODEL    DMOT    D(IS=2.2E-15 BV=1200V CJO=0 TT=0) ; Diode model parameters
.MODEL    SMOD    VSWITCH (RON=0.01 ROFF=10E+6 VON=10V VOFF=5V)
.ENDS    STMOD                        ; Ends subcircuit definition
```

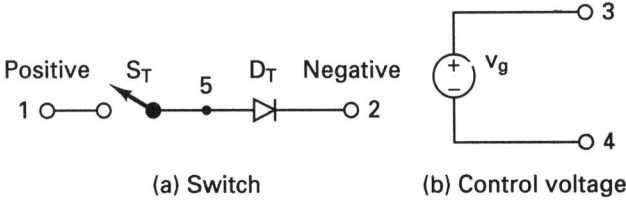

(a) Switch   (b) Control voltage

Figure 12-5 Switched transistor model.

The conduction angles are generated from two reference signals as shown in Fig. 12-6. Assume that reference voltage $V_r = 10$ V. For $M = 0.4$, the carrier voltage is $V_c = MV_r = 4$ V. We can use the subcircuit PWM in Fig. 12-1 for the generation of control signals. The list of the circuit file for the inverter is as follows:

■ ■

Example 12-1   Single-phase inverter with PWM control

```
SOURCE   ■ VS    1    0    DC    100V
           Vg1   11   0    PULSE (0  -6V  0  1NS  1NS  8333.33US  16666.67US)
           Rg1   7    0    10MEG
           Vg2   12   0    PULSE (0  -6V  8333.33US  1NS  1NS  8333.33US  16666.67US)
           Rg2   8    0    10MEG
           Vc    9    0    PULSE (10V  0  0  1041.67US  1041.67US  1NS  2083.34US)
CIRCUIT  ■ ■ R   4    6    2.5
          *L     5    6    10MH          ; Inductor L is excluded
           VX    3    4    DC    0V      ; Measures load current
           VY    1    2    DC    0V      ; Voltage source to measure supply current
           D1    3    2    DMOD          ; Diode
           D2    0    6    DMOD          ; Diode
           D3    6    2    DMOD          ; Diode
           D4    0    3    DMOD          ; Diode
          .MODEL  DMOD  D(IS=2.2E-15  BV=1200V  TT=0  CJO=0)   ; Diode model parameters
```

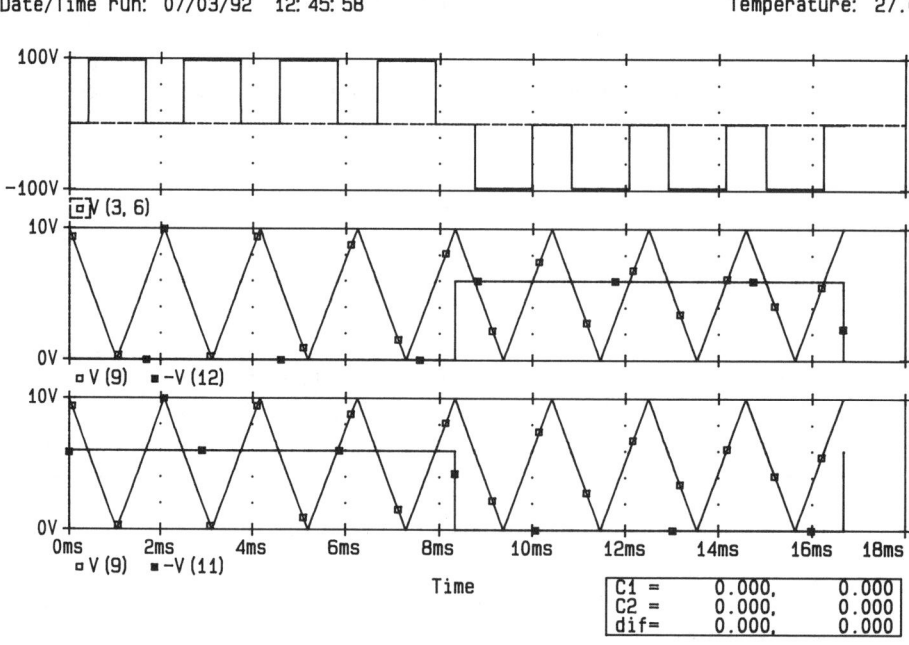

Example 12-1   Single-Phase Inverter with PWM Control
Date/Time run: 07/03/92   12:45:58   Temperature: 27.0

Figure 12-6 Plots for Example 12-1.

```
*    Subcircuit call for switched transistor model:
XT1    2    3    7    0    STMOD     ; Transistor T1
XT2    6    0    7    0    STMOD     ; Transistor T2
XT3    2    6    8    0    STMOD     ; Transistor T3
XT4    3    0    8    0    STMOD     ; Transistor T4
*    Subcircuit call for PWM control:
XPW1   11   9    7    0    PWM       ; Control voltage for transistor T1
XPW2   12   9    8    0    PWM       ; Control voltage for transistor T2
*    Subcircuit STMOD, which is missing, must be inserted.
*    Subcircuit PWM, which is missing, must be inserted.
```

ANALYSIS ■ ■ ■
```
.TRAN    10US    16.67MS              ; Transient analysis
.PROBE                               ; Graphics post-processor
.OPTIONS ABSTOL = 1.00N   RELTOL = 0.01   VNTOL = 0.1   ITL5=40000
.FOUR    60HZ    V(3,6)              ; Fourier analysis
```
`.END`
■ ■

(a) The PSpice plots of the instantaneous output voltage V(3,6), the instantaneous carrier voltages, and the instantaneous reference voltages are shown in Fig. 12-6.
(b) The Fourier coefficients of the output voltage are as follows:

```
FOURIER COMPONENTS OF TRANSIENT RESPONSE V(3,6)
DC COMPONENT =   8.339770E-04
```

HARMONIC NO	FREQUENCY (HZ)	FOURIER COMPONENT	NORMALIZED COMPONENT	PHASE (DEG)	NORMALIZED PHASE (DEG)
1	6.000E+01	7.449E+01	1.000E+00	−2.945E−02	0.000E+00
2	1.200E+02	1.720E−03	2.310E−05	1.596E+02	1.597E+02
3	1.800E+02	2.871E+01	3.855E−01	−1.179E−01	−8.846E−02
4	2.400E+02	1.852E−03	2.486E−05	−1.307E+02	−1.307E+02
5	3.000E+02	2.472E+01	3.318E−01	2.240E−02	5.185E−02
6	3.600E+02	1.994E−03	2.677E−05	−5.965E+01	−5.962E+01
7	4.200E+02	4.608E+01	6.186E−01	6.028E−01	6.323E−01
8	4.800E+02	2.076E−03	2.787E−05	1.365E+01	1.368E+01
9	5.400E+02	3.132E+01	4.205E−01	−1.791E+02	−1.790E+02

```
TOTAL HARMONIC DISTORTION =   9.044873E+01 PERCENT
```

Example 12-2

The PWM inverter in Example 12-1 has a load resistance $R = 2.5\ \Omega$, and a load inductance $L = 10$ mH. Use PSpice (a) to plot the instantaneous output voltage v_o, the instantaneous output current i_o, and the instantaneous supply current i_s, and (b) to calculate the Fourier coefficients of the output voltage v_o and the output current i_o.

Solution   The list of the circuit file is identical to that for Example 12-1, except that the statements for R and L are changed as follows:

```
R    4    5    2.5
L    5    6    10MH       ; Inductor L is included
```

(a) The PSpice plots of the instantaneous output voltage V(3,6), the output current I(VX), and the supply current I(VY) are shown in Fig. 12-7.

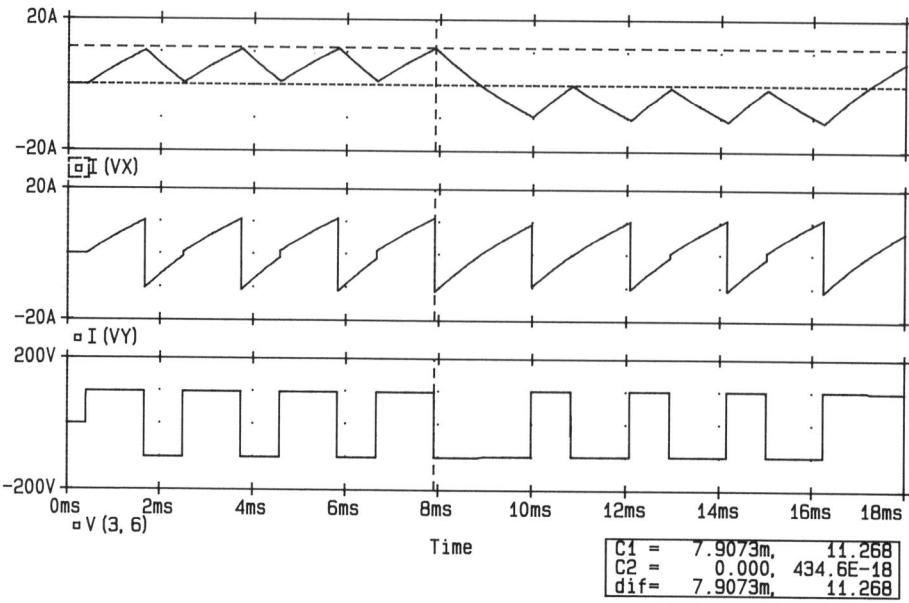

C1 =	7.9073m,	11.268
C2 =	0.000,	434.6E-18
dif=	7.9073m,	11.268

Figure 12-7 Plots for Example 12-2.

(b) The Fourier coefficients of the output voltage and the output current are as follows:

FOURIER COMPONENTS OF TRANSIENT RESPONSE V(3,6)

DC COMPONENT =   −7.146290E−02

HARMONIC NO	FREQUENCY (HZ)	FOURIER COMPONENT	NORMALIZED COMPONENT	PHASE (DEG)	NORMALIZED PHASE (DEG)
1	6.000E+01	3.267E+01	1.000E+00	6.813E+01	0.000E+00
2	1.200E+02	1.267E−01	3.879E−03	−1.364E+02	−2.045E+02
3	1.800E+02	2.885E+01	8.831E−01	1.298E+02	6.168E+01
4	2.400E+02	1.155E−01	3.535E−03	1.703E+02	1.021E+02
5	3.000E+02	3.777E+01	1.156E+00	1.736E+02	1.055E+02
6	3.600E+02	1.202E−01	3.679E−03	1.140E+02	4.591E+01
7	4.200E+02	8.850E+01	2.709E+00	−1.471E+02	−2.153E+02
8	4.800E+02	1.341E−01	4.103E−03	6.515E+01	−2.979E+00
9	5.400E+02	6.730E+01	2.060E+00	6.753E+01	−5.927E−01

TOTAL HARMONIC DISTORTION =   3.700917E+02 PERCENT

FOURIER COMPONENTS OF TRANSIENT RESPONSE I(VX)

DC COMPONENT =   3.288993E−01

HARMONIC NO	FREQUENCY (HZ)	FOURIER COMPONENT	NORMALIZED COMPONENT	PHASE (DEG)	NORMALIZED PHASE (DEG)
1	6.000E+01	7.469E+00	1.000E+00	−1.394E+01	0.000E+00
2	1.200E+02	2.804E−01	3.754E−02	4.549E+01	5.942E+01

HARMONIC NO	FREQUENCY (HZ)	FOURIER COMPONENT	NORMALIZED COMPONENT	PHASE (DEG)	NORMALIZED PHASE (DEG)
3	1.800E+02	2.535E+00	3.395E-01	-2.891E+01	-1.497E+01
4	2.400E+02	1.832E-01	2.453E-02	4.369E+01	5.762E+01
5	3.000E+02	1.946E+00	2.605E-01	-4.735E+01	-3.341E+01
6	3.600E+02	1.489E-01	1.994E-02	4.922E+01	6.316E+01
7	4.200E+02	3.264E+00	4.370E-01	-7.088E+01	-5.694E+01
8	4.800E+02	1.471E-01	1.969E-02	5.020E+01	6.414E+01
9	5.400E+02	2.088E+00	2.796E-01	8.048E+01	9.442E+01

TOTAL HARMONIC DISTORTION =   6.745586E+01 PERCENT

Example 12-3

Repeat Example 12-1 with sinusoidal pulse-width modulation (SPWM) control.
Solution $P = 4$, $M = 0.6$. Let us assume that $A_r = 10$ V. $A_c = 10 \times 0.6 = 6$ V. The two reference voltages are generated with a PWL function as shown in Fig. 12-8. A precision rectifier converts a sine-wave input signal to two half sine-wave pulses, and a comparator generates the PWM waveform. This is implemented in Fig. 12-3. We can use the subcircuit SPWM for generation of the control signals. The list of the circuit file for the inverter is as follows:

■ ■

Example 12-3   Single-phase inverter with sinusoidal PWM control

```
SOURCE   ■ VS    1    0    DC    100V
         Vr1  11    0    PWL (0 10V  1041.67US 0V 2083.33US 10V  3125US 0V
         +  4166.67US 10V  5208.33US 0V  6250US 10V  7291.67US 0V  8333.33US 10V
```

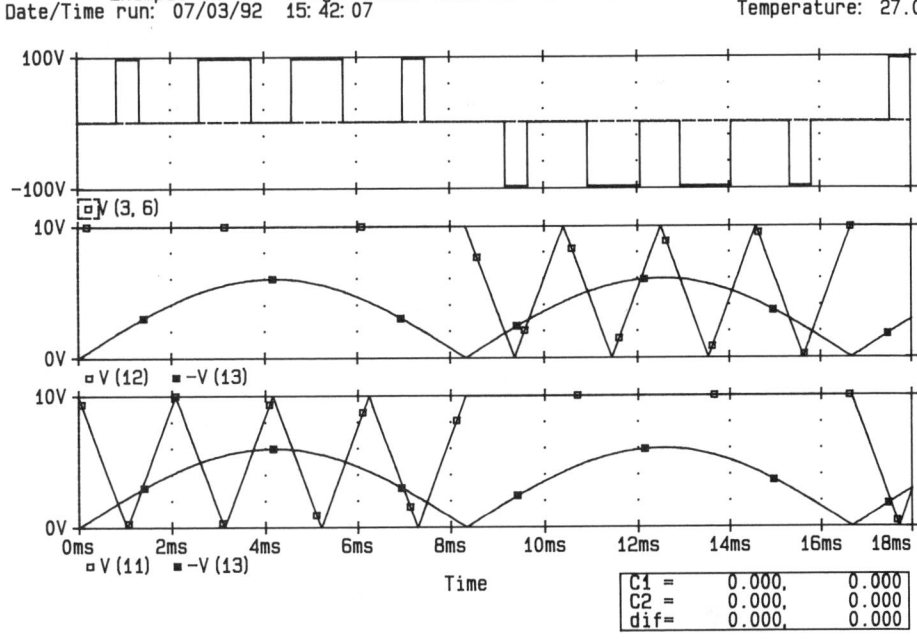

Example 12-3   Single-Phase Inverter with Sinusoidal PWM Control
Date/Time run: 07/03/92   15:42:07   Temperature: 27.0

Figure 12-8   Plots for Example 12-3.

```
+   16666.67US 10V 17708.33US 0V 18750US 10V  19791.67US 0V 20833.33US 10V
+   21875US 0V  22916.67US 10V  23958.33US 0V  25000US 10V
+   33333.33US 10V 34375US 0V  35416.67US 10V 36458.33US 0V 37500US 10V
+   38541.67US 0V 39583.33US 10V 40625US 0V 41666.67US 10V
+   50000US 10V  51041.67US 0V 52083.33US 10V 53125US 0V  54166.67US 10V
+   55208.33US 0V  56250US 10V  57291.67US 0V 58333.33US 10V 66666.67US 10V)
Vc1   9    0    SIN (1MV   6V   60HZ)
Rg1   7    0    10MEG
Vr3   12   0    PWL (0   10V   8333.33US 10V  9375US 0V  10416.67US 10V
+   11458.33US 0V  12500US 10V  13541.67US 0V 14583.33US 10V
+   15625US 0V  16666.67US   10V
+   25000US 10V  26041.67US 0V 27083.33US 10V 28125US 0V  29166.67US 10V
+   30208.33US 0V  31250US 10V  32291.67US 0V 33333.33US 10V
+   41666.67US 10V 42708.33US 0V 43750US 10V 44791.67US 0V 45833.33US 10V
+   46875US 0V  47916.67US 10V 48958.33US 0V 50000US 10V
+   58333.33US 10V  59375US 0V  60416.67US  10V 61458.33US 0V 62500US 10V
+   63541.67US 0V 64583.33US 10V  65625US 0V 66666.67US 10V)
Rg2   8    0    10MEG
Rc    13   0    10MEG
```

CIRCUIT ■ ■
```
R       4    6    2.5
*L      5    6    10MH          ; Inductor L is excluded
VX      3    4    DC    0V      ; Measures load current
VY      1    2    DC    0V      ; Voltage source to measure supply current
D1      3    2    DMOD          ; Diode
D2      0    6    DMOD          ; Diode
D3      6    2    DMOD          ; Diode
D4      0    3    DMOD          ; Diode
.MODEL   DMOD   D(IS=2.2E-15 BV=1800V TT=0)  ; Diode model parameters
*   Subcircuit call for switched transistor model:
XT1     2    3    7    0    STMOD     ; Transistor T1
XT2     6    0    7    0    STMOD     ; Transistor T2
XT3     2    6    8    0    STMOD     ; Transistor T3
XT4     3    0    8    0    STMOD     ; Transistor T4
*   Subcircuit call for SPWM control:
XPW1   11    9    7    0    13   SPWM  ; Control voltage for thyristor T1
XPW2   12    9    8    0    13   SPWM  ; Control voltage for thyristor T2
*   Subcircuit TMOD, which is missing, must be inserted.
*   Subcircuit SPWM, which is missing, must be inserted.
```

ANALYSIS ■ ■ ■
```
.TRAN    10US    18MS                   ; Transient analysis
.PROBE                                  ; Graphics post-processor
.OPTIONS ABSTOL = 1.00N  RELTOL = 0.01  VNTOL = 0.1  ITL5=40000
.FOUR    60HZ   V(3,6)                  ; Fourier analysis
.END
```
■■■

(a) The PSpice plots of the instantaneous output voltage V(3,6), the carrier voltages, and the reference voltages are shown in Fig. 12-8.

(b) The Fourier coefficients of the output voltage are as follows:

FOURIER COMPONENTS OF TRANSIENT RESPONSE V(3,6)

DC COMPONENT = -4.386737E-03

HARMONIC NO	FREQUENCY (HZ)	FOURIER COMPONENT	NORMALIZED COMPONENT	PHASE (DEG)	NORMALIZED PHASE (DEG)
1	6.000E+01	5.734E+01	1.000E+00	2.895E+01	0.000E+00
2	1.200E+02	1.233E-02	2.150E-04	-5.973E+01	-8.868E+01
3	1.800E+02	2.328E-02	4.061E-04	-1.778E+01	-4.673E+01
4	2.400E+02	1.630E-02	2.842E-04	-5.811E+01	-8.706E+01
5	3.000E+02	6.256E+00	1.091E-01	1.427E+02	1.137E+02
6	3.600E+02	1.604E-02	2.798E-04	-6.504E+01	-9.399E+01
7	4.200E+02	3.599E+01	6.278E-01	-1.583E+02	-1.873E+02
8	4.800E+02	1.197E-02	2.087E-04	-7.286E+01	-1.018E+02
9	5.400E+02	3.616E+01	6.307E-01	7.955E+01	5.060E+01

TOTAL HARMONIC DISTORTION = 8.965353E+01 PERCENT

Example 12-4

A three-phase bridge inverter is shown in Fig. 12-9(a). The dc input voltage is 100 V. The control voltages are shown in Fig. 12-9(b). The output frequency is f_o = 1 kHz. The load resistance is R = 10 Ω and the load inductance is L = 5 mH. Use PSpice (a) to plot the instantaneous output line–line voltage v_L, the instantaneous output phase voltage v_p, and the instantaneous output current i_a for phase a, and (b) to calculate the Fourier coefficients of the phase voltage v_p and the phase current i_a. The model parameters of the BJTs are IS=2.33E−27, BF=13, CJE=1PF, CJC=607.3PF, and TF=26.5NS.

Solution The list of the circuit file for the inverter is as follows:

■ ■

Example 12-4 Three-phase inverter

```
SOURCE  ■ VS    1    0    DC   100V
          RB1   22   6    50
          Rg1   22   3    10MEG
          Vg1   22   3    PULSE (0   40V   0         1NS   1NS   0.5MS   1MS)
          Rb2   16   15   50
          Rg2   16   0    10MEG
          Vg2   16   0    PULSE (0   40V   166.67US  1NS   1NS   0.5MS   1MS)
          Rb3   8    7    50
          Rg3   8    4    10MEG
          Vg3   8    4    PULSE (0   40V   333.33US  1NS   1NS   0.5MS   1MS)
          Rb4   12   11   50
          Rg4   12   0    10MEG
          Vg4   12   0    PULSE (0   40V   500US     1NS   1NS   0.5MS   1MS)
          Rb5   10   9    50
          Rg5   10   5    10MEG
          Vg5   10   5    PULSE (0   40V   666.67US  1NS   1NS   0.5MS   1MS)
          Rb6   14   13   50
          Rg6   14   0    10MEG
          Vg6   14   0    PULSE (0   40V   833.33US  1NS   1NS   0.5MS   1MS)
CIRCUIT ■ ■ VY   1    2    DC   0V     ; Voltage source to measure supply current
          VX    3    20   DC   0V     ; Measures load phase current
```

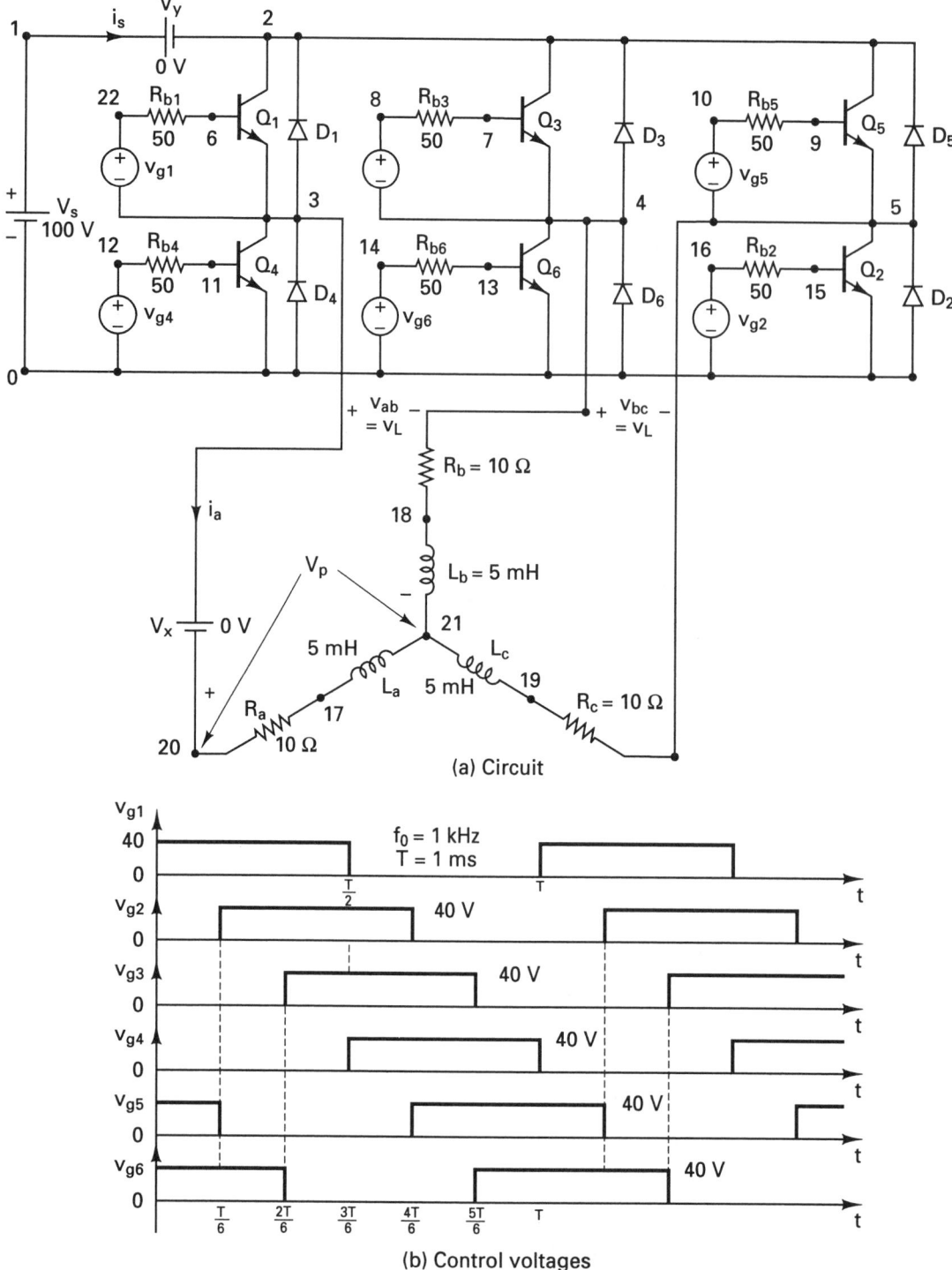

(a) Circuit

(b) Control voltages

Figure 12-9 Three-phase bridge inverter.

```
RA      20    17    10
LA      17    21    5MH              ; LA is included
RB       4    18    10
LB      18    21    5MH              ; LB is included
RC       5    19    10
LC      19    21    5MH              ; LC is included
D1       3     2    DMOD             ; Diode
D3       4     2    DMOD             ; Diode
D5       5     2    DMOD             ; Diode
D2       0     5    DMOD             ; Diode
D4       0     3    DMOD             ; Diode
D6       0     4    DMOD             ; Diode
.MODEL  DMOD  D(IS=2.22E-15 BV=1200V IBV=13E-3 CJO=0 TT=0) ; Diode model
Q1       2     6     3     3    2N6546      ; BJT switch
Q3       2     7     4     4    2N6546      ; BJT switch
Q5       2     9     5     5    2N6546      ; BJT switch
Q2       5    15     0     0    2N6546      ; BJT switch
Q4       3    11     0     0    2N6546      ; BJT switch
Q6       4    13     0     0    2N6546      ; BJT switch
.MODEL  2N6546 NPN(IS=2.33E-27 BF=13 CJE=1PF CJC=607.3PF TF=26.5NS)
```
ANALYSIS ■ ■ ■
```
.TRAN     5US    2.5MS   1.0MS             ; Transient analysis
.PROBE                                     ; Graphics post-processor
.OPTIONS ABSTOL = 1.00N  RELTOL = 0.01  VNTOL = 0.1  ITL5=20000
.FOUR    1KHZ    I(VX)    V(3,21)          ; Fourier analysis
```
`.END` ■

(a) The PSpice plots of the instantaneous output line–line voltages V(3,4), V(4,5), and V(5,3) are shown in Fig. 12-10. The output phase voltages V(3,21) and V(4,21) and the output phase current I(VX) are shown in Fig. 12-11. The phase current is almost sinusoidal with THD = 4.72%.

(b) The Fourier coefficients of the output phase voltage and phase current are as follows:

FOURIER COMPONENTS OF TRANSIENT RESPONSE V(3,21)

DC COMPONENT =   −3.689844E−03

HARMONIC NO	FREQUENCY (HZ)	FOURIER COMPONENT	NORMALIZED COMPONENT	PHASE (DEG)	NORMALIZED PHASE (DEG)
1	1.000E+03	6.349E+01	1.000E+00	1.796E+02	0.000E+00
2	2.000E+03	2.118E−02	3.336E−04	1.098E+02	−6.981E+01
3	3.000E+03	6.643E−01	1.046E−02	8.981E+01	−8.979E+01
4	4.000E+03	2.138E−02	3.368E−04	−6.049E+01	−2.401E+02
5	5.000E+03	1.306E+01	2.057E−01	1.775E+02	−2.135E+00
6	6.000E+03	6.534E−03	1.029E−04	1.216E+02	−5.799E+01
7	7.000E+03	9.015E+00	1.420E−01	1.759E+02	−3.683E+00
8	8.000E+03	1.352E−02	2.129E−04	−1.845E+01	−1.980E+02
9	9.000E+03	6.873E−01	1.083E−02	8.654E+01	−9.305E+01

TOTAL HARMONIC DISTORTION =   2.503806E+01 PERCENT

Example 12-4   Three-Phase Inverter

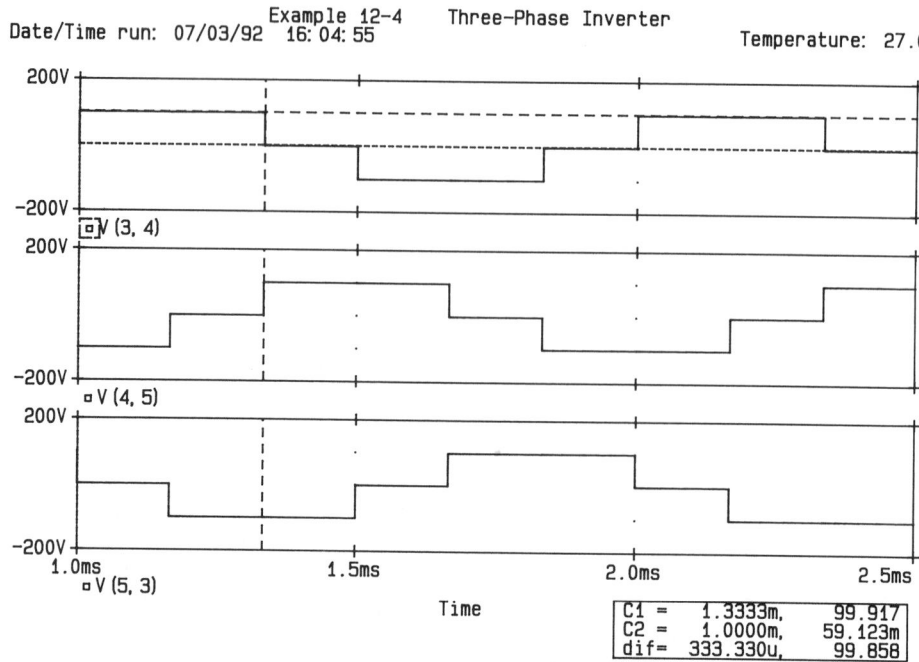

C1 =	1.3333m,	99.917
C2 =	1.0000m,	59.123m
dif=	333.330u,	99.858

Figure 12-10 Plots for Example 12-4.

Example 12-4   Three-Phase Inverter

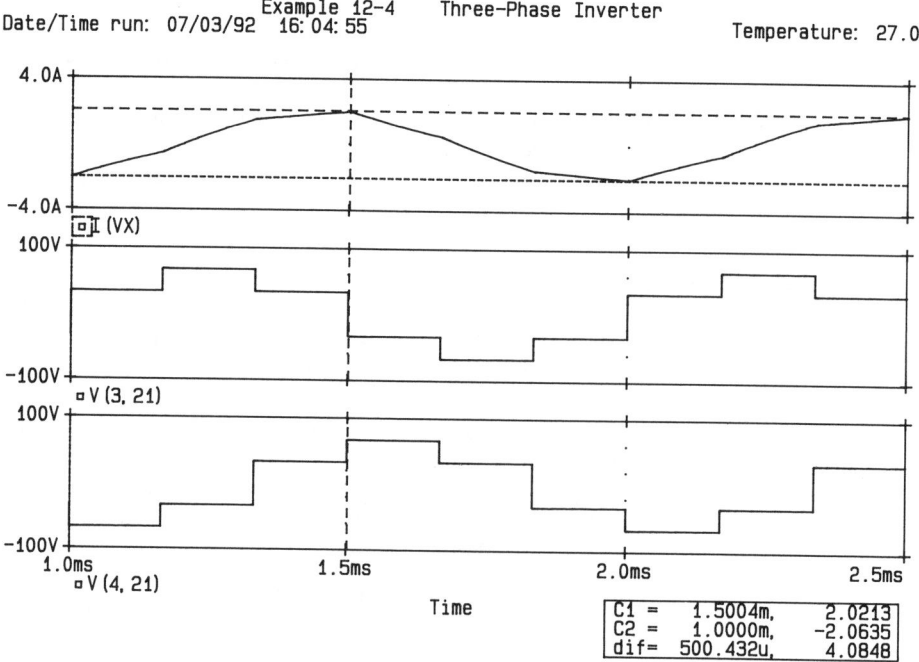

C1 =	1.5004m,	2.0213
C2 =	1.0000m,	-2.0635
dif=	500.432u,	4.0848

Figure 12-11 Plots for Example 12-4.

FOURIER COMPONENTS OF TRANSIENT RESPONSE I(VX)

DC COMPONENT = −4.969840E−03

HARMONIC NO	FREQUENCY (HZ)	FOURIER COMPONENT	NORMALIZED COMPONENT	PHASE (DEG)	NORMALIZED PHASE (DEG)
1	1.000E+03	1.931E+00	1.000E+00	1.078E+02	0.000E+00
2	2.000E+03	8.719E−04	4.516E−04	−1.507E+02	−2.586E+02
3	3.000E+03	4.645E−04	2.406E−04	1.384E+02	3.053E+01
4	4.000E+03	8.019E−04	4.153E−04	1.741E+02	6.625E+01
5	5.000E+03	8.075E−02	4.182E−02	9.345E+01	−1.439E+01
6	6.000E+03	4.428E−04	2.293E−04	−1.476E+02	−2.555E+02
7	7.000E+03	4.190E−02	2.170E−02	9.405E+01	−1.380E+01
8	8.000E+03	2.887E−04	1.495E−04	1.744E+02	6.659E+01
9	9.000E+03	2.040E−04	1.057E−04	1.227E+02	1.488E+01

TOTAL HARMONIC DISTORTION = 4.711973E+00 PERCENT

12-3 CURRENT-SOURCE INVERTERS

The input current of a current-source inverter is maintained approximately constant by having a large inductor at the input side. The magnitude of this current is normally varied by a chopper with an output filter. The time during which the input source current flows through the load is controlled by varying the turn-on and turn-off times of the inverter switches. The control can also use PWM, SPWM, and other advanced modulation techniques.

Example 12-5

A single-phase current-source inverter is shown in Fig. 12-12(a). The control voltages are shown in Fig. 12-12(b). The dc input voltage is 100 V. The output frequency is f_o = 1 kHz. The chopping frequency is f_s = 2 kHz, and its duty cycle is k = 0.6. The load resistive is R = 10 Ω, and the load inductance is L = 6.5 mH. Use PSpice (a) to plot the instantaneous output current i_o, the instantaneous source current i_s, and the instantaneous current i_1 through inductor L_e, and (b) to calculate the Fourier coefficients of the output current i_o. The model parameters of the BJTs are IS=2.33E−27, BF=13, CJE=1PF, CJC=607.3PF, and TF=26.5NS. The model parameters of the MOSFETs are VTO=2.83, KP=31.2U, L=1U, W=3.0M, CGDO=1.359N, and CGSO=2.032N.

Solution f_o = 1 kHz, f_s = 2 kHz, and k = 0.6. The list of the circuit file for the inverter is as follows:

■ ■

Example 12-5   Single-phase current-source inverter

```
SOURCE  ■ VS    1    0    DC    100V
          Vg    22   2    PULSE (0V  40V  0   1NS  1NS  300US  500US)
          Rg    22   2    10MEG
          RB    22   21   250        ; Transistor base resistance
          Rb1   8    7    50
          Rg1   8    9    10MEG
          Vg1   8    9    PULSE (0  40V  0  1NS  1NS  0.5MS  1MS)
          Rb2   17   16   50
```

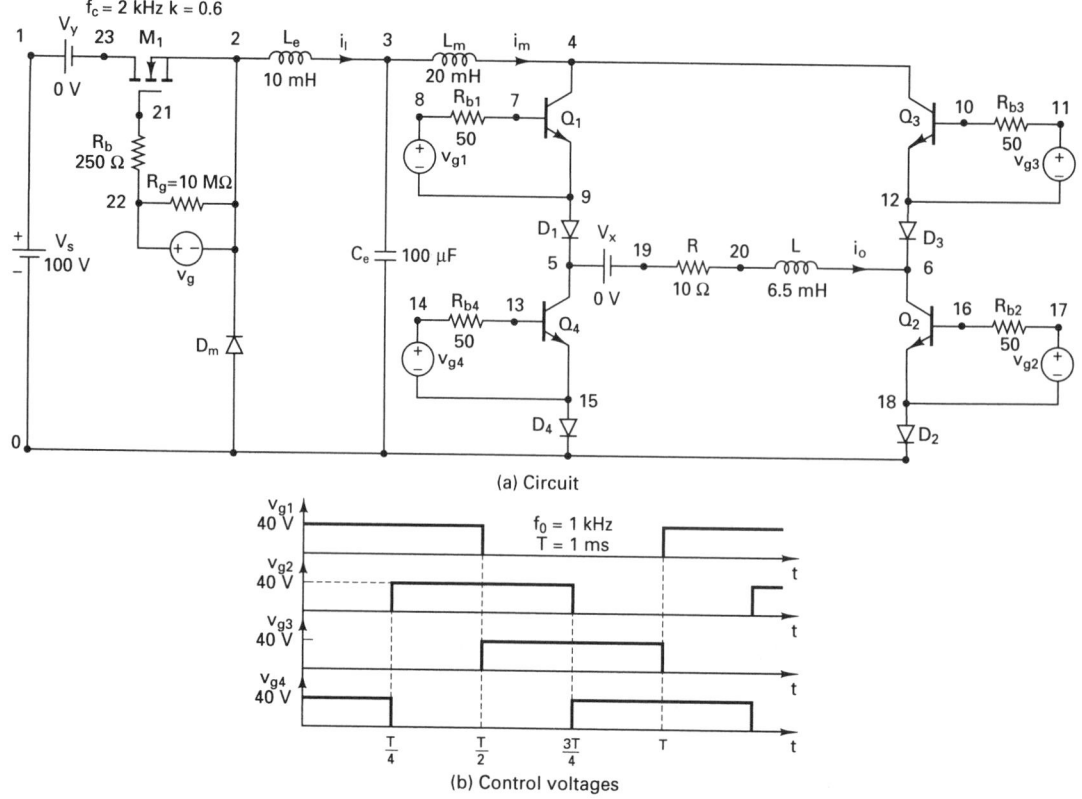

(a) Circuit

(b) Control voltages

Figure 12-12   Single-phase current-source inverter.

```
              Rg2   17   18   10MEG
              Vg2   17   18   PULSE (0   40V   250US   1NS   1NS   0.5MS   1MS)
              Rb3   11   10   50
              Rg3   11   12   10MEG
              Vg3   11   12   PULSE (0   40V   500US   1NS   1NS   0.5MS   1MS)
              Rb4   14   13   50
              Rg4   14   15   10MEG
              Vg4   14   15   PULSE (0   40V   750US   1NS   1NS   0.5MS   1MS)
CIRCUIT ■ ■   VY     1   23   DC    0V    ; Voltage source to measure supply current
              VX     5   19   DC    0V    ; Measures load current
              R     19   20   10
              L     20    6   6.5MH       ; L is included
              Le     2    3   10MH   IC=1A
              Ce     3    0   100UF
              Lm     3    4   20MH   IC=3A
              D1     9    5   DMOD        ; Diode
              D2    18    0   DMOD        ; Diode
              D3    12    6   DMOD        ; Diode
```

```
D4     15    0       DMOD              ; Diode
DM      0    2       DMOD              ; Diode
.MODEL DMOD D(IS=2.22E-15 BV-1200V IBV=13E-3 CJO=1PF TT=0) ; Diode model
M1     23   21    2    2    IRF150            ; MOSFET switch
.MODEL IRF150  NMOS (VTO=2.83 KP=31.2U L=1U W=3.0M
+      CGDO=1.359N CGSO=2.032N)             ; MOSFET parameters
Q1      4    7    9    9    2N6546    ; BJT switch
Q2      6   16   18   18    2N6546    ; BJT switch
Q3      4   10   12   12    2N6546    ; BJT switch
Q4      5   13   15   15    2N6546    ; BJT switch
.MODEL 2N6546  NPN(IS=2.33E-27 BF=13 CJE=1PF CJC=607.3PF TF=26.5NS)
```

ANALYSIS ■ ■ ■ .TRAN   10US   5MS   3MS   UIC   ; Transient analysis

```
              .PROBE                              ; Graphics post-processor
              .OPTIONS ABSTOL = 1.00U  RELTOL = 0.02  VNTOL = 0.1  ITL5=50000
              .FOUR    1KHZ    I(VX)               ; Fourier analysis
.END
```

(a) The PSpice plots of the instantaneous output current I(VX), the current source I(Lm), and the inductor current I(Le) are shown in Fig. 12-13. The output voltage is a square wave, as expected. It should be noted that the currents have not reached steady state.

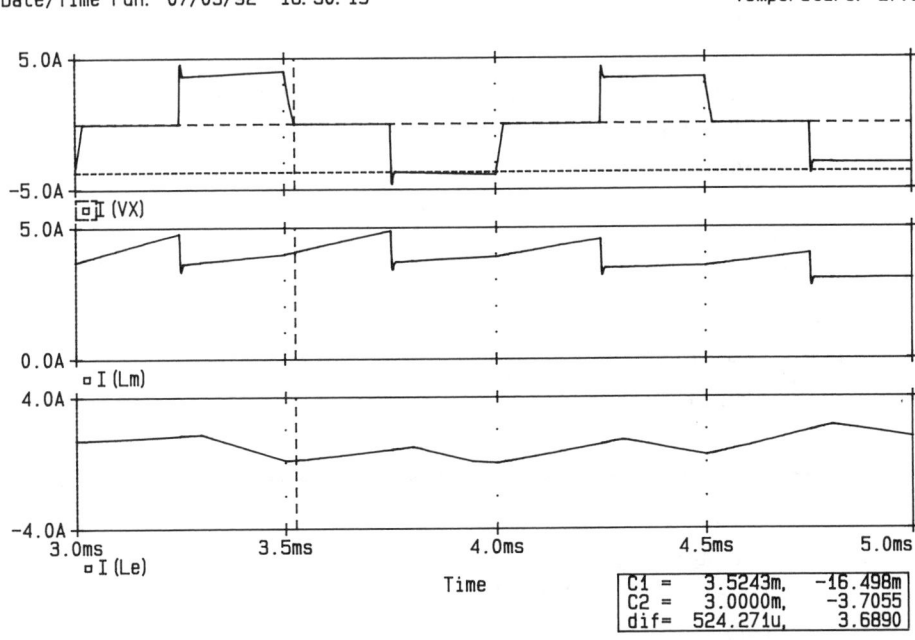

Figure 12-13 Plots for Example 12-4.

(b) The Fourier coefficients of the output current are as follows:

FOURIER COMPONENTS OF TRANSIENT RESPONSE I(VX)
DC COMPONENT = 9.959469E−02

HARMONIC NO	FREQUENCY (HZ)	FOURIER COMPONENT	NORMALIZED COMPONENT	PHASE (DEG)	NORMALIZED PHASE (DEG)
1	1.000E+03	3.009E+00	1.000E+00	−4.814E+01	0.000E+00
2	2.000E+03	1.437E−01	4.775E−02	−1.759E+02	−1.278E+02
3	3.000E+03	9.189E−01	3.054E−01	3.520E+01	8.334E+01
4	4.000E+03	2.512E−02	8.347E−03	−1.064E+02	−5.831E+01
5	5.000E+03	6.507E−01	2.162E−01	−5.957E+01	−1.143E+01
6	6.000E+03	5.224E−02	1.736E−02	−1.633E+02	−1.152E+02
7	7.000E+03	3.570E−01	1.186E−01	2.344E+01	7.158E+01
8	8.000E+03	2.363E−02	7.853E−03	−1.062E+02	−5.803E+01
9	9.000E+03	3.878E−01	1.289E−01	−7.128E+01	−2.314E+01

TOTAL HARMONIC DISTORTION = 4.164380E+01 PERCENT

12-4 LABORATORY EXPERIMENTS

It is possible to develop many experiments to demonstrate the operation and characteristics of inverters. The following experiments are suggested:

Single-phase half-bridge inverter
Single-phase full-bridge inverter
Single-phase full-bridge inverter with PWM control
Single-phase full-bridge inverter with sinusoidal PWM control
Three-phase bridge inverter
Single-phase current-source inverter
Three-phase current-source inverter

12-4.1 Experiment PW-1

SINGLE-PHASE HALF-BRIDGE INVERTER

Objective   The objective is to study the operation and characteristics of a single-phase half-bridge (transistor) inverter under various load conditions.

Applications   The single-phase half-bridge inverter is used to control power flow in many applications (e.g., ac and dc power supplies, and input stages of other converters).

Apparatus
1. Two BJTs/MOSFETs with ratings of at least 50 A and 400 V, mounted on heat sinks
2. Two fast-recovery diodes with ratings of at least 50 A and 400 V, mounted on heat sinks
3. A firing pulse generator with isolating signals for gating transistors
4. An *RL* load
5. One dual-beam oscilloscope with floating or isolating probes
6. Ac and dc voltmeters and ammeters and one noninductive shunt

Warning   Before making any circuit connection, switch the dc power OFF. **Do not** switch the power ON unless the circuit is checked and approved by your lab instructor. **Do not** touch the transistor/diode heat sinks, which are connected to live terminals.

Experimental procedure
1. Set up the circuit as shown in Fig. 12-14. Use the load resistance *R* only.
2. Connect the measuring instruments as required.
3. Connect the firing pulses to the appropriate transistors.
4. Set one pulse per half-cycle with duty cycle $k = 0.5$.
5. Observe and record the waveforms of the load voltage v_o and the load current i_o.
6. Measure the rms load voltage $V_{o(rms)}$, the rms load current $I_{o(rms)}$, the average input current $I_{s(dc)}$, the dc input voltage $V_{s(dc)}$, and the total harmonic distortion THD of the output voltage and current.
7. Measure the conduction angles of transistor Q_1 and diode D_1.
8. Repeat steps 2 to 7 with both load resistance *R* and load inductance *L*.

Report
1. Present all recorded waveforms and discuss all significant points.
2. Compare the waveforms generated by SPICE with the experimental results, and comment.

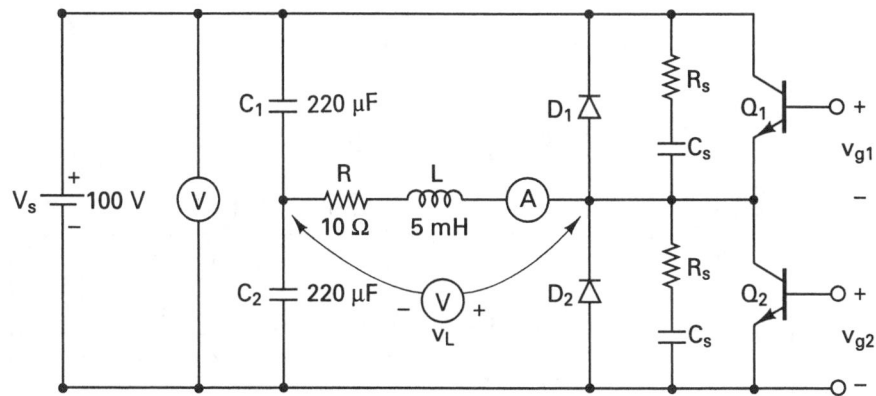

Figure 12-14   Single-phase half-bridge inverter.

3. Compare the experimental results with the results predicted.

4. Discuss the advantages and disadvantages of this type of inverter.

5. Discuss the effects of the diodes on the performance of the inverter.

12-4.2 Experiment PW-2

SINGLE-PHASE FULL-BRIDGE INVERTER

Objective   The objective is to study the operation and characteristics of a single-phase full-bridge (transistor) inverter under various load conditions.

Applications   See Experiment PW-1.

Textbook   See Ref. 1, Sec. 10-4.

Apparatus   Similar to Experiment PW-1, except four BJTs/MOSFETs and four fast-recovery diodes are required, as shown in Fig. 12-15.

Warning   See Experiment PW-1.

Experimental   See Experiment PW-1.
procedure

Report   See Experiment PW-1.

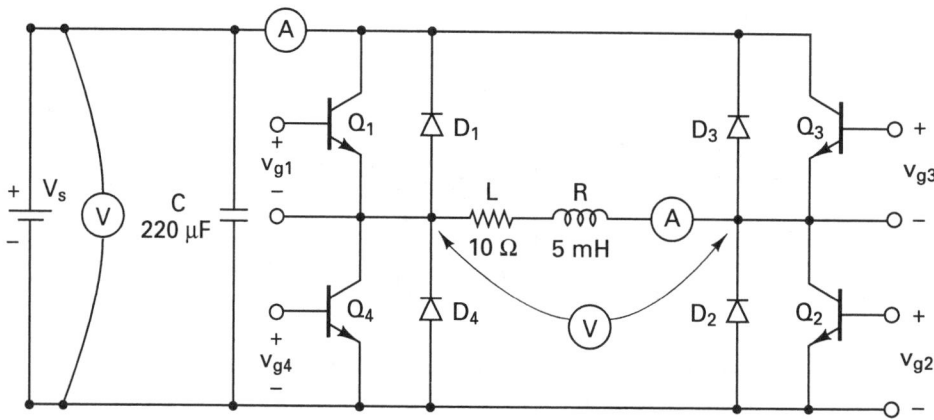

Figure 12-15   Single-phase full-bridge inverter.

12-4.3 Experiment PW-3

SINGLE-PHASE FULL-BRIDGE INVERTER WITH PWM CONTROL

Objective   The objective is to study the effects of PWM control on the total harmonic distortion (THD) of the output voltage for a single-phase full-bridge (transistor) inverter with a resistive load.

Applications   Similar to Experiment PW-1.

Textbook   See Ref. 1, Secs. 10-4 and 10-6.

Apparatus   Similar to Experiment PW-1, except that four BJTs/MOSFETs and four fast-recovery diodes are required.

Warning   See Experiment PW-1.

Experimental procedure
1. Set up the circuit as shown in Fig. 12-15. Use the load resistance R only.
2. Set one pulse per half cycle, $p = 1$, and the modulation index $M = 0.5$.
3. Measure the rms load voltage $V_{o(rms)}$, and the total harmonic distortion of the output voltage.
4. Repeat step 3 for $p = 2, 3, 4, 5$.
5. Repeat steps 2 and 4 for $M = 0.1$ to 1 with an increment of 0.1.

Report
1. Plot the rms output voltage and THD of the output voltage against the modulation M for various values of p.
2. Use SPICE or MathCAD to predict the rms output voltage and THD.
3. Compare the experimental results with the results predicted, and comment.

12-4.4 Experiment PW-4

SINGLE-PHASE FULL-BRIDGE INVERTER WITH SINUSOIDAL PWM CONTROL

Objective   The objective is to study the effects of sinusoidal PWM control on the total harmonic distortion (THD) of the output voltage for a single-phase full-bridge (transistor) inverter under a resistive load.

Applications   See Experiment PW-1.

Apparatus See Experiment PW-2.

Warning See Experiment PW-1

Experimental procedure See Experiment PW-3.

Report See Experiment PW-3.

12-4.5 Experiment PW-5

THREE-PHASE BRIDGE INVERTER

Objective The objective is to study the operation and characteristics of a three-phase bridge (transistor) inverter under various load conditions.

Applications The three-phase bridge inverter is used to control power flow under many applications (e.g., ac power supplies and ac motor drives).

Textbook See Ref. 1, Secs. 10-5 and 10-6.

Apparatus Similar to Experiment PW-1, except that six BJTs/MOSFETs and six fast-recovery diodes are required.

Warning See Experiment PW-1.

Experimental procedure
1. Set up the circuit as shown in Fig. 12-16. Use a wye-connected resistive load R only.
2. Connect the measuring instruments as required.
3. Connect the firing pulses to the appropriate transistors.
4. Set one pulse per half-cycle with duty cycle $k = 0.5$.
5. Observe and record the waveforms of the output phase voltage, the output line–line voltage, and the load phase current i_o.
6. Measure the rms load phase voltage $V_{p(rms)}$, the rms load phase current $I_{p(rms)}$, the average input current $I_{s(dc)}$, the average input voltage $V_{s(dc)}$, and the total harmonic distoriton of the output phase voltage and phase current.
7. Measure the conduction angles of transistor Q_1 and diode D_1.
8. Repeat steps 2 to 7 with both load resistance R and load inductance L.
9. Repeat steps 1 to 8 with a delta-connected load.

Report See Experiment PW-1.

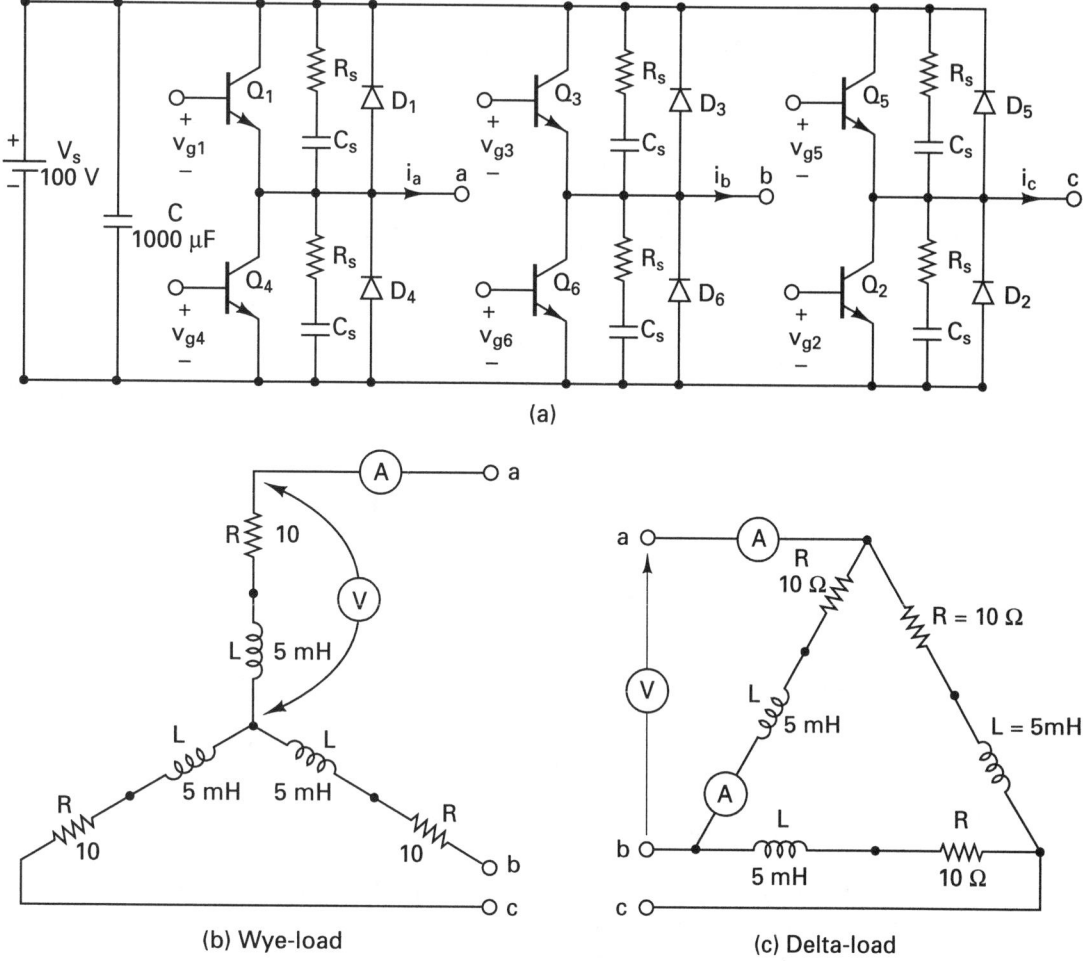

Figure 12-16   Three-phase bridge inverter.

12-4.6 Experiment PW-6

SINGLE-PHASE CURRENT-SOURCE INVERTER

Objective   The objective is to study the operation and characteristics of a single-phase current-source (transistor) inverter under various load conditions.

Applications   The current-source inverter is used to control power flow in many applications (e.g., ac power supplies).

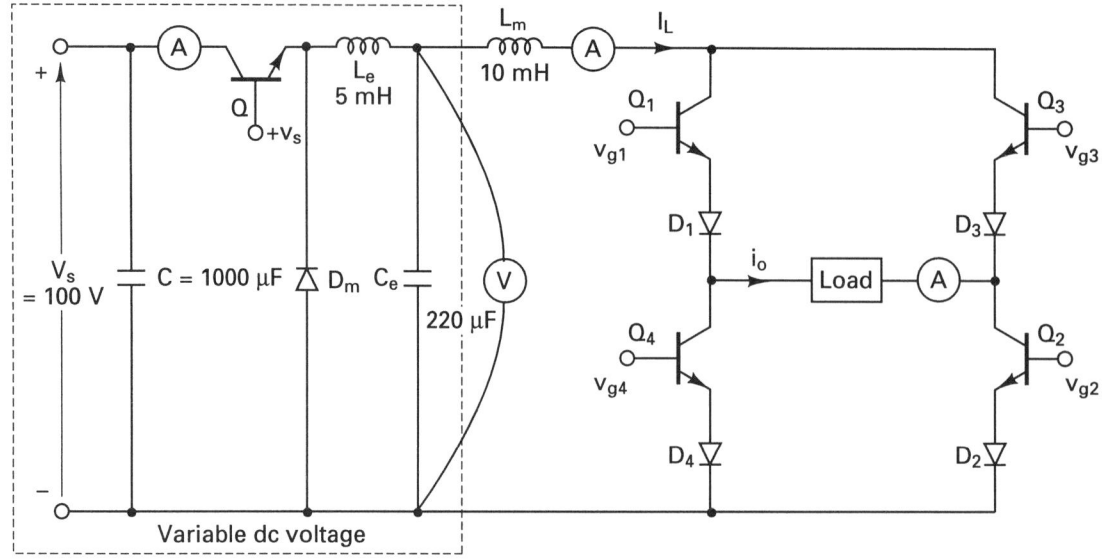

Figure 12-17 Single-phase current-source inverter.

Textbook See Ref. 1, Sec. 10-11.

Apparatus Similar to Experiment PW-1, except that four BJTs/MOSFETs and four fast-recovery diodes are required.

Warning See Experiment PW-1.

Experimental procedure
1. Set up the circuit as shown in Fig. 12-17. Use a resistive load R only.
2. Connect the measuring instruments as required.
3. Connect the firing pulses to the appropriate transistors.
4. Set the duty cycle of the chopper at $k = 0.5$. Set one pulse per half-cycle with duty cycle $k = 0.5$ of the inverter.
5. Observe and record the waveforms of the output voltage and the output current.
6. Measure the rms load voltage $V_{o(rms)}$, the rms load current $I_{o(rms)}$, the average input current $I_{s(dc)}$, the dc input voltage $V_{s(dc)}$, and the total harmonic distortion of the output current.
7. Measure the conduction angles of the transistor Q_1.
8. Repeat steps 2 to 7 with both load resistance R and load inductance L.

Report See Experiment PW-1.

Objective The objective is to study the operation and characteristics of a three-phase current-source (transistor) inverter under resistive load.

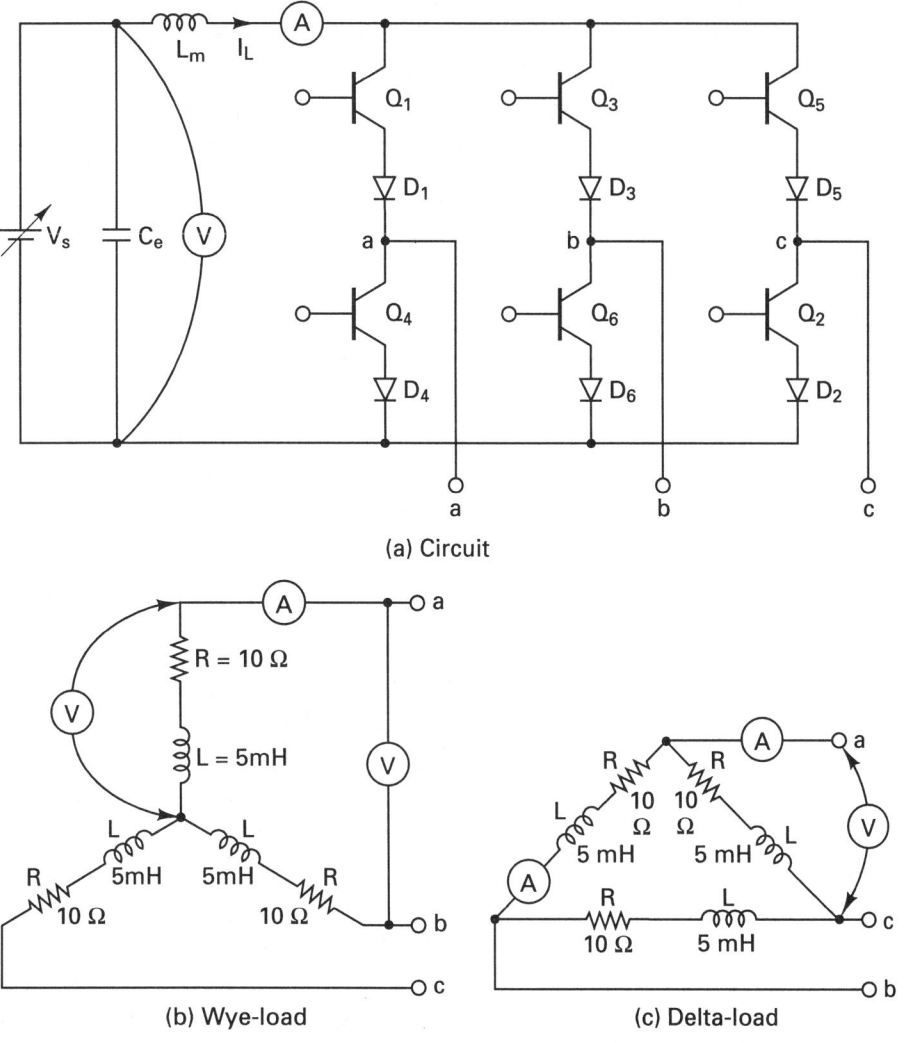

(a) Circuit

(b) Wye-load

(c) Delta-load

Figure 12-18 Three-phase current-source inverter.

Applications	The three-phase current-source inverter is used to control power flow in many applications (e.g., ac power supplies and ac motor drives).

Textbook	See Ref. 1, Sec. 10-11.

Apparatus	Similar to Experiment PW-1, except that six BJTs/MOSFETs and six fast-recovery diodes are required.

Warning	See Experiment PW-1.

Experimental procedure	**1.** Set up the circuit as shown in Fig. 12-18. Use a wye-connected resistive load R only.

2. Connect the measuring instruments as required.

3. Connect the firing pulses to the appropriate transistors.

4. Set one pulse per half-cycle with duty cycle $k = 0.5$.

5. Observe and record the waveforms of the output phase current i_o.

6. Measure the rms load phase current, average input current, the dc input voltage, and the total harmonic distortion of the output phase current.

7. Measure the conduction angles of the transistor Q_1 and diode D_1.

Report	See Experiment PW-1.

SUMMARY

The statements for an ac thyristor are:

```
*    Subcircuit call for switched transistor model:
XT1    +N        -N        +NG       -NG       QM
       positive  negative +control  -control   model
*                          voltage   voltage    name
*    Subcircuit call for PWM control:
XPWM    VR        VC        +NG       -NG       PWM
*       ref.     carrier   +control  -control  model
*       input     input     voltage   voltage    name
*    Subcircuit call for sinusoidal PWM control:
XSPWM    VR        VS        +NG       -NG           VC            SPWM
*        ref.    sine-wave  +control  -control   rectified        model
*        input     input     voltage   voltage   carrier sine wave  name
```

SUGGESTED READING

1. M. H. Rashid, *Power Electronics: Circuits, Devices, and Applications*, 2nd ed. Englewood Cliffs, N.J.: Prentice Hall, 1993, Chap. 14.

2. M. H. Rashid, *SPICE for Circuits and Electronics Using PSpice*. Englewood Cliffs, N.J.: Prentice Hall, 1990, Chap. 4.

3. J. G. Kassakian, M. F. Schlecht, and G. C. Verghese, *Principles of Power Electronics.* Reading, Mass.: Addison-Wesley, 1991.

4. V. Voperian, R. Tymerski, and F. C. Y. Lee, "Equivalent circuit models for resonant and PWM switches," *IEEE Transactions on Power Electronics*, Vol. 4, No. 2, 1990, pp. 205–214.

DESIGN PROBLEMS

12-1. It is required to design the single-phase half-bridge inverter of Fig. 12-14 with the following specifications:

Dc supply voltage, $V_s = 100$ V

Load resistance, $R = 5\ \Omega$

Load inductance, $L = 5$ mH

Output frequency, $f_o = 1$ kHz

(a) Determine the ratings of all components and devices under worst-case conditions.
(b) Use SPICE to verify your design.
(c) Provide a cost estimate of the circuit.

12-2. It is required to design the single-phase full-bridge inverter of Fig. 12-15 with the following specifications:

Dc supply voltage, $V_s = 100$ V

Load resistance, $R = 5\ \Omega$

Load inductance, $L = 5$ mH

Output frequency, $f_o = 1$ kHz

(a) Determine the ratings of all components and devices under worst-case conditions.
(b) Use SPICE to verify your design.
(c) Provide a cost estimate of the circuit.

12-3. (a) Design an output C filter for the single-phase full-bridge inverter of Problem 12-3. The harmonic content of the load current should be less than 5% of the value without the filter.
(b) Use SPICE to verify your design in part (a).

12-4. It is required to design the three-phase bridge inverter of Fig. 12-16 with the following specifications:

Dc supply voltage, $V_s = 100$

Load resistance per phase, $R = 5\ \Omega$

Load inductance per phase, $L = 5$ mH

Output frequency, $f_o = 1$ kHz

(a) Determine the ratings of all components and devices under worst-case conditions.
(b) Use SPICE to verify your design.
(c) Provide a cost estimate of the circuit.

12-5. It is required to design the single-phase current-source inverter of Fig. 12-17 with the following specifications:

Dc supply voltage, $V_s = 100$ v

Average value dc current source,
$$I_s = 10\ \text{A}$$

Load resistance, $R = 5\ \Omega$

Load inductance, $L = 5$ mH

Output frequency, $f_o = 400$ Hz

(a) Determine the ratings of all components and devices under worst-case conditions.
(b) Use SPICE to verify your design.
(c) Provide a cost estimate of the circuit.

12-6. (a) Design an output C filter for the single-phase current-source inverter of Problem 12-5. The harmonic content of the load current should be less than 5% of the value without the filter.
(b) Use SPICE to verify your design in part (a).

12-7. It is required to design the three-phase current-source inverter of Fig. 12-18 with the following specifications:

Dc supply voltage, $V_s = 100$ V

Average value dc current source,
$$I_s = 10 \text{ A}$$

Load resistance per phase, $R = 5 \, \Omega$

Load inductance per phase, $L = 5$ mH

Output frequency, $f_o = 400$ Hz

(a) Determine the ratings of all components and devices under worst-case conditions.

(b) Use SPICE to verify your design.

(c) Provide a cost estimate of the circuit.

12-8. (a) Design an output C filter for the three-phase current-source inverter of Problem 12-7. The harmonic content of the load current should be less than 5% of the value without the filter.

(b) Use SPICE to verify your design in part (a).

Chapter 13

■■■■■■■■■

Resonant Pulse Inverters

13-1 INTRODUCTION

The input to a resonant inverter is a dc voltage or current source, and the output is a resonant pulse of voltage or current. The power semiconductor devices perform the switching action and the desired output is obtained by varying their turn-on and turn-off times. The commonly used devices are BJTs, MOSFETs, IGBTs, MCTs, GTOs, and SCRs. We shall use PSpice switches and BJTs to simulate the characteristics of the following inverters:

> Resonant pulse inverters
> Zero-current switching converters
> Zero-voltage switching converter

13-2 RESONANT PULSE INVERTERS

The switches of resonant inverters are turned on to initiate resonant oscillation and are maintained in an on-state condition to complete resonant oscillation. The output waveform depends mostly on circuit parameters and the input source. The on-time and switching frequency of power devices must match the resonant frequency of the circuit.

Example 13-1

A half-bridge resonant inverter is shown in Fig. 13-1(a). The control voltages are shown in Fig. 13-1(b). The dc input voltage is 100 V. The output frequency is $f_o = 5$ kHz. The load resistance is $R = 1$ Ω, and the load inductance is $L = 50$ μH. Use PSpice (a) to plot the instantaneous output current i_o and the instantaneous input supply current i_s, and (b) to calculate the Fourier coefficients of the output current

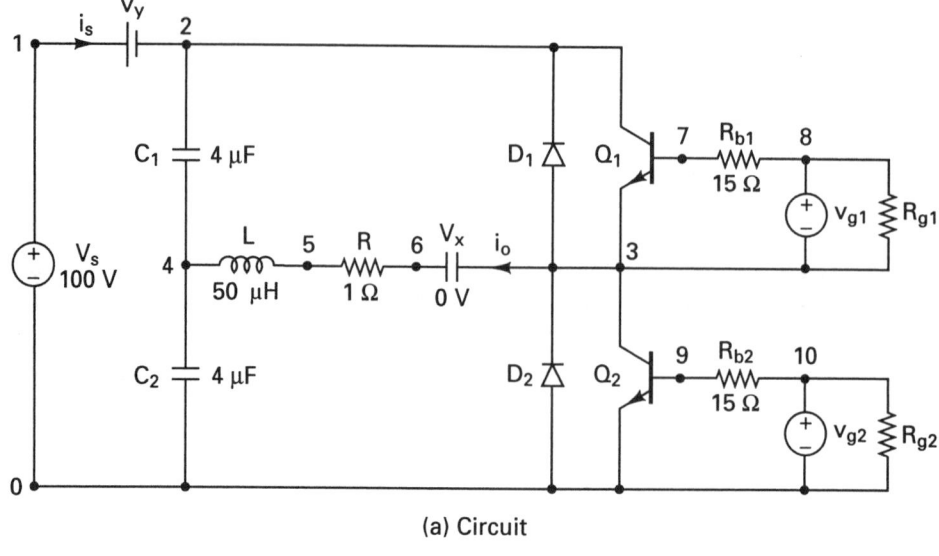

(a) Circuit

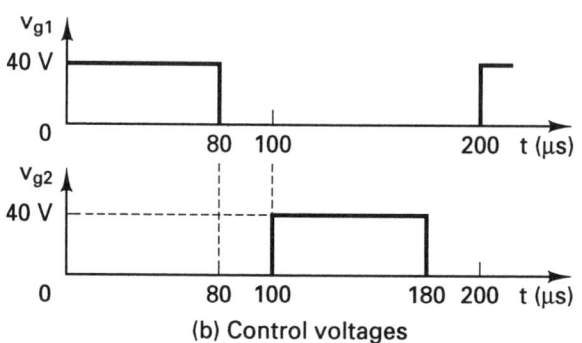

(b) Control voltages

Figure 13-1 Half-bridge resonant inverter.

i_o. The BJT parameters are IS=2.33E−27, BF=13, CJE=1PF, CJC=607.3PF, and TF=26.5NS.

Solution The values of gate voltage V_g and base resistance R_b must be such that the transistors are driven into saturation at the load current expected. The list of the circuit file is as follows:

■■■

Example 13-1 Half-bridge resonant inverter

SOURCE	■ VS	1	0	DC	100V					
	Rg1	8	3	10MEG						
	Vg1	8	3	PULSE (0	40V	0	1NS	1NS	80US	200US)
	Rg2	10	0	10MEG						
	Vg2	10	0	PULSE (0	40V	100US	1NS	1NS	80US	200US)
CIRCUIT	■■ Rb1	8	7	15						
	Rb2	10	9	15						

```
VY      1   2   DC   0V      ; Voltage source to measure supply current
VX      3   6   DC   0V      ; Measures load current
R       6   5   1
L       5   4   50UH
C1      2   4   4UF
C2      4   0   4UF
D1      3   2   DMOD           ; Diode
D2      0   3   DMOD           ; Diode
.MODEL  DMOD   D(IS=2.2E-15 BV=1200V CJO=0 TT=0) ; Diode model parameters
Q1      2   7   3   3   2N6546        ; BJT switch
Q2      3   9   0   0   2N6546        ; BJT switch
.MODEL 2N6546   NPN(IS=2.33E-27 BF=13 CJE=1PF CJC=607.3PF TF=26.5NS)
```

ANALYSIS ■ ■ ■
```
.TRAN   1US   400US                  ; Transient analysis
.PROBE                               ; Graphics post-processor
.OPTIONS ABSTOL = 1.00N  RELTOL = 0.01  VNTOL = 0.1  ITL5=50000
.FOUR   5KHZ   I(VX)                 ; Fourier analysis
```
`.END`

(a) The PSpice plots of the instantaneous output current I(VX) and the current source I(VY) are shown in Fig. 13-2. For $f_o = 4$ kHz, the switching period is changed to 250 μs, and the plots are shown in Fig. 13-3. As expected, the output voltage can be varied by changing the switching frequency.

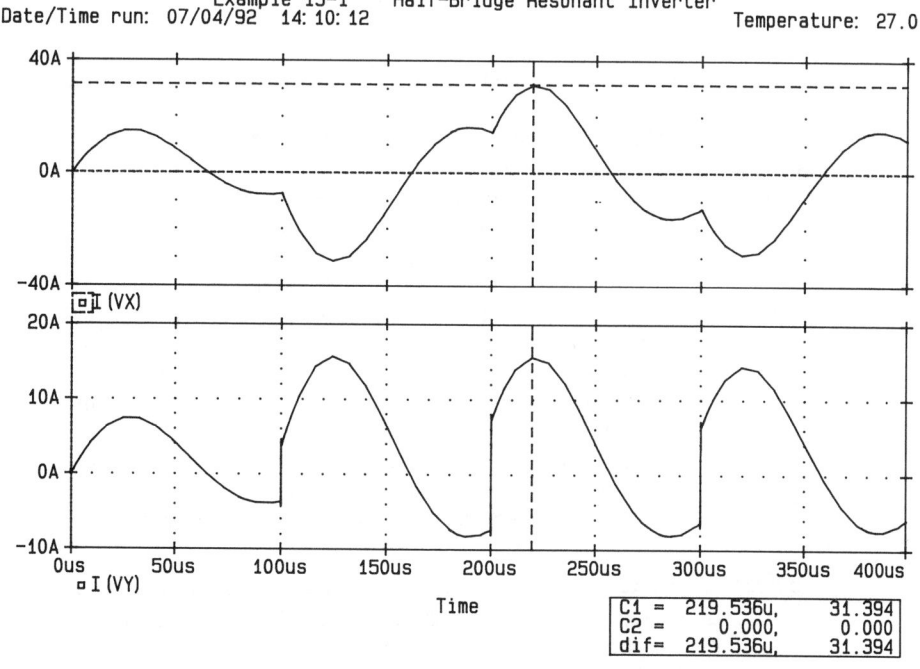

Example 13-1   Half-Bridge Resonant Inverter
Date/Time run: 07/04/92 14:10:12   Temperature: 27.0

Figure 13-2 Plots for Example 13-1 at $f_0 = 5$ kHz.

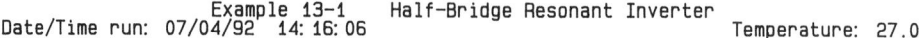

Example 13-1   Half-Bridge Resonant Inverter
Date/Time run: 07/04/92 14:16:06   Temperature: 27.0

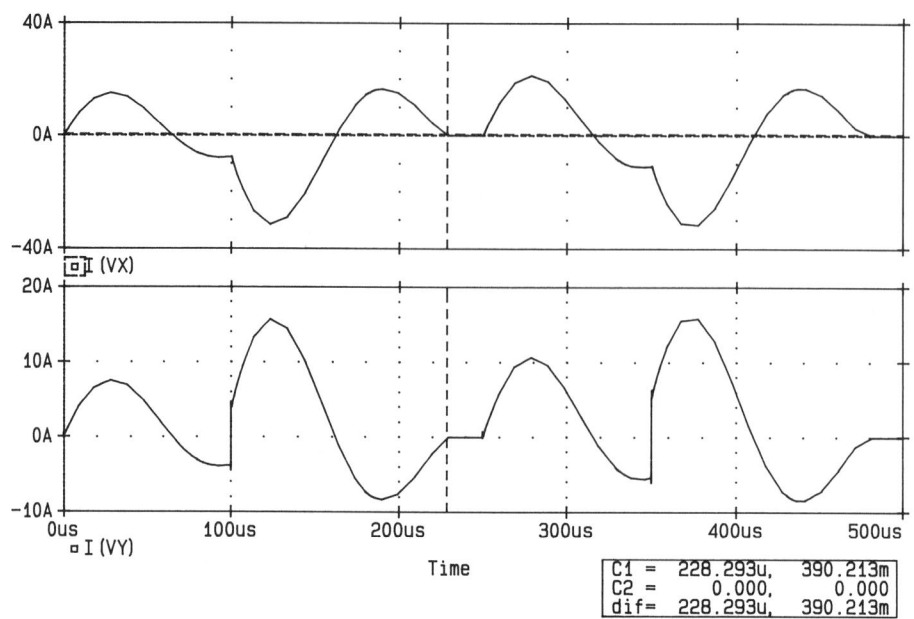

C1 =	228.293u,	390.213m
C2 =	0.000,	0.000
dif=	228.293u,	390.213m

Figure 13-3   Plots for Example 13-1 at f_0 = 4 kHz.

(b) The Fourier coefficients of the output current are as follows:

FOURIER COMPONENTS OF TRANSIENT RESPONSE I(VX)
DC COMPONENT =   2.185143E-01

HARMONIC NO	FREQUENCY (HZ)	FOURIER COMPONENT	NORMALIZED COMPONENT	PHASE (DEG)	NORMALIZED PHASE (DEG)
1	5.000E+03	2.566E+01	1.000E+00	6.723E+01	0.000E+00
2	1.000E+04	9.637E-01	3.756E-02	1.020E+01	-5.703E+01
3	1.500E+04	6.027E+00	2.349E-01	-7.231E+01	-1.395E+02
4	2.000E+04	2.216E-01	8.636E-03	-3.107E-01	-6.754E+01
5	2.500E+04	1.781E+00	6.943E-02	-7.669E+01	-1.439E+02
6	3.000E+04	1.362E-01	5.308E-03	-5.628E+00	-7.286E+01
7	3.500E+04	8.531E-01	3.325E-02	-7.579E+01	-1.430E+02
8	4.000E+04	1.193E-01	4.649E-03	1.046E+00	-6.619E+01
9	4.500E+04	5.075E-01	1.978E-02	-7.477E+01	-1.420E+02

TOTAL HARMONIC DISTORTION =   2.510610E+01 PERCENT

Example 13-2

A parallel resonant inverter is shown in Fig. 13-4(a). The control voltages are shown in Fig. 13-4(b). The dc input voltage is 100 V. The output frequency is f_o = 29.3 kHz. Use PSpice (a) to plot the instantaneous output current i_o and the instantaneous input supply current i_s, and (b) to calculate the Fourier coefficients of the output current i_o. Use voltage-controlled switches to perform the switching action.

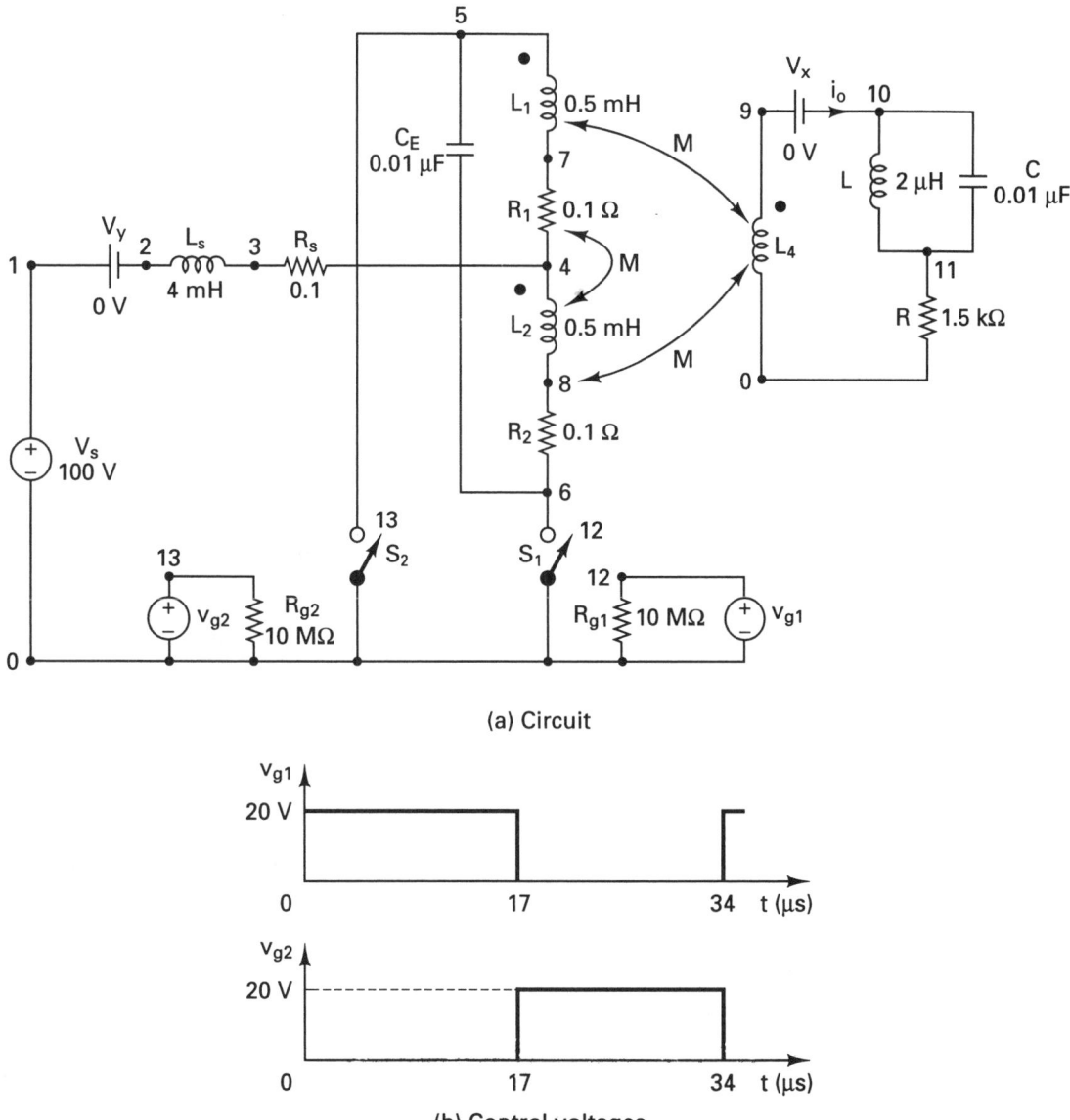

(a) Circuit

(b) Control voltages

Figure 13-4 Parallel resonant inverter.

Solution The inductor L_s acts as a current source. It is generally necessary to adjust the on-time of the switches to match the resonant frequency of the circuit. The list of the circuit file is as follows:

■ ■

Example 13-2   Push–pull resonant inverter

```
SOURCE  ■ VS    1    0    DC    100V
          Vg1   12   0    PULSE (0  20V  0  1NS  1NS  17US  34.11US)
```

```
              Rg1   12   0    10MEG
              Vg2   13   0    PULSE (0  20V  17US  1NS  1NS  17US  34.1US)
              Rg2   13   0    10MEG
CIRCUIT  ■ ■  VX     9  10    DC   0V    ; Measures load current
              VY     1   2    DC   0V    ; Voltage source to measure supply current
              RS     3   4    0.1
              LS     2   3    4MH
              CE     5   6    0.01UF
              L1     5   7    0.5MH
              R1     7   4    0.1
              L2     4   8    0.5MH
              R2     8   6    0.1
              L4     9   0    3.5MH
              K12   L1  L2    0.9999
              K14   L1  L4    0.9999
              K24   L2  L4    0.9999
              L     10  11    2UH
              C     10  11    0.01UF
              R     11   0    1.5K
              S1     6   0    12   0    SMOD   ; Voltage-controlled switch
              S2     5   0    13   0    SMOD   ; Voltage-controlled switch
              .MODEL  SMOD  VSWITCH (RON=0.01 ROFF=10E+6 VON=0.2V VOFF=10MV)
ANALYSIS ■ ■ ■ .TRAN  0.5US    120US                ; Transient analysis
              .PROBE                                ; Graphics post-processor
```

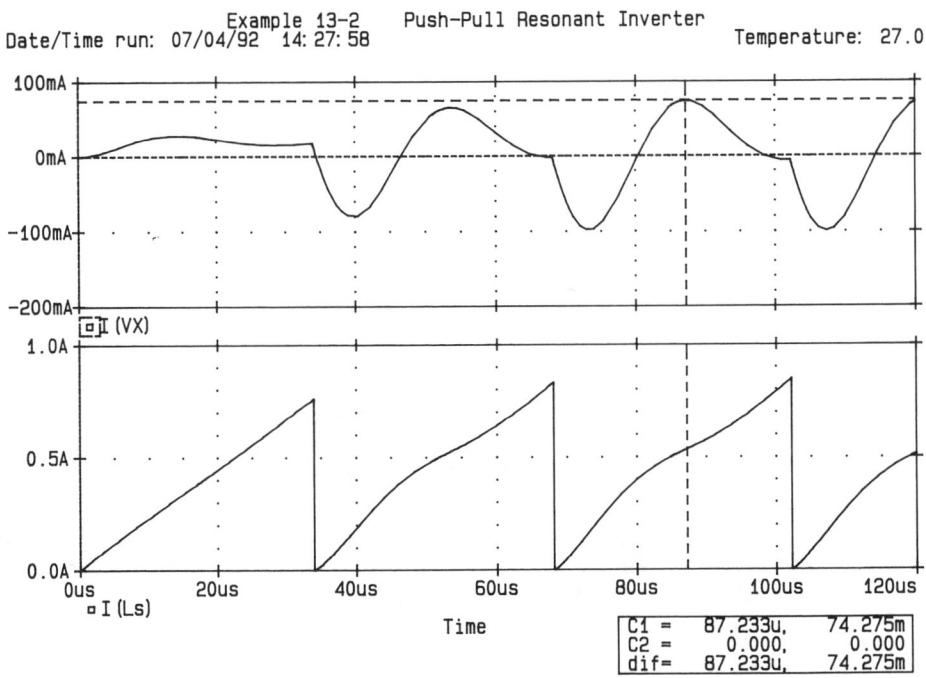

Figure 13-5 Plots for Example 13-2 at f_0 = 29.3 kHz.

```
.OPTIONS ABSTOL = 1.00N  RELTOL = 0.01  VNTOL = 0.1  ITL5=50000
.FOUR  29.3KHZ  I(VX)                       ; Fourier analysis
.END
```
■ ■

(a) The PSpice plots of the instantaneous output current I(VX) and the current source I(VY) are shown in Fig. 13-5.

(b) The Fourier coefficients of the output current are as follows:

FOURIER COMPONENTS OF TRANSIENT RESPONSE I(VX)

DC COMPONENT =   9.019181E-05

HARMONIC NO	FREQUENCY (HZ)	FOURIER COMPONENT	NORMALIZED COMPONENT	PHASE (DEG)	NORMALIZED PHASE (DEG)
1	2.930E+04	7.433E-02	1.000E+00	5.243E+01	0.000E+00
2	5.860E+04	2.302E-02	3.098E-01	1.345E+02	8.202E+01
3	8.790E+04	8.330E-03	1.121E-01	-4.895E+01	-1.014E+02
4	1.172E+05	4.527E-03	6.091E-02	1.299E+02	7.745E+01
5	1.465E+05	2.641E-03	3.553E-02	-4.305E+01	-9.548E+01
6	1.758E+05	1.825E-03	2.456E-02	1.424E+02	8.993E+01
7	2.051E+05	1.300E-03	1.749E-02	-2.515E+01	-7.759E+01
8	2.344E+05	1.083E-03	1.457E-02	1.535E+02	1.011E+02
9	2.637E+05	8.259E-04	1.111E-02	-1.467E+01	-6.710E+01

TOTAL HARMONIC DISTORTION =   3.387136E+01 PERCENT

13-3 ZERO-CURRENT SWITCHING CONVERTERS

A power device of converters is turned on and off at zero current by creating an *LC*-resonant circuit. The device remains on and provides a path for completing the resonant oscillation. When the device is off, it has to withstand the peak voltage, but at a zero current, thereby reducing the switching loss of the device. The output waveforms depend primarily on the circuit parameters and the input supply voltage. The switching period must be long enough to complete the resonant oscillation.

Example 13-3

A zero-current switching converter (ZCSC) is shown in Fig. 13-6(a). The control voltage is shown in Fig. 13-6(b). The dc input voltage is 15 V. The output frequency is $f_o = 8.33$ kHz. Use PSpice (a) to plot the instantaneous capacitor current i_c, the instantaneous capacitor voltage v_c, and the diode voltage v_{Dm}, and (b) to calculate the instantaneous voltage, current, and power of the switch. Use voltage- and current-controlled switches to perform the switching action.

Solution A current-controlled switch W_1 is required to break the circuit when the resonant current falls to zero. When the supply is turned on, the capacitor will be charged to V_s and have an initial voltage. The list of the circuit file is as follows:

■ ■

Example 13-3   Zero-current switching converter

```
SOURCE  ■ VS   1   0   DC    15V
          Rg   9   0   10MEG
          Vg   9   0   PULSE (0  20V  0  1NS  1NS  75US  120US)
```

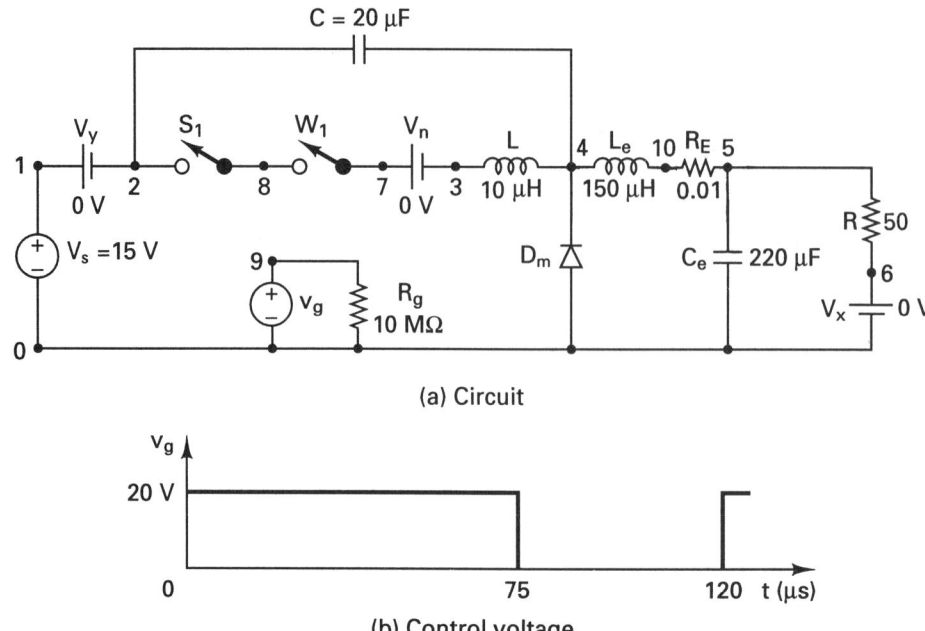

(a) Circuit

(b) Control voltage

Figure 13-6 Zero-current switching converter.

CIRCUIT ■ ■ VY   1   2   DC   0V   ; Voltage source to measure supply current
         VX   6   0   DC   0V   ; Measures load current
         VN   7   3   DC   0V   ; Measures the current-controlled switch
         R   5   6   50
         LE   4   10   150UH
         RE   10   5   0.01
         CE   5   0   220UF
         L   3   4   10UH
         C   2   4   20UF   IC=15V ; Initial condition
         DM   0   4   DMOD   ; Diode
         .MODEL DMOD D(IS=2.2E−15 BV=1200V CJO=0 TT=0) ; Diode model parameters
         S1   2   8   9   0   SMOD   ; Voltage-controlled switch
         .MODEL SMOD VSWITCH (RON=0.001 ROFF=10E+6 VON=10V VOFF=5V)
         W1   8   7   VN   IMOD   ; Current-controlled switch
         .MODEL IMOD ISWITCH (RON=1E+6 ROFF=0.01 ION=0 IOFF=1UA) ; Model parameters
ANALYSIS ■ ■ ■ .TRAN 1US   400US UIC   ; Transient analysis
         .PROBE   ; Graphics post-processor
         .OPTIONS ABSTOL = 1.00N RELTOL = 0.1 VNTOL = 0.1 ITL5=50000
         .END

(a) The PSpice plots of the gate voltage V(9), the instantaneous capacitor voltage V(2,4), the capacitor current I(C), and the diode voltage V(4) are shown in Fig. 13-8.

(b) The instantaneous voltage, current, and power of the switch are shown in Fig. 13-7, where its peak instantaneous power is 12.2 W.

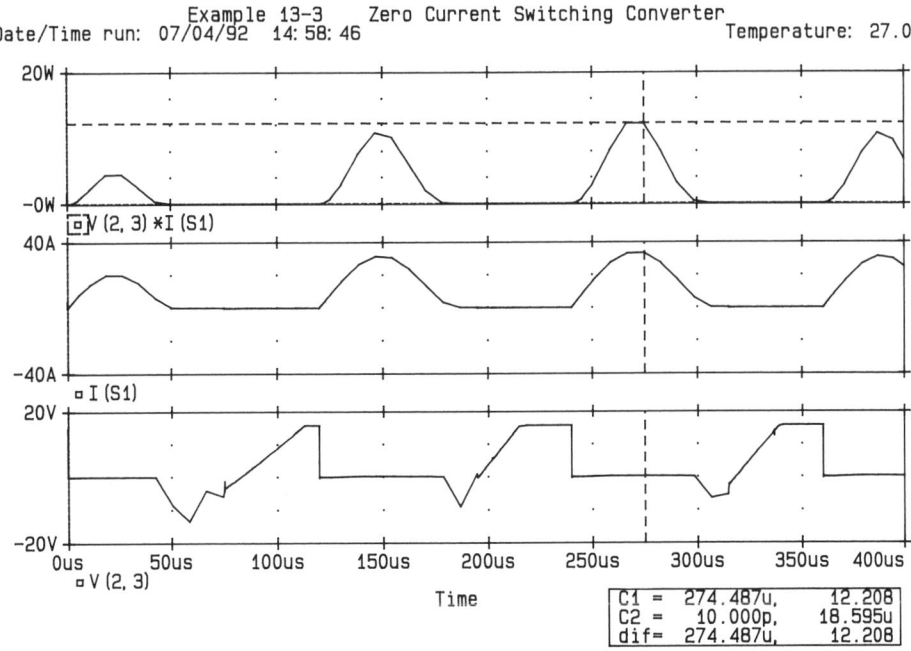

Figure 13-7 Plots for Example 13-3.

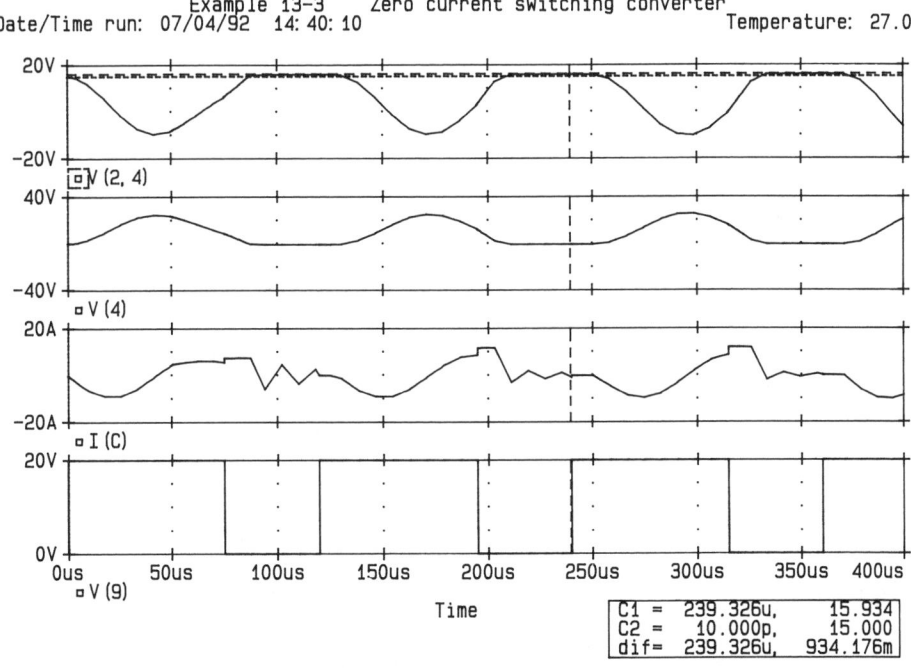

Figure 13-8 Plots for Example 13-3.

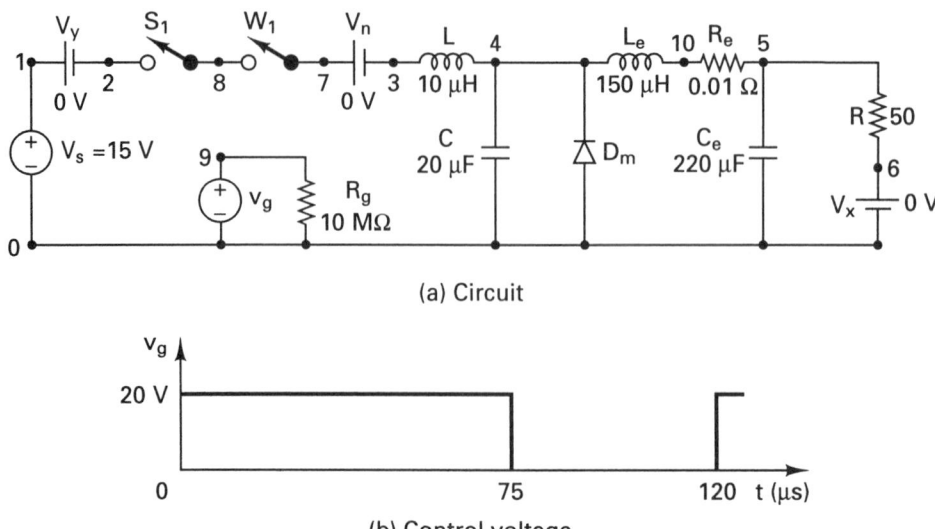

(a) Circuit

(b) Control voltage

Figure 13-9 Zero-current switching converter.

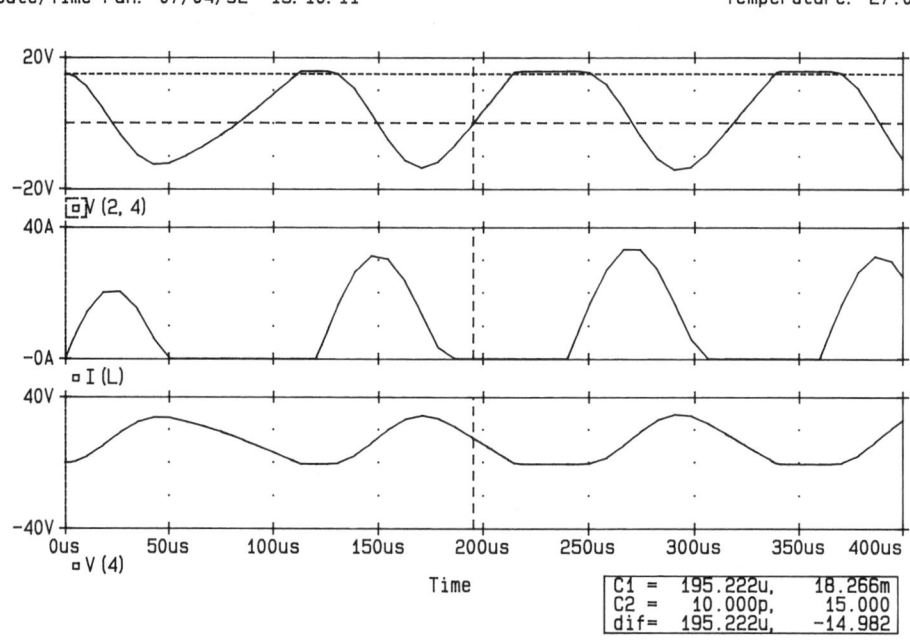

Figure 13-10 Plots for Example 13-4.

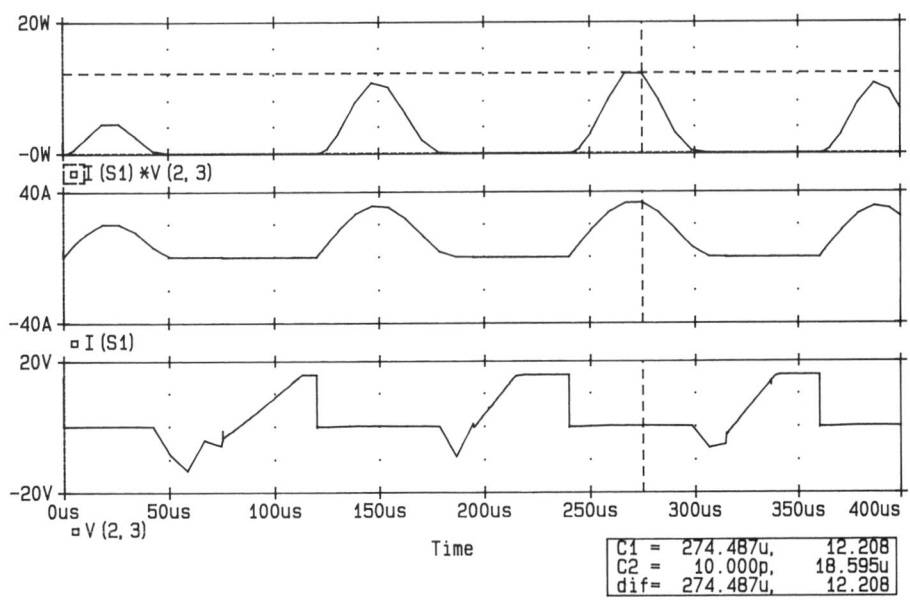

Figure 13-11 Plots for Example 13-4.

Example 13-4

Repeat Example 13-3 for the converter shown in Fig. 13-9(a). The control voltage is shown in Fig. 13-9(b).

Solution   The circuit file is similar to that for Example 13-3, except that the capacitor C is connected across the diode D_m. The statement for C is changed to

```
C    4    0    20UF        ; No initial condition
```

(a) The PSpice plots of the instantaneous capacitor voltage V(2,4), the inductor current I(L), and the diode voltage V(4) are shown in Fig. 13-10.

(b) The instantaneous voltage, current, and power of the switch are shown in Fig. 13-11. The instantaneous power, which is the same as Example 13-3, is 12.2 W.

13-4 ZERO-VOLTAGE SWITCHING CONVERTER

A power device of a zero-voltage switching converter (ZVSC) is turned on when its voltage becomes zero due to the resonant oscillation. At zero voltage, the resonant current becomes maximum. The device remains on and supplies the load current. The switching period must be long enough to complete the resonant oscillation.

Example 13-5

A zero-voltage switching converter is shown in Fig. 13-12(a). The control voltage is shown in Fig. 13-12(b). The dc input voltage is $V_s = 15$ V. The switching frequency is $f_s = 2.5$ kHz. Use PSpice to plot the instantaneous capacitor voltage v_c, the inductor current i_L, the diode current v_{Dm}, and the load voltage v_L. Use a voltage-controlled switch to perform the switching action.

Solution $V_s = 1/2.5$ kHz $= 400$ μs. The list of the circuit file is as follows:

■ ■

Example 13-5   Zero-voltage switching converter

SOURCE ■	VS	1	0	DC	15V	
	Rg	9	0	10MEG		
	Vg	9	0	PULSE (0 20V 0S 1NS 1NS 100US 400US)		
CIRCUIT ■ ■	VY	1	2	DC	0V	; Voltage source to measure supply current
	VX	6	0	DC	0V	; Measures load current
	R	5	6	50		
	LE	4	10	150UH		
	RE	10	5	0.01		
	CE	5	0	220UF		
	L	3	4	20UH		
	C	2	3	20UF		
	D1	3	2	DMOD	; Diode	
	DM	0	4	DMOD	; Diode	

.MODEL DMOD D(IS=2.2E-15 BV=1200V CJO=0 TT=0) ; Diode model parameters

	S1	2	3	9	0	SMOD ; Voltage-controlled switch

.MODEL SMOD VSWITCH (RON=0.01 ROFF=10E+6 VON=10V VOFF=5V)

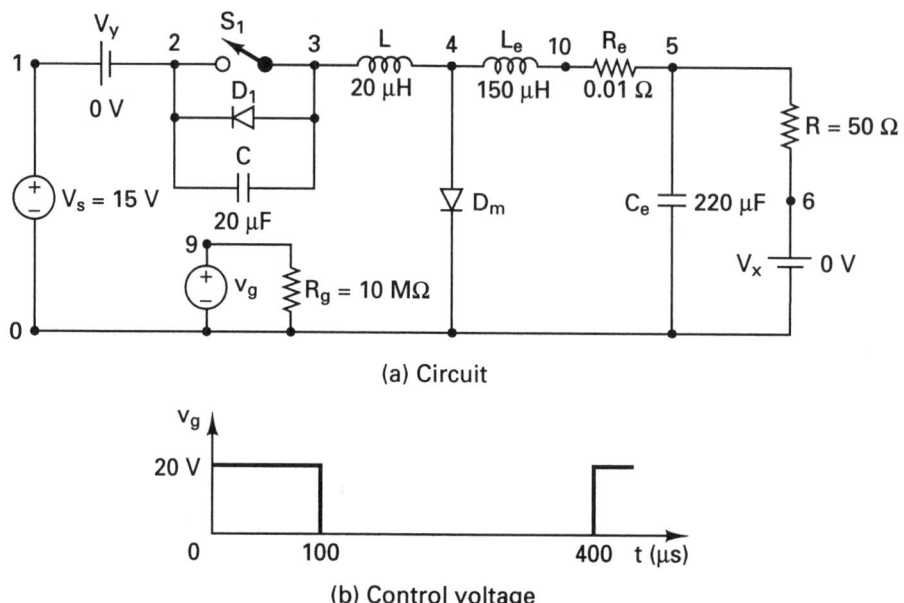

(a) Circuit

(b) Control voltage

Figure 13-12   Zero voltage switching converter.

```
ANALYSIS  ■ ■ ■ .TRAN    1US   1.6MS   0.401MS      ; Transient analysis
               .PROBE                               ; Graphics post-processor
               .OPTIONS ABSTOL = 1.00N  RELTOL = 0.1  VNTOL = 0.1  ITL5=50000
          .END
          ■ ■ ■ ■ ■ ■ ■ ■ ■ ■ ■ ■ ■ ■ ■ ■ ■ ■ ■ ■ ■ ■ ■ ■ ■ ■ ■ ■ ■ ■ ■ ■ ■ ■ ■ ■ ■ ■ ■ ■ ■ ■ ■ ■
```

The PSpice plots of the instantaneous capacitor voltage V(2,4), the inductor current I(L), the diode voltage V(4), and the load voltage V(5) are shown in Fig. 13-13.

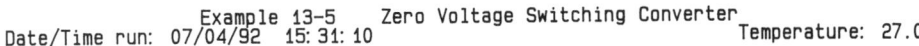

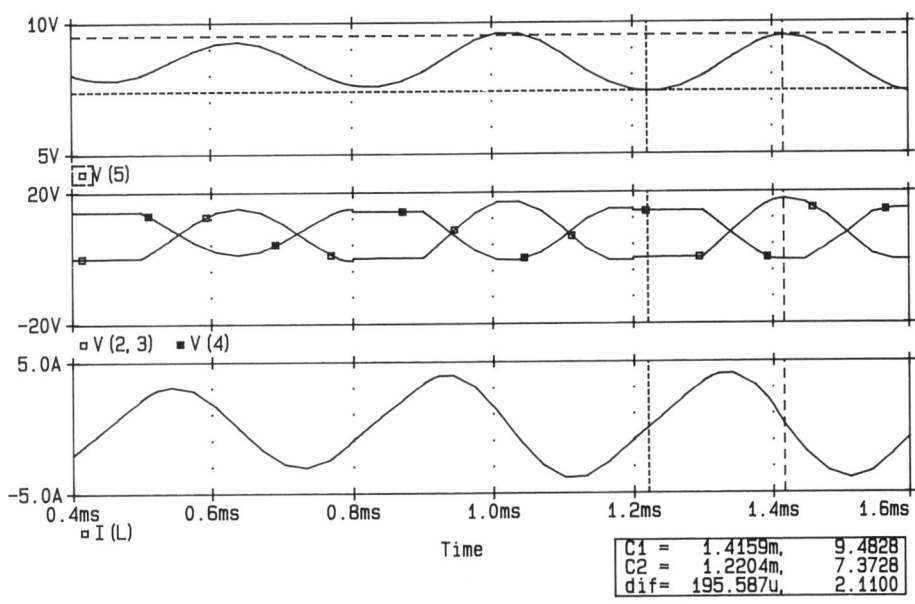

Figure 13-13 Plots for Example 13-5.

13-5 LABORATORY EXPERIMENTS

It is possible to develop many experiments to demonstrate the operation and characteristics of inverters. The following experiments are suggested:

Single-phase half-bridge resonant inverter

Single-phase full-bridge resonant inverter

Push–pull inverter

Parallel resonant inverter

Zero-current switching converter

Zero-voltage switching converter

13-5.1 Experiment RI-1

SINGLE-PHASE HALF-BRIDGE RESONANT INVERTER

Objective   The objective is to study the operation and characteristics of a single-phase half-bridge resonant (transistor) inverter.

Applications   The resonant inverter is used to control power flow in many applications (e.g., ac and dc power supplies and input stages of other converters).

Textbook   See Ref. 1, Sec. 11-2.

Apparatus   1. Two BJTs/MOSFETs with ratings of at least 50 A and 400 V, mounted on heat sinks

2. Two fast-recovery diodes with ratings of at least 50 A and 400 V, mounted on heat sinks

3. A firing pulse generator with isolating signals for gating transistors

4. An *RL* load

5. One dual-beam oscilloscope with floating or isolating probes

6. Ac and dc voltmeters and ammeters and one noninductive shunt

Warning   Before making any circuit connection, switch the dc power OFF. **Do not** switch the power ON unless the circuit is checked and approved by your lab instructor. **Do not** touch the transistor/diode heat sinks, which are connected to live terminals.

Experimental procedure   1. Set up the circuit as shown in Fig. 13-14. Use an *RLC* load. Design suitable values of snubbers.

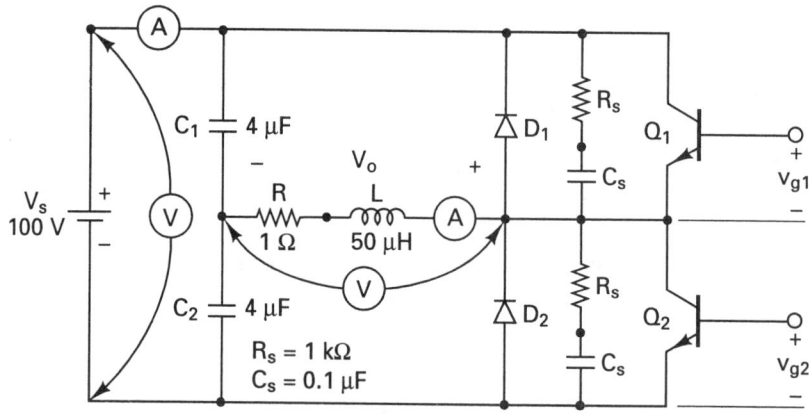

Figure 13-14   Single-phase half-bridge resonant inverter.

2. Connect the measuring instruments as required.

3. Connect the firing pulses to the appropriate transistors.

4. Set the duty cycle $k = 0.5$.

5. Observe and record the waveforms of the load voltage v_o and the load current i_o.

6. Measure the rms load voltage, the rms load current, the average input current, the average input voltage, and the total harmonic distortion of the output voltage and current.

7. Measure the conduction angles of the transistor Q_1 and diode D_1.

Report
1. Present all recorded waveforms and discuss all significant points.

2. Compare the waveforms generated by SPICE with the experimental results, and comment.

3. Compare the experimental results with the results predicted.

4. Discuss the advantages and disadvantages of this type of inverter.

5. Discuss the effects of the diodes on the performance of the inverter.

13-5.2 Experiment RI-2

SINGLE-PHASE FULL-BRIDGE RESONANT INVERTER

Objective   The objective is to study the operation and characteristics of a single-phase full-bridge resonant (transistor) inverter.

Applications   The resonant inverter is used to control power flow in many applications (e.g., high-frequency applications, ac and dc power supplies, and input stages of other converters).

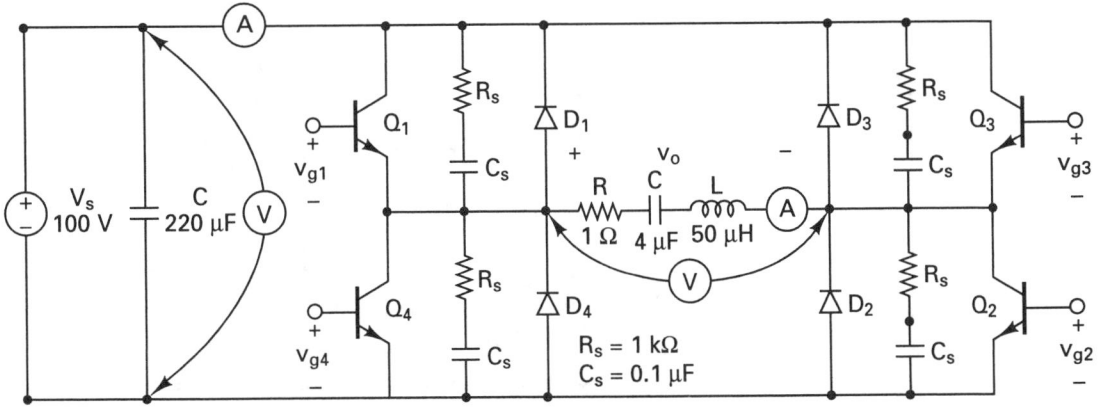

Figure 13-15   Single-phase full-bridge resonant inverter.

Textbook   See Ref. 1, Sec. 11-2.

Apparatus   See Experiment RI-1.

Warning   See Experiment RI-1.

Experimental procedure   Set up the circuit as shown in Fig. 13-15. Repeat the steps in Experiment RI-1.

Report   See Experiment RI-1.

13-5.3 Experiment RI-3

PUSH–PULL INVERTER

Objective   The objective is to study the operation and characteristics of a push-pull (transistor) inverter.

Applications   The push-pull inverter is used to control power flow in many applications (e.g., high-frequency applications, ac and dc power supplies, and input stages of other converters).

Textbook   See Ref. 1, Sec. 11-2.

Apparatus   See Experiment RI-1.

Warning   See Experiment RI-1.

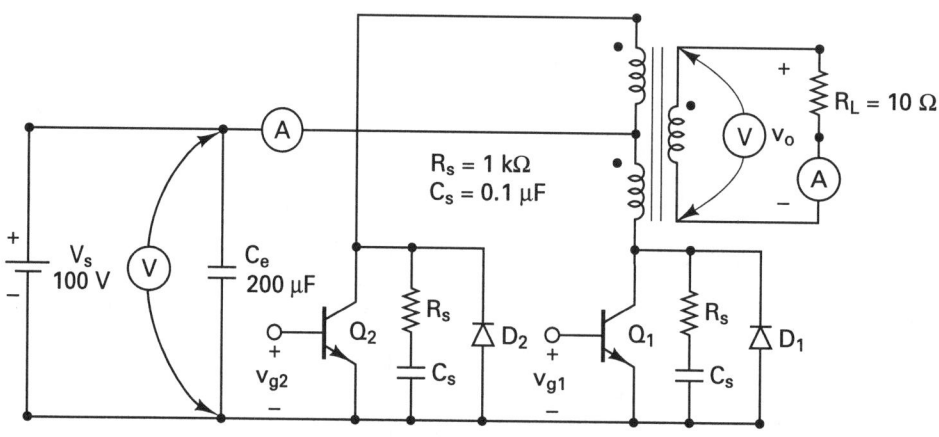

Figure 13-16   Push-pull inverter.

Experimental
procedure
Set up the circuit as shown in Fig. 13-16. Repeat the steps in Experiment RI-1.

Report   See Experiment RI-1.

13-5.4 Experiment RI-4
PARALLEL RESONANT INVERTER

Objective   The objective is to study the operation and characteristics of a single-phase push-pull parallel resonant (transistor) inverter.

Applications   The parallel resonant inverter is used to control power flow in many applications (e.g., high-frequency applications, ac and dc power supplies, and input stages of other converters).

Textbook   See Ref. 1, Sec. 11-3.

Apparatus   See Experiment RI-1.

Warning   See Experiment RI-1.

Experimental   Set up the circuit as shown in Fig. 13-17. Repeat the steps in Experiment RI-1.
procedure

Report   See Experiment RI-1.

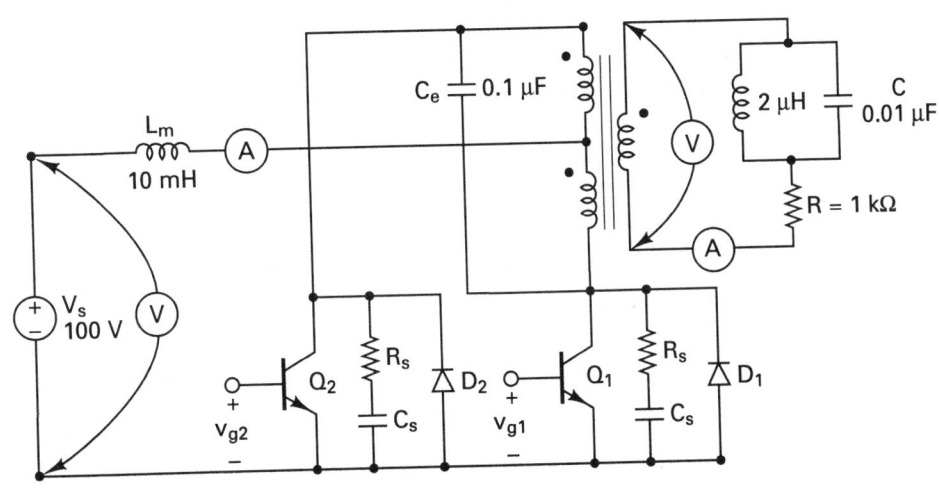

Figure 13-17 Parallel resonant inverter.

Objective The objective is to study the operation and characteristics of a zero-current switching converter (ZCSC).

Applications The zero-current switching converter is used to control power flow in many applications (e.g., ac and dc power supplies).

Textbook See Ref. 1, Sec. 11-6.

Apparatus
1. One BJT/MOSFET with ratings of at least 50 A and 400 V, mounted on a heat sink
2. One fast-recovery diode with ratings of at least 50 A and 400 V, mounted on a heat sink
3. A firing pulse generator with isolating signals for gating transistor
4. An R load, capacitors, and chokes
5. One dual-beam oscilloscope with floating or isolating probes
6. Ac and dc voltmeters and ammeters and one noninductive shunt

Warning See Experiment RI-1.

Experimental procedure
1. Set up the circuit as shown in Fig. 13-18.
2. Connect the measuring instruments as required.
3. Connect the firing pulses to the appropriate transistors.
4. Set the duty cycle $k = 0.5$ cycle.
5. Observe and record the waveforms of the load voltage v_o, the currents through C and L and Q_1, and the voltage across diode D_m.

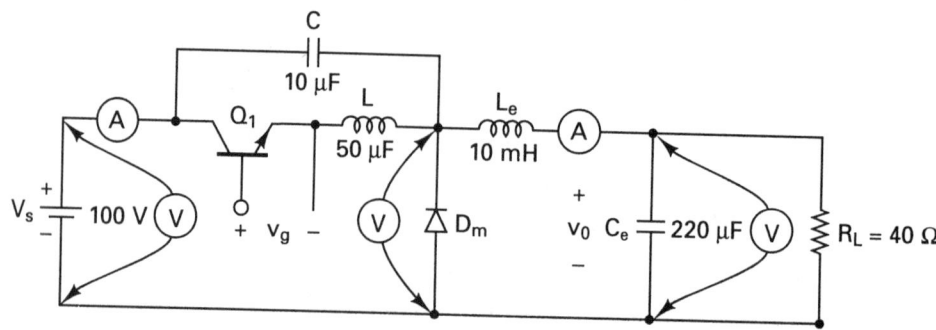

Figure 13-18 Zero-current switching converter.

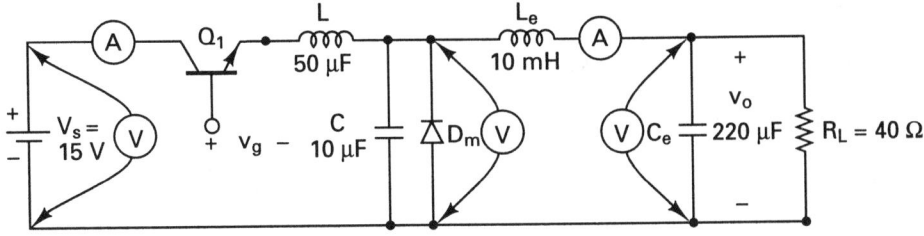

Figure 13-19 Zero-current switching converter.

6. Measure the average output voltage, the average output current, the average input current, and the average input voltage.

7. Measure the conduction angles of transistor Q_1.

8. Repeat steps 2 to 7 for the circuit of Fig. 13-19.

Report   See Experiment RI-1.

13-5.6 Experiment RI-6
ZERO-VOLTAGE SWITCHING CONVERTER

Objective   The objective is to study the operation and characteristics of a zero-voltage switching converter (ZVSC).

Applications   The zero-voltage switching converter is used to control power flow in many applications (e.g., ac and dc power supplies).

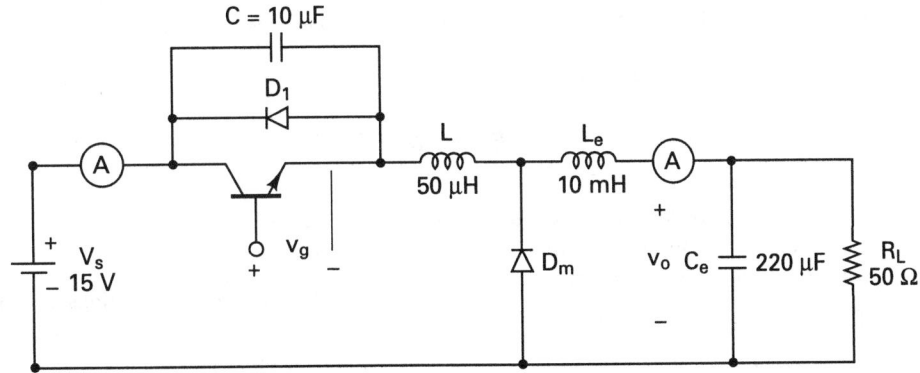

Figure 13-20 Zero-voltage switching converter.

SUMMARY

The statements for an ac thyristor are:

```
*     Subcircuit call for switched transistor model:
XT1    +N        −N        +NG         −NG         QM
       positive  negative  +control    −control    model
*                          voltage     voltage     name
*     Subcircuit call for PWM control:
XPWM   VR        VC        +NG         −NG         PWM
*      ref.      carrier   +control    −control    model
*      input     input     voltage     voltage     name
*     Subcircuit call for sinusoidal PWM control:
XSPWM  VR        VS        +NG         −NG         VC             SPWM
*      ref.      sine-wave +control    −control    rectified      model
*      input     input     voltage     voltage     carrier sine wave  name
```

SUGGESTED READING

1. M. R. Rashid, *Power Electronics: Circuits, Devices, and Applications*, 2nd ed. Englewood Cliffs, N.J.: Prentice Hall, 1993, Chap. 11.

2. N. Mohan, T. M. Undeland, and W. P. Robbins, *Power Electronics: Converters, Applications, and Design*. New York: Wiley, 1989.

DESIGN PROBLEMS

13-1. It is required to design the single-phase half-bridge resonant inverter of Fig. 13-14 with the following specifications:

Dc supply voltage, $V_s = 100$ V

Load resistance, $R = 1\ \Omega$

Load inductance, $L = 100\ \mu$H

Output frequency should be as high as possible

(a) Determine the ratings of all components and devices under worst-case conditions.

(b) Use SPICE to verify your design.

(c) Provide a cost estimate of the circuit.

13-2. It is required to design the single-phase full-bridge resonant inverter of Fig. 13-15 with the following specifications:

Dc supply voltage, $V_s = 100$ V

Load resistance, $R = 1$ Ω

Load inductance, $L = 50$ μH

Load capacitance, $C = 4$ μH

Output frequency should be as high as possible

(a) Determine the ratings of all components and devices under worst-case conditions.
(b) Use SPICE to verify your design.
(c) Provide a cost estimate of the circuit.

13-3. It is required to design the push-pull inverter of Fig. 13-16 with the following specifications:

Dc supply voltage, $V_s = 100$ V

Load resistance, $R = 10$ Ω

Peak value of load voltage, $V_p = 140$ V

Output frequency, $f_o = 1$ kHz

(a) Determine the ratings of all components and devices under worst-case conditions.
(b) Use SPICE to verify your design.
(c) Provide a cost estimate of the circuit.

13-4. It is required to design the parallel resonant inverter of Fig. 13-17 with the following specifications:

Dc supply voltage, $V_s = 100$ V

Load resistance, $R = 1$ kΩ

Load inductance, $L = 2$ μH

Load capacitance, $C = 0.1$ μF

Peak value of load voltage, $V_p = 140$ V

(a) Determine the ratings of all components and devices under worst-case conditions.

(b) Use SPICE to verify your design.
(c) Provide a cost estimate of the circuit.

13-5. It is required to design the zero-current switching converter of Fig. 13-18 with the following specifications:

Dc supply voltage, $V_s = 20$ V

Load resistance, $R = 100$ Ω

Average output voltage,
$V_{(dc)} = 10$ V with $\pm 5\%$ ripple

(a) Determine the ratings of all components and devices under worst-case conditions.
(b) Use SPICE to verify your design.
(c) Provide a cost estimate of the circuit.

13-6. It is required to design the zero-current switching converter of Fig. 13-19 with the following specifications:

Dc supply voltage, $V_s = 20$ V

Load resistance, $R = 100$ Ω

Average output voltage,
$V_{(dc)} = 10$ V with $\pm 5\%$ ripple

(a) Determine the ratings of all components and devices under worst-case conditions.
(b) Use SPICE to verify your design.
(c) Provide a cost estimate of the circuit.

13-7. It is required to design the zero-voltage switching converter of Fig. 13-20 with the following specifications:

Dc supply voltage, $V_s = 20$ V

Load resistance, $R = 100$ Ω

Average output voltage,
$V_{(dc)} = 10$ V with $\pm 5\%$ ripple

(a) Determine the ratings of all components and devices under worst-case conditions.
(b) Use SPICE to verify your design.
(c) Provide a cost estimate of the circuit.

Chapter 14

∎∎∎∎∎∎∎∎

Control Applications

14-1 INTRODUCTION

In practical applications, power converters are normally operated under closed-loop control, which requires comparing the output desired with the actual output. The control implementation requires summing, differentiating, and integrating signals to obtain the control strategy desired. SPICE can be used in modeling:

    Op-amp circuits
    Control systems
    Signal conditioning

14-2 OP-AMP CIRCUITS

An operational amplifier (op-amp) may be modeled as a linear amplifier to simplify the design and analysis of op-circuits. The linear models give reasonable results, especially for determining the approximate design values of op-amp circuits. The simulation of the actual behavior of op-amps is required in many applications to obtain accurate response of electronic circuits. An op-amp can be simulated from the circuit arrangement of the particular type of op-amp. The μ741 type of op-amp consists of 24 transistors and is beyond the capability of the student (or demo) version of PSpice. However, a macromodel, which is a simplified version of the op-amp and requires only two transistors, is quite accurate for many applications and can be simulated as a subcircuit or library file. Some op-amp manufacturers supply macromodels of their op-amps [1]. In the absence of a complex op-amp model, the characteristic of op-amp circuits may be represented approximately by one of the following models:

Dc linear models
Ac linear model
Nonlinear macromodel

14-2.1 DC Linear Models

An op-amp may be modeled as a voltage-controlled voltage source, as shown in Fig. 14-1(a). The input resistance is high, typically 2 MΩ, and the output resistance is very low, typically 75 Ω. For an ideal op-amp, the model of Fig. 14-1(a) can be reduced to Fig. 14-1(b). These models do not take into account the saturation effect and slew rate that exist in practical op-amps. The voltage gain is also assumed independent of the frequency, but in practical op-amps the gain falls with the frequency. These simple models are normally suitable for dc or low-frequency applications.

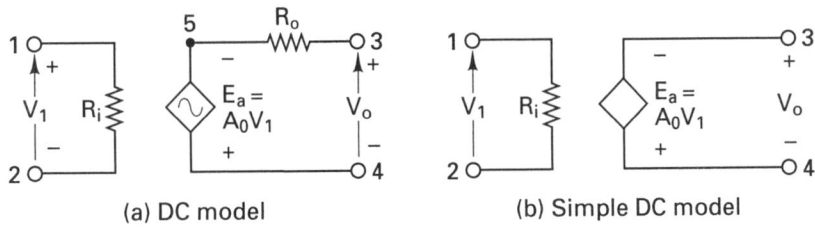

(a) DC model   (b) Simple DC model

Figure 14-1 Dc linear models.

14-2.2 AC Linear Model

The frequency response of an op-amp can be approximated by a single-break frequency, as shown in Fig. 14-2(a). This characteristic can be modeled by the circuit of Fig. 14-2(b), a high-frequency op-amp model. If an op-amp has more than one break frequency, it can be represented by using as many capacitors as the number of breaks. R_{in} is the input resistance, and R_o is the output resistance.

The dependent sources in the op-amp model of Fig. 14-2(b) have a common node. Without this, PSpice will give an error message because there is no dc path from the nodes of the dependent current source. The common node could be in either the input stage or output stage. This model does not take into account the saturation effect, and is suitable only if the op-amp operates within the linear region.

The output voltage can be expressed as

$$V_o = A_o V_2 = \frac{A_o V_i}{1 + R_1 C_1 s}$$

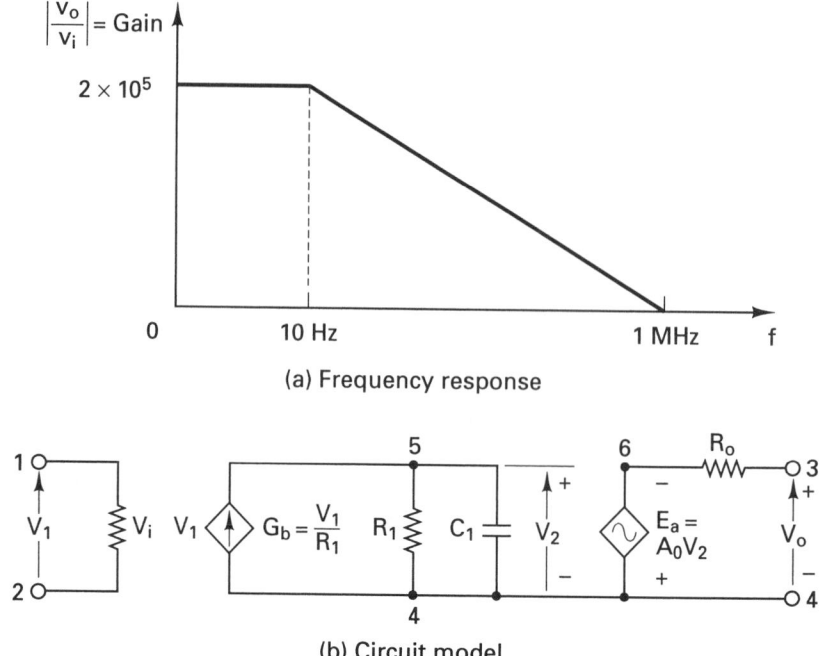

(a) Frequency response

(b) Circuit model

Figure 14-2 Ac linear-model with single break frequency.

Substituting $s = j2\pi f$ yields

$$V_o = \frac{A_o V_i}{1 + j2\pi f R_1 C_1} = \frac{A_o V_i}{1 + j(f/f_b)}$$

where $f_b = 1/(2\pi R_1 C_1)$ is called the *break frequency* (in hertz) and A_o is the *large-signal* (or *dc*) *gain* of the op-amp. Thus the open-loop voltage gain is

$$A(f) = \frac{V_o}{V_i} = -\frac{A_o}{1 + j(f/f_b)}$$

For $\mu741$ op-amps, $f_b = 10$ Hz, $A_o = 2 \times 10^5$, $R_i = 2$ MΩ, and $R_o = 75$ Ω. Letting $R_1 = 10$ kΩ, $C_1 = 1/(2\pi \times 10 \times 10 \times 10^3) = 1.15619$ μF.

14-2.3 Nonlinear Macromodel

The circuit arrangement of the op-amp macromodel is shown in Fig. 14-3 [2,3]. The macromodel can be used as a subcircuit with a .SUBCKT command. However, if an op-amp is used in various circuits, it is convenient to have the macromodel as a library file (i.e., EVAL.LIB), and it is not necessary to type the statements of the macromodel in every circuit where the macromodel is employed. The library file EVAL.LIB that comes with the student version of PSpice has macromodels for op-amps, comparators, diodes, MOSFETs, BJTs, and SCRs. The professional version of PSpice supports library files for many de-

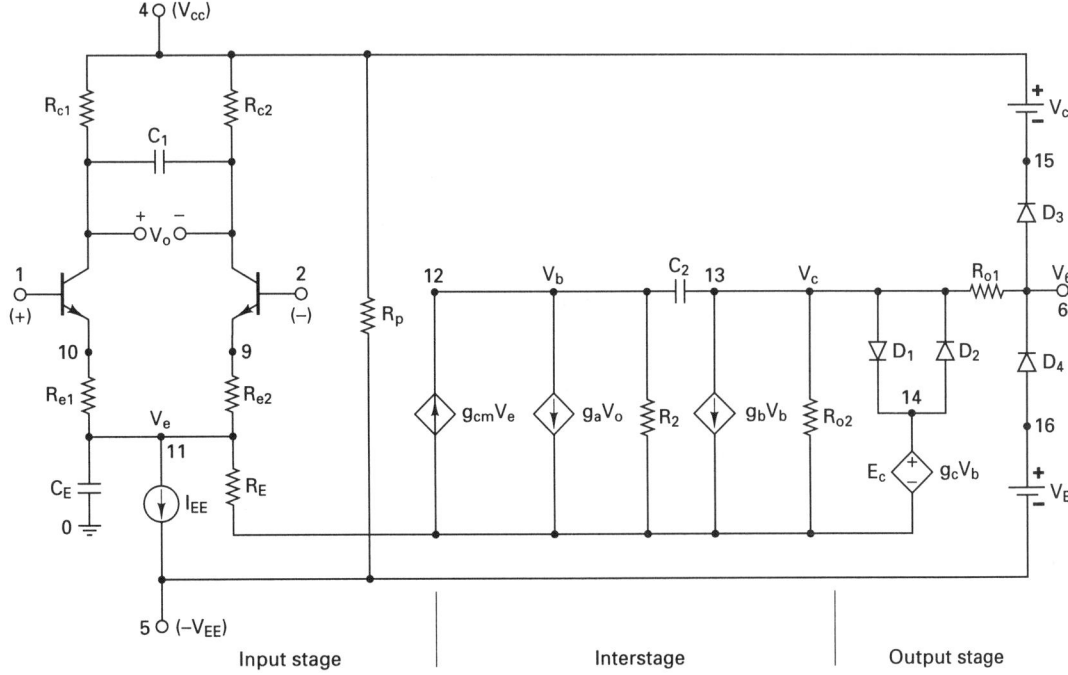

Figure 14-3 Circuit diagram of op-amp macromodel.

vices. Check the name of the current library file by listing the files of the PSPice programs (by the DOS command DIR).

The macromodel of the μ741 op-amp is simulated at room temperature. The library file EVAL.LIB contains the op-amp macromodel model as a subcircuit definition UA741 with a set of .MODEL statements. This op-amp model contains nominal, not worst-case devices, and does not consider the effects of temperature.

The listing of the subcircuit UA741 in library file EVAL.LIB is as follows:

```
* Subcircuit for μ741 op-amp
* connections:    noninverting input
*                 :   inverting input
*                 :   :
*                 :   :   positive power supply
*                 :   :   :   negative power supply
*                 :   :   :   :   output
*                 :   :   :   :   :
.SUBCKT   UA741   1   2   4   5   6
*                 Vi+ Vi- Vp+ Vp- Vout
Q1    7   1   10   UA741QA
Q2    8   2   9    UA741QB
RC1   4   7   5.305165D+03
```

```
RC2     4    8         5.305165D+03
C1      7    8         5.459553D-12
RE1    10   11         2.151297D+03
RE2     9   11         2.151297D+03
IEE    11    5         1.666000D-05
CE     11    0         3.000000D-12
RE     11    0         1.200480D+07
GCM     0   12        11    0     5.960753D-09
GA     12    0         8    7     1.884955D-04
R2     12    0         1.000000D+05
C2     12   13         3.000000D-11
GB     13    0        12    0     2.357851D+02
RO2    13    0         4.500000D+01
D1     13   14         UA741DA
D2     14   13         UA741DA
EC     14    0         6    0     1.0
RO1    13    6         3.000000D+01
D3      6   15         UA741DB
VC      4   15         2.803238D+00
D4     16    6         UA741DB
VE     16    5         2.803238D+00
RP      4    5         18.16D+03
*   Models for diodes and transistors:
.MODEL    UA741DA    D  (IS=9.762287D-11)
.MODEL    UA741DB    D  (IS=8.000000D-16)
.MODEL    UA741QA    NPN  (IS=8.000000D-16 BF=9.166667D+01)
.MODEL    UA741QB    NPN  (IS=8.309478D-16 BF=1.178571D+02)
.ENDS     UA741                         ; End of subcircuit definition
```

Example 14-1

An inverting amplifier is shown in Fig. 14-4. Use PSpice to plot the dc transfer characteristic if the input is varied from -1 to $+1$ V with an 0.2-V increment. (a) Use the op-amp of Fig. 14-1(a) as a subcircuit; its parameters are $A_o = 2 \times 10^5$, $R_i = 2\,\text{M}\Omega$, and $R_o = 75\,\Omega$. (b) Use the op-amp of Fig. 14-2 as a subcircuit; its parameters are $R_i = 2\,\text{M}\Omega$, $R_o = 75\,\Omega$, $C_1 = 1.5619\,\mu\text{F}$, and $R_1 = 10\,\text{k}\Omega$. (c) Use the macromodel of Fig. 10-3 for the UA741.

Solution   The list of the subcircuit OPAMP-DC for Fig. 14-1(a) is as follows:

```
*    Subcircuit definition for OPAMP-DC:
.SUBCKT     OPAMP-DC         1     2     3     4
*          model name       Vi-   Vi+   Vo+   Vo-
RIN    1    2    2MEG
RO     5    3    75
EA     5    4    2    1    2E+5       ; Voltage-controlled voltage source
*   End of subcircuit definition:
.ENDS    OPAMP-DC                     ; End of subcircuit definition
```

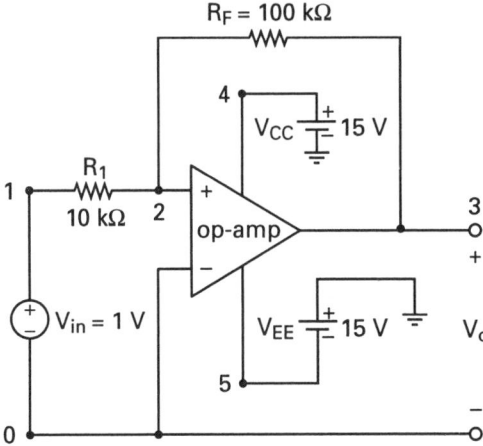

Figure 14-4 Inverting amplifier.

The list of the subcircuit OPAMP-AC for Fig. 14-2 is as follows:

```
*     Subcircuit definition for OPAMP-AC:
.SUBCKT     OPAMP-AC        1     2     3     4
*           model name      Vi-   Vi+   Vo+   Vo-
RI    1     2     2MEG
GB    4     5     1     2     0.1M   ; Voltage-controlled current source
R1    5     4     10K
C1    5     4     1.5619UF
EA    4     6     5     4     2E+5   ; Voltage-controlled voltage source
RO    6     3     75
.ENDS     OPAMP-AC                  ; End of subcircuit definition
```

(a) The list of the circuit file with the dc op-amp model is as follows:

■■

Example 14-1(a)   Inverting amplifier with dc op-amp model

SOURCE ■ VIN   1   0   DC   1V

CIRCUIT ■■ R1   1   2   10K

          RF   2   3   100K

          * Calling subcircuit OPAMP-DC:

          XA1   2   0   3   0   OPAMP-DC

          *   Vi-   Vi+   Vo+   Vo-   model name

          * Subcircuit definition OPAMP-DC <u>must</u> be inserted.

ANALYSIS ■■■ .DC   VIN   -1V   1V   0.1V   ; DC sweep

         .PROBE   ; Graphics post-processor

 .END

■■

The plot of the transfer characteristic using the dc op-amp model of Fig. 14-1(a) is shown in Fig. 14-5. The gain expected is $R_F/R_1 = 100/10 = 10$.

(b) The list of the circuit file with the ac op-amp model is as follows:

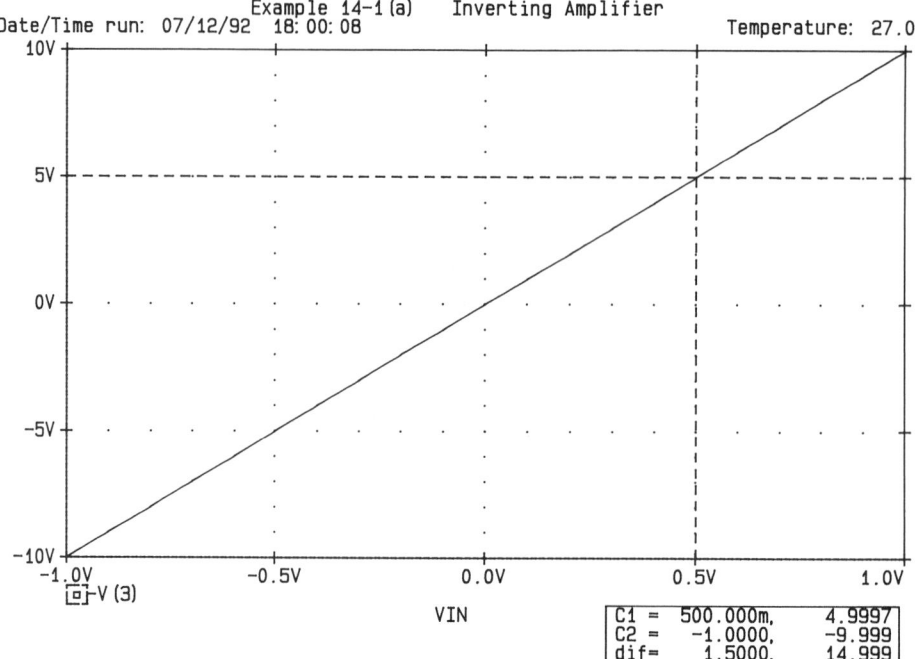

C1 = 500.000m, 4.9997
C2 = −1.0000, −9.999
dif=   1.5000, 14.999

Figure 14-5 Plot for Example 14-1 with dc op-amp model.

■ ■

Example 14-1(b)   Inverting amplifier with ac op-amp model

```
SOURCE   ■ VIN     1     0     DC     1V
CIRCUIT  ■ ■ R1    1     2     10K
            RF     2     3     100K
            *   Calling subcircuit OPAMP-AC:
            XA1    2     0     3     0     OPAMP-AC
            *      Vi−   Vi+   Vo+   Vo−   model name
            *   Subcircuit definition OPAMP-AC must be inserted.
ANALYSIS ■ ■ ■ .DC  VIN   −1V   1V   0.1V        ; Dc sweep
            .PROBE                                ; Graphics post-processor
         .END
```

■ ■

The plot of the transfer characteristic using the ac op-amp model of Fig. 14-2 is shown in Fig. 14-6. No difference is expected with a dc signal. However, the output will be dependent on the frequency of the input signal.

(c) The list of the circuit file with the UA741 macromodel is as follows:

■ ■

Example 14-1(c)   Inverting amplifier with μA741 macromodel

```
SOURCE   ■ VIN    1     0     DC     1V
CIRCUIT  ■ ■ R1   1     2     10K
            RF    2     3     100K
```

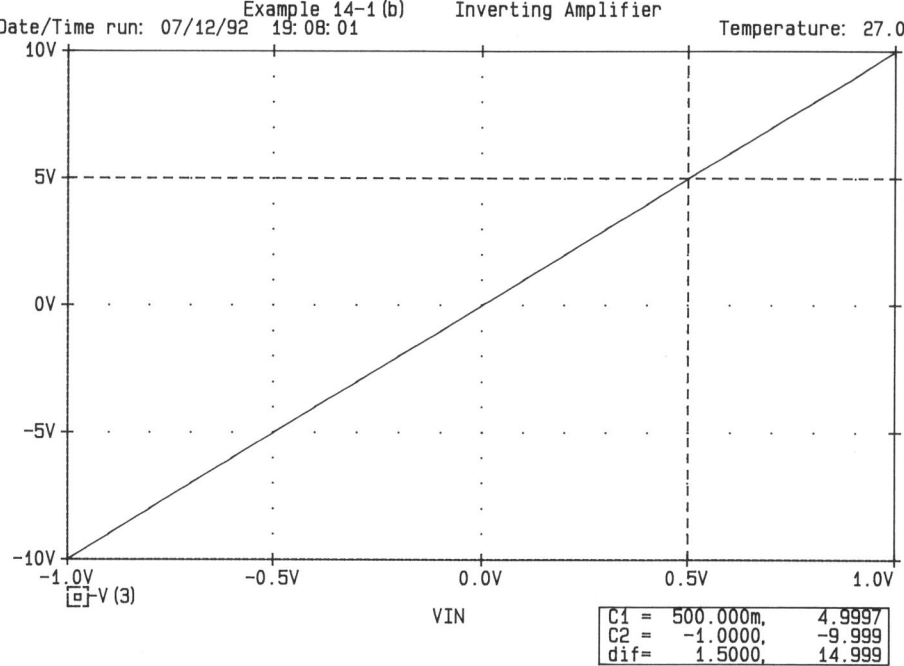

Figure 14-6   Plot for Example 14-1 with ac op-amp model.

```
VCC      4    0    DC    15V
VEE      0    5    DC    15V
*  Calling subcircuit op-amp UA741:
XA1      2    0    4    5    3    UA741
*        Vi-  Vi+  Vp+  Vp-  Vout  model name
*  Subcircuit definition UA741 must be inserted.
```
ANALYSIS ■ ■ ■ `.DC   VIN   -1V   1V   0.1V        ; Dc sweep`
`.PROBE                         ; Graphics post-processor`
`.END`

The plot of the transfer characteristic using the UA741 macromodel of Fig. 14-3 is shown in Fig. 14-7. A macromodel affects the characteristics because it has saturation limits. The voltage gain is 4.98.

Example 14-2

An op-amp integrator circuit is shown in Fig. 14-8(a). The input voltage is shown in Fig. 14-8(b). Use PSpice to plot the transient response of the output voltage for a duration of 0 to 4 ms in steps of 50 μs. Use the op-amp model OPAMP-DC.
Solution   The list of the circuit file with the op-amp dc model is as follows:

Example 14-2   Integrator circuit with op-amp dc model
SOURCE ■ `VIN   1    0    PULSE (-1V  1V  1MS  1NS  1NS  1MS  2MS) ; Pulse waveform`

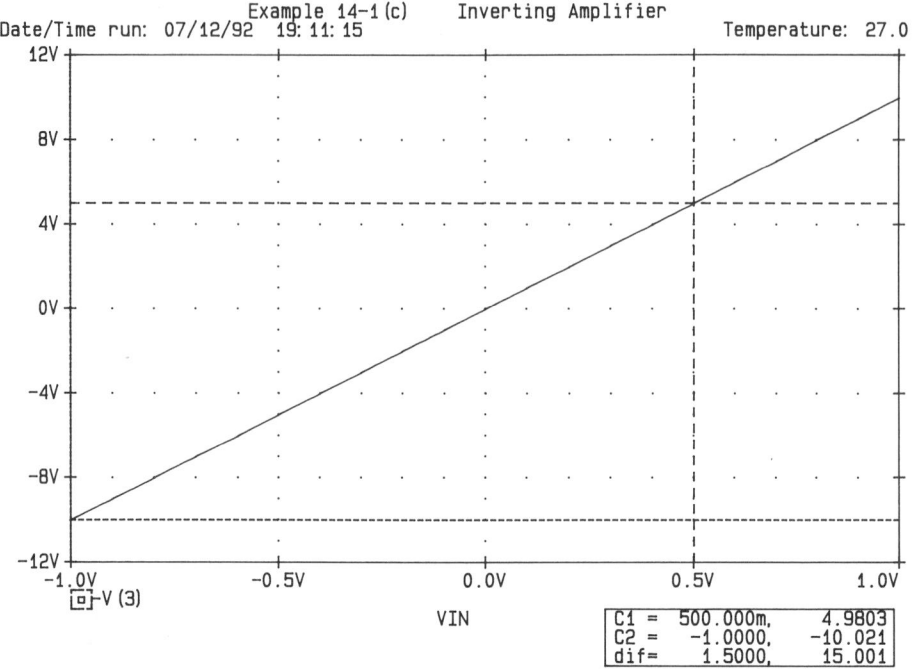

	C1 =	500.000m,	4.9803
	C2 =	−1.0000,	−10.021
	dif=	1.5000,	15.001

Figure 14-7   Plot for Example 14-1 with the UA741 macromodel.

```
CIRCUIT  ■ ■ R1      1    2    2.5K
           RF      2    3    1MEG
           C1      2    3    0.1UF   IC=0V      ; Set initial condition
           *  Calling subcircuit OPAMP-DC:
           XA1     2    0    3    0      OPAMP-DC
           *       Vi-   Vi+   Vo+   Vo-   model name
           *  Subcircuit definition OPAMP-DC must be inserted.
ANALYSIS ■ ■ ■ .TRAN   10US   4MS    UIC ; Use initial condition in transient analysis
           .PLOT    TRAN   V(3)  V(1)           ; Prints on the output file
           .PROBE                               ; Graphics post-processor
         .END
         ■ ■ ■ ■ ■ ■ ■ ■ ■ ■ ■ ■ ■ ■ ■ ■ ■ ■ ■ ■ ■ ■ ■ ■ ■ ■ ■ ■ ■ ■ ■ ■ ■ ■ ■ ■ ■ ■ ■ ■ ■ ■ ■
```

The plot of the transient response using the op-amp dc model is shown in Fig. 14-9. The output voltage is triangular in response to a square-wave input. Equating the areas under two curves, the expected height h is given by 0.5×1 ms $h = (1\ \text{V} + 1\ \text{V}) \times 1$ ms, or $h = 4$ V, which is verified by SPICE simulation.

Example 14-3

A practical differentiator circuit is shown in Fig. 14-10(a). The input voltage is shown in Fig. 14-10(b). Use PSpice to plot the transient response of output voltage for a duration of 0 to 4 ms in steps of 50 μs. Use the op-amp model OPAMP-DC.

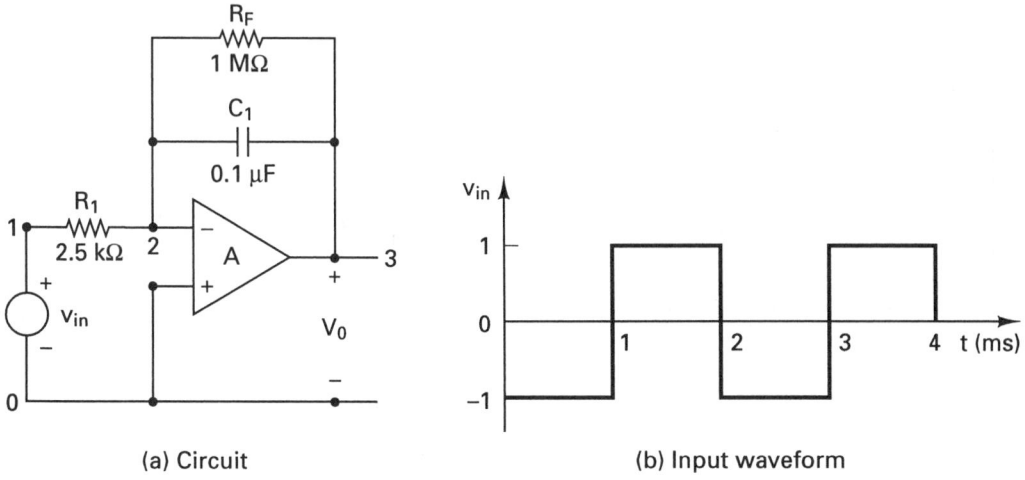

(a) Circuit   (b) Input waveform

Figure 14-8 Integrator circuit.

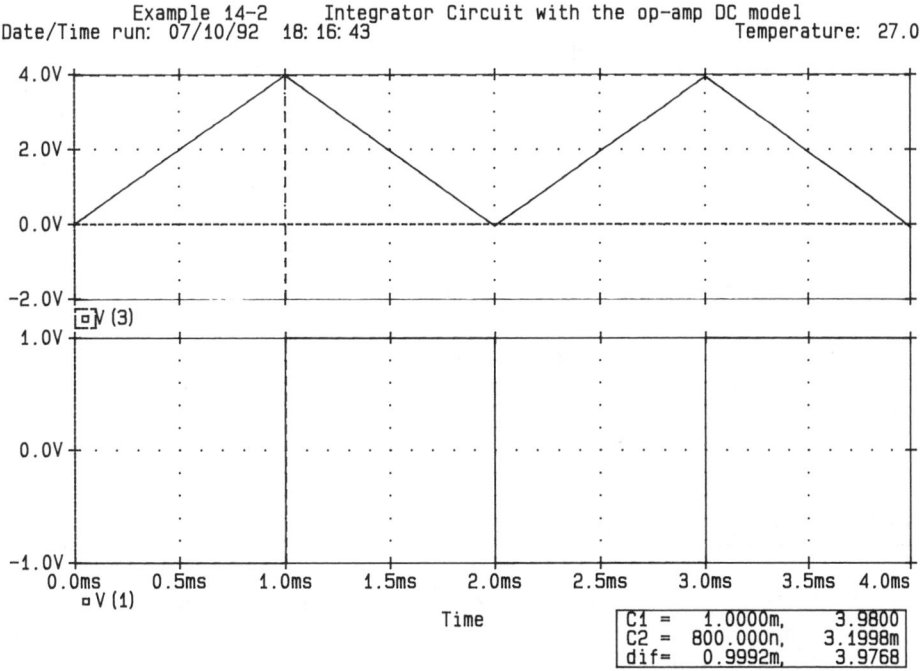

Figure 14-9 Transient response for Example 14-2 with the op-amp dc model.

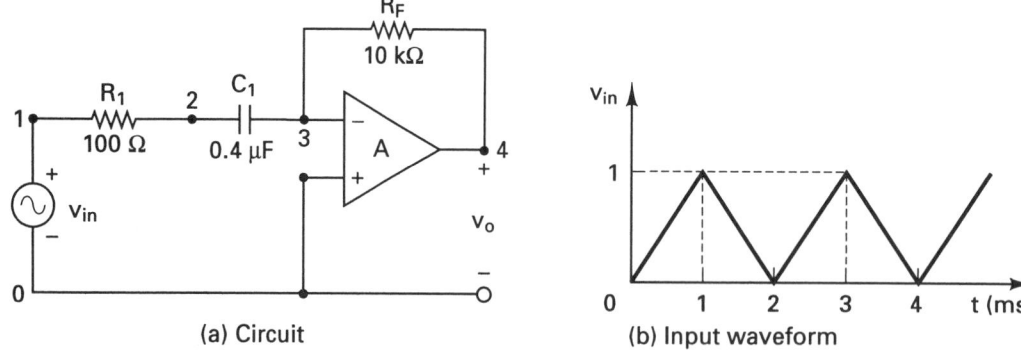

(a) Circuit　　　　　　　　(b) Input waveform

Figure 14-10　Differentiator circuit.

Solution　The list of the circuit file is as follows:

■ ■

Example 14-3　Differentiator circuit

SOURCE	■ VIN	1	0	PULSE (0 1V 0 1MS 1MS 1NS 2MS) ; Pulse waveform
CIRCUIT	■ ■ R1	1	2	100
	C1	2	3	0.4UF
	RF	3	4	10K

```
        *   Calling subcircuit OPAMP-DC:
        XA1      3      0      4      0      OPAMP-DC
        *        Vi-    Vi+    Vo+    Vo-    model name
        *   Subcircuit definition OPAMP-DC must be inserted.
```

ANALYSIS	■ ■ ■ .TRAN	10US 4MS	; Transient analysis
	.PLOT	TRAN V(4) V(1)	; Prints on the output file
	.PROBE		; Graphics post-processor

```
.END
```
■ ■

　　　　The plot of the transient response is shown in Fig. 14-11. The output voltage is a square wave in response to a triangular input. The time constant of the circuit limits the sharp rise and fall of the output voltage.

14-3 CONTROL SYSTEMS

The simulation of control systems requires integrators, multipliers, summing amplifiers, and function generation. These features can easily be simulated by PSpice. Additional PSpice features, such as Polynomial, Table, Frequency, Laplace, Parameter, Value, and Step, make PSpice a versatile tool to simulate complex control system.

Example 14-4

A unity feedback control system is shown in Fig. 14-12. The reference input v_r is a step voltage of 1 V. Use PSpice to plot the transient response of the output voltage

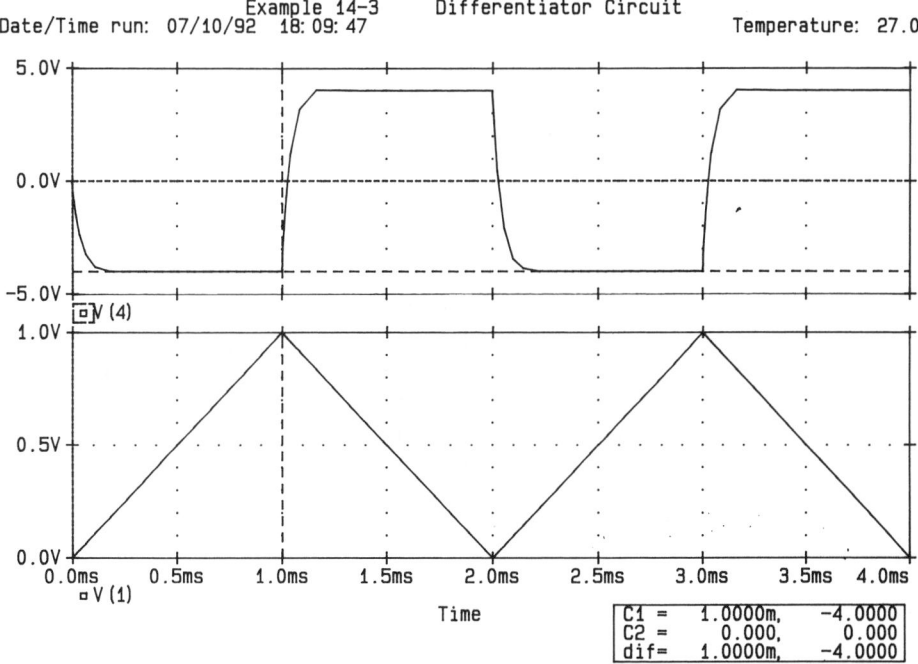

Figure 14-11 Transient response for Example 14-3.

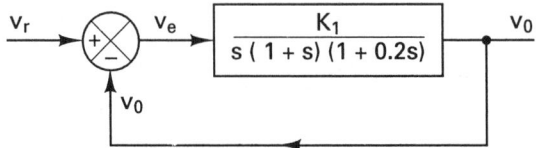

Figure 14-12 Unity feedback control system.

v_o for a duration of 0 to 10 s in steps of 10 ms. The gain K is to be varied from 0.5 to 2 with an increment of 0.5. Assume that all initial conditions are zero.

Solution The relations among V_r, V_o, and V_e, in Laplace's s domain, are

$$V_e(s) = V_r(s) - V_o(s)$$

$$\frac{V_o(s)}{V_e(s)} = \frac{K}{s(1 + s)(1 + 0.2s)}$$

which gives

$$[s(1 + s)(1 + 0.2s)]V_o(s) = KV_e(s)$$

$$[s + 1.2s^2 + 0.2s^3]V_o(s) = KV_e(s)$$

which can be written in the time domain as

$$0.2\frac{d^3v_o}{dt^3} + 1.2\frac{d^2v_0}{dt^2} + \frac{dv_o}{dt} = Kv_e = K(v_r - v_o) \qquad (14\text{-}1)$$

Dividing both sides by 0.2 gives

$$\frac{d^3v_o}{dt^3} + 6\frac{d^2v_o}{dt^2} + 5\frac{dv_o}{dt} = 5Kv_e$$

where $v_e = v_r - v_o$, or

$$\frac{d^3v_o}{dt^3} = -6\frac{d^2v_o}{dt^2} - 5\frac{dv_o}{dt} + 5Kv_e \qquad (14\text{-}2)$$

which can be denoted by

$$\dddot{v}_o = -6\ddot{v}_o - 5\dot{v}_o + 5Kv_e \qquad (14\text{-}3)$$

Integrating the third derivative three times should yield the second derivative of v_o, the first derivative of v_o, and the output v_o. The third derivative on the left-hand side of Eq. (14-3) must equal the sum of the terms on the right-hand side. The circuit for PSpice simulation of Eq. (14-2) is shown in Fig. 14-13. If $R_1 = R_2 = R_3 = 1 \text{ k}\Omega$ and $C_1 = C_2 = C_3 = 0.001$ F, the time constants of the integrators are equal and $\tau = \tau_1 = \tau_2 = \tau_3 = R_1C_1 = 1 \text{ k}\Omega \times 0.001 = 1$ s.

The gain, which is

$$K = \frac{R_F}{R} \qquad (14\text{-}4)$$

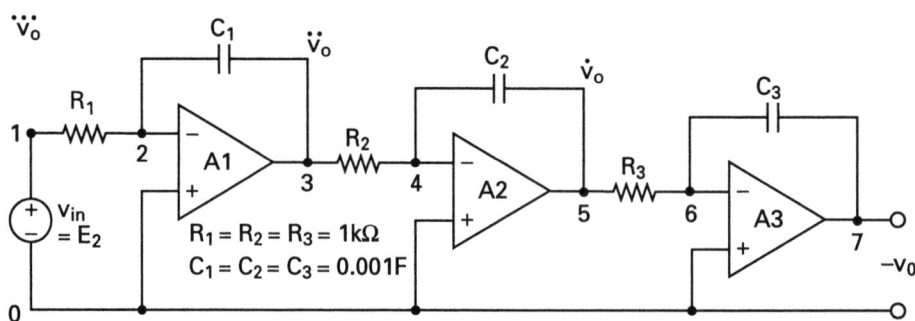

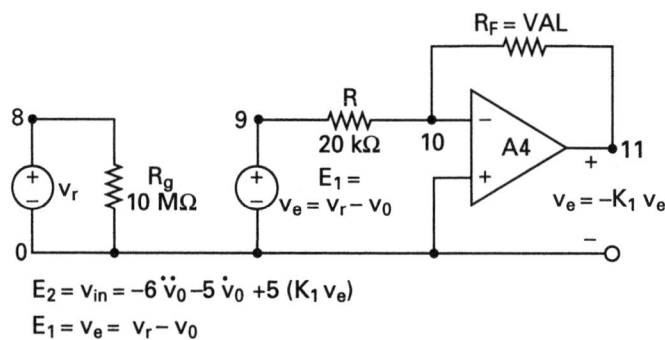

$$E_2 = v_{in} = -6\,\ddot{v}_0 - 5\,\dot{v}_0 + 5\,(K_1\,v_e)$$
$$E_1 = v_e = v_r - v_0$$

Figure 14-13 Unity-feedback control system for PSpice Psimulation.

can be varied by varying R_F. The input voltage v_{in} is obtained from

$$v_{in} = \ddot{v}_o = -6\ddot{v}_o - 5\dot{v}_o + 5Kv_e$$

It should be noted that the output signal of an integrator is inverted. This sign change should be taken into account in summing signals. v_{in} and v_e are represented by polynomial sources.

The list of the circuit file is as follows:

■ ■

Example 14-4 Unity feedback control system with a step input

SOURCE ■ Vr   8   0   PWL (0   0V   1NS   1V   10MS   1V)   ; Reference voltage

           Rg   8   0   10MEG

CIRCUIT ■ ■ .PARAM   VAL = 10K   ; Parameter VAL

         .STEP   PARAM   VAL   10K   40K   10K   ; Step change of parameter VAL

         R1   1   2   1K

         C1   2   3   0.001   IC=0V   ; Set initial condition

         R2   3   4   1K

         C2   4   5   0.001   IC=0V   ; Set initial condition

         R3   5   6   1K

         C3   6   7   0.001   IC=0V   ; Set initial condition

         R   9   10   20K

         RF   10   11   {VAL}

         *   Calling subcircuit OPAMP-DC:

         XA1   2   0   3   0   OPAMP-DC

         XA2   4   0   5   0   OPAMP-DC

         XA3   6   0   7   0   OPAMP-DC

         XA4   10   0   11   0   OPAMP-DC

         E1   9   0   POLY(2)   8   0   7   0   0   1   1

         E2   1   0   POLY(3)   3   0   5   0   11   0   0   6.0   −5.0   −5.0

         *   Subcircuit definition OPAMP-DC <u>must</u> be inserted.

ANALYSIS ■ ■ ■ .TRAN   0.01S   10S   UIC   ; Use initial condition in transient analysis

         .PLOT   TRAN   V(7)   ; Prints on the output file

         .OPTIONS ABSTOL = 1.00N   RELTOL = 0.01   VNTOL = 0.1   ITL5=0

         .PROBE   ; Graphics post-processor

.END

■ ■

The plot of the transient response for the feedback control system is shown in Fig. 14-14. A higher value of K gives more overshoot and the system tends to be unstable. The transient should settle to the input signal level.

Example 14-5

The well-known Van der Pol equation is represented by

$$\ddot{y} - \mu(1 - y^2)\dot{y} + y = 0$$

Use PSpice to plot the transient response of the output signal for a duration of 0 to 20 s in steps of 10 ms, and the phase plane (dy/dt against y). Assume a constant $\mu = 2$ and an initial disturbance of 0.1 unit.

Solution Van der Pol's equation can be written as

$$\frac{d^3y}{dt^2} = \mu(1 - y^2)\frac{dy}{dt} - y \tag{14-5}$$

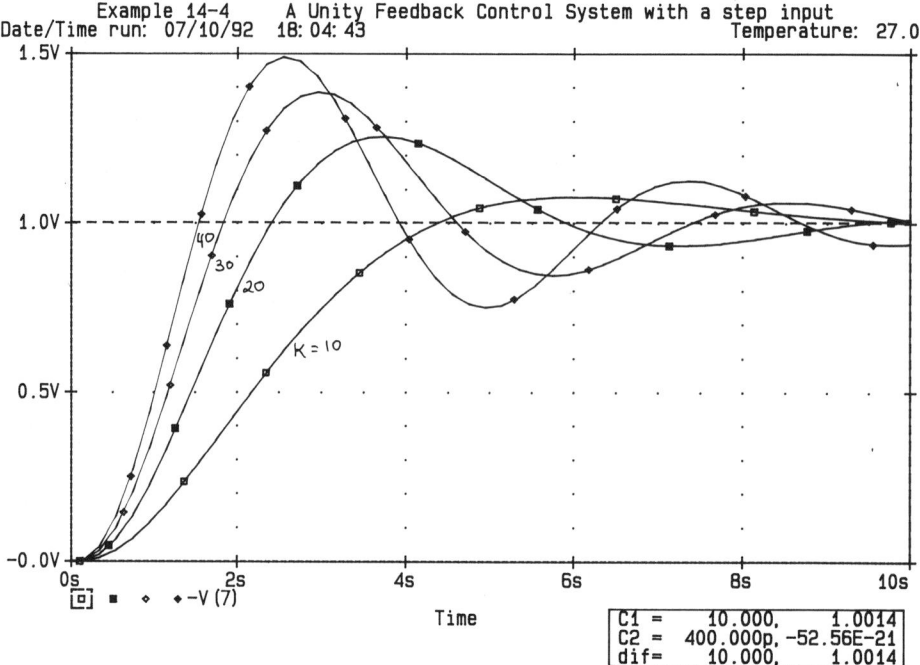

Figure 14-14 Transient response for Example 14-4.

This is identical to that for the LC circuit of Fig. 14-15. Let us assume that $L = 1$ H and $C = 1$ F. The equation describing the LC circuit of Fig. 14-15 is

$$v = L\frac{di}{dt} + \frac{1}{C}\int i\,dt = \frac{di}{dt} + \int i\,dt$$

Differentiating both sides yields

$$\frac{dv}{dt} = \frac{d^2i}{dt^2} + i$$

Solving for d^2i/dt^2, we get

$$\frac{d^2i}{dt^2} = \frac{dv}{dt} - i \tag{14-6}$$

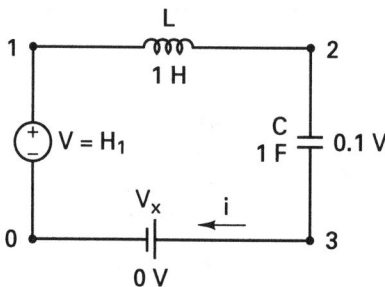

Figure 14-15 LC circuit for representing Van der Pol's equation.

Equation (14-5) will be identical to Eq. (14-6) if the current i represents y and dv/dt equals

$$\frac{dv}{dt} = \mu(1 - y^2)\frac{dy}{dt} = \mu(1 - i^2)\frac{di}{dt} \tag{14-7}$$

which can be integrated to give

$$v = \int \mu(1 - i^2)di = \mu\left(i - \frac{i^3}{3}\right) \tag{14-8}$$

Equation (14-8) is a polynomial of the form

$$v = P_o + P_1 i + P_2 i^2 + P_3 i^3$$

where $P_o = P_2 = 0$, $P_1 = 2$, and $P_3 = -\mu/3$.

The voltage across the inductor $v_L = di/dt$ represents dy/dt. The list of the circuit file is as follows:

■■■

Example 14-5   Van Der Pol's equation

CIRCUIT
```
■ H1    1    0    POLY(1)  VX   0   2   0   -2/3
  L     1    2    1
  C     2    3    1    IC=0.1              ; Set initial condition
  VX    3    0    DC    0V         ; Senses the circuit current
```
ANALYSIS
```
■ ■ .TRAN  0.01S  20S  UIC  ; Use initial condition in transient analysis
    .PLOT   TRAN   I(VX)  V(1)       ; Prints on the output file
```

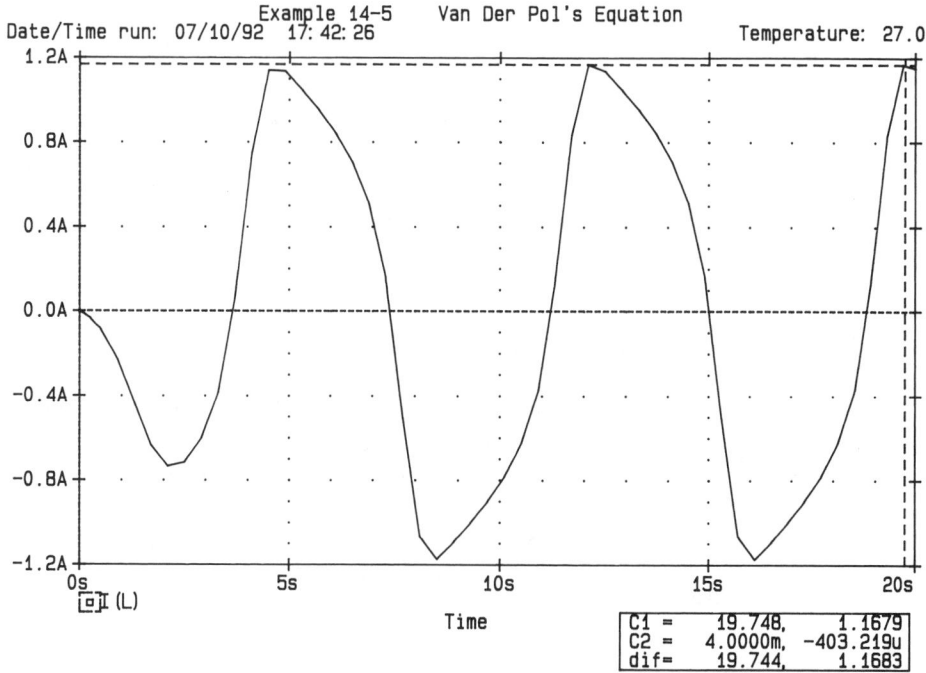

Figure 14-16   Transient response of Van Der Pol's equation in Example 14-5.

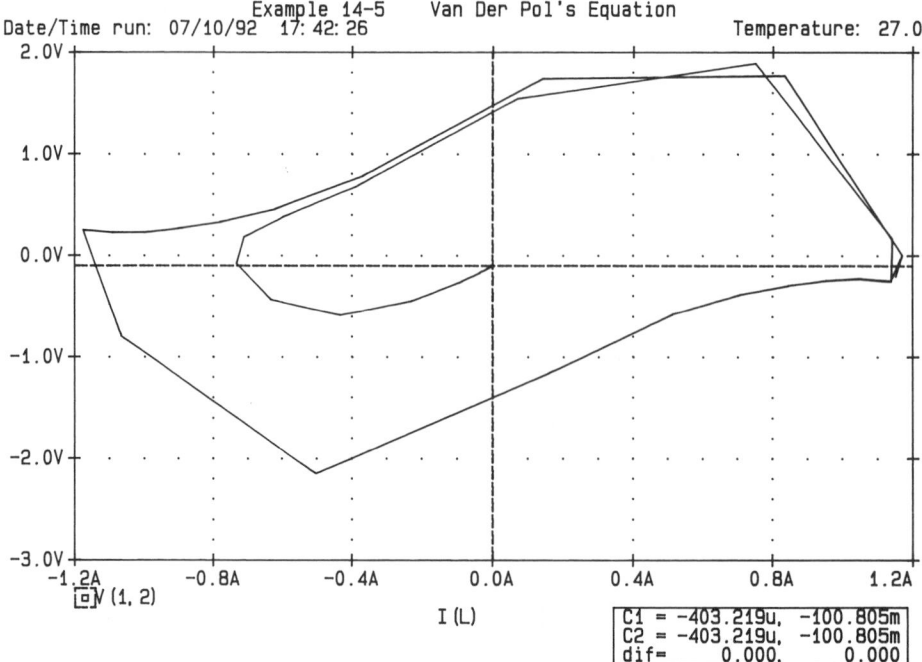

```
C1 = −403.219u,   −100.805m
C2 = −403.219u,   −100.805m
dif=       0.000,        0.000
```

Figure 14-17   Phase plane plot of Van Der Pol's equation in Example 14-5.

```
.OPTIONS ABSTOL = 1.00N   RELTOL = 0.01   VNTOL = 0.1   ITL5=0
 .PROBE                              ; Graphics post-processor
.END
```
■ ■

The plots of the transient response and the phase plane for the Van Der Pol equation is shown in Figs. 14-16 and 14-17, respectively. The system is unstable and oscillates between limit cycles. A smoother curve can be obtained by reducing the printing time of 0.01 s.

14-4 SIGNAL CONDITIONING

An op-amp can be used for wave shaping to yield a desired control characteristic. The student version of PSpice allows only one op-amp macromodel; it also increases the computation time. Simple *RC* circuits can be used to perform the functions of an op-amp, thereby reducing the computation time. We illustrate the applications of *RC* circuits to perform the following operations:

Integration
Averaging
Rms
Hysteresis

Example 14-6

The input signal $v_{in} = \sin(120\pi t + 90°)$ is to be integrated and then fed to a load resistance $R_L = 1$ GΩ. This is shown in Fig. 14-18(a). Use PSpice to plot the transient response of the input and output voltages for a duration of 0 to 33.33 ms in steps of 10 μs.

Solution The integration is accomplished by the *RC* circuit of Fig. 14-18(b). The input frequency is $f = 60$ Hz. The list of the circuit file is as follows:

■ ■

Example 14-6   Integrating circuit

```
SOURCE    ■ Vin   1    0     SIN (0   1V    60HZ   0   0    90DEG)
CIRCUIT   ■ ■ .PARAM   FREQ = 60HZ                    ; Input frequency in hertz
             .PARAM   TWO_PI = 6.2832
             RL    2    0    1G
             * Calling subcircuit INTG:
             X1    1    2    0      INTG
             *     Vi+   Vo+   Vo-    model name
             *   Subcircuit definition for INTG:
             .SUBCKT     INTG     1    3    2    PARAMS: FREQ = 60HZ
             *          model name   Vi+   Vo+   Vo-
             RI    1    0    10G
             GINTG  2    4    1    0    1
             C     4    2    1    IC=0V
             R     4    2    10G
             EOUT   3    2    VALUE = {V(4,2)*TWO_PI*FREQ}
             RO    3    2    10G
             .ENDS   INTG                          ; End of subcircuit definition
ANALYSIS  ■ ■ ■ .TRAN  10US   33.33MS   UIC  ; Use initial condition in transient analysis
             .PLOT   TRAN   V(2)   V(1)             ; Prints on the output file
```

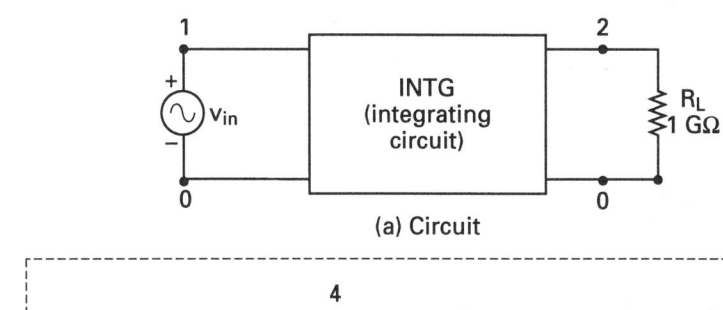

(a) Circuit

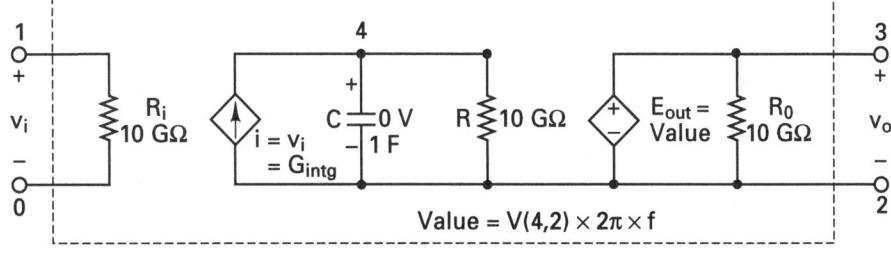

(c) Subcircuit

Figure 14-18 Integrating circuit.

```
              .OPTIONS ABSTOL = 1.00N   RELTOL = 0.01   VNTOL = 0.1   ITL5=0
              .PROBE                                    ; Graphics post-processor
       .END
```
■■

The plot of the transient response for the input and output voltages are shown in Fig. 14-19. The output is a sine wave in response to a cosine signal.

Example 14-7

An input signal is to be averaged and then fed to a load resistance $R_L = 1$ GΩ. This is shown in Fig. 14-20(a). The input signal is a triangular wave, as shown in Fig. 14-20(b). Use PSpice to plot the transient response of input and output voltages for a duration of 0 to 33.33 ms in steps of 10 μs.

Solution   The averaging is accomplished by the *RC* circuit of Fig. 14-20(c). This is similar to Example 14-6, except that the value is divided by time. The input frequency is $f = 1$ kHz. The list of the circuit file is as follows:

■■

Example 14-7   Averaging circuit

SOURCE
CIRCUIT
```
■ Vin   1    0     PULSE (0  1V  0  0.5MS  0.5MS  1NS  1MS)
■ ■ RL   2    0     1G
   *   Calling subcircuit AVRG:
   X1     1     2     0      AVRG
   *      Vi+   Vo+   Vo−    model name
```

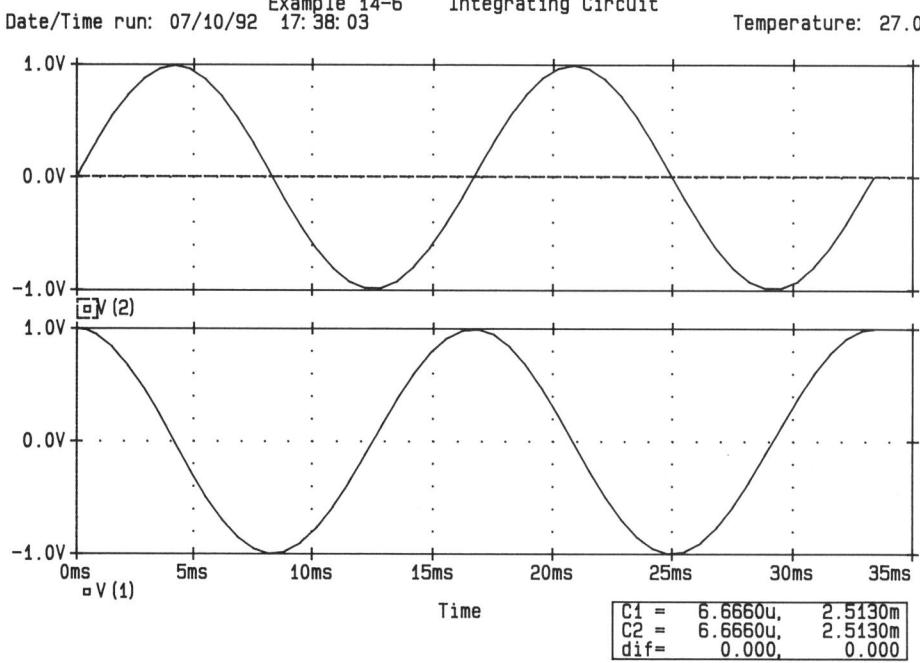

Figure 14-19   Transient response for the integrating circuit of Example 14-6.

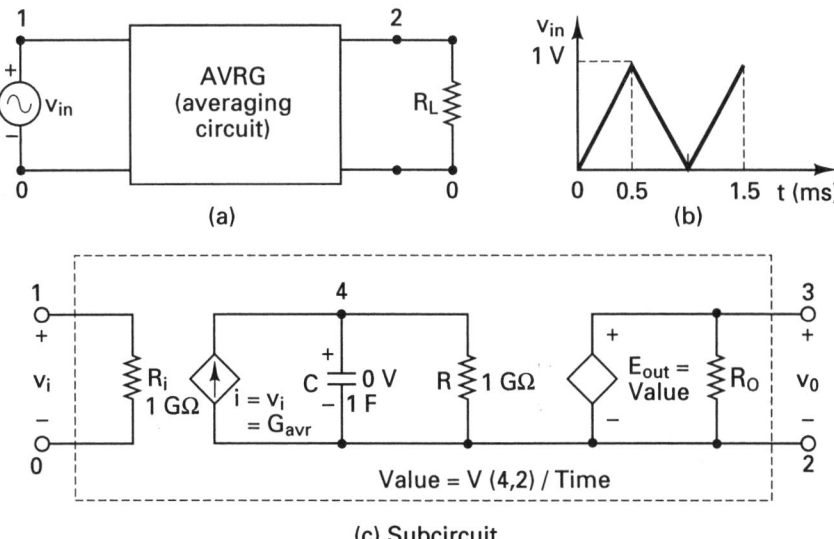

(a)

(b)

(c) Subcircuit

Figure 14-20 Averaging circuit.

```
        *    Subcircuit definition for AVRG:
        .SUBCKT      AVRG        1      3      2
        *           model name   Vi+    Vo+    Vo−
        RI      1    0    1G
        GAVR    2    4    1     0      1
        C       4    2    1     IC=0V
        R       4    2    1G
        EOUT    3    2    VALUE = {V(4,2)/TIME}
        RO      3    2    1G
        .ENDS   AVRG                              ; End of subcircuit definition
ANALYSIS ■ ■ ■ .TRAN    10US   4MS   UIC   ; Use initial condition in transient analysis
        .PLOT    TRAN    V(2)    V(1)              ; Prints on the output file
        .OPTIONS ABSTOL = 1.00N   RELTOL = 0.01   VNTOL = 0.1   ITL5=0
        .PROBE                                ; Graphics postprocessor
.END
```

The plots of the transient response for the input and output voltages are shown in Fig. 14-21. Under steady-state conditions, the average value of a triangular wave remains constant at $0.5bh/T = 0.5 \times 1$ ms $\times 1$ V/1 ms $= 0.5$ V.

Example 14-8

The rms value of an input signal $v_{in} = \sin(120\pi t)$ is to be fed to a load resistance $R_L = 1$ GΩ. This is shown in Fig. 14-22(a). Use PSpice to plot the transient response of the input and output voltages for a duration of 0 to 33.33 ms in steps of 10 μs.
Solution The rms value is determined by the RC circuit of Fig. 14-22(b). The input frequency is $f = 60$ Hz. The list of the circuit file is as follows:

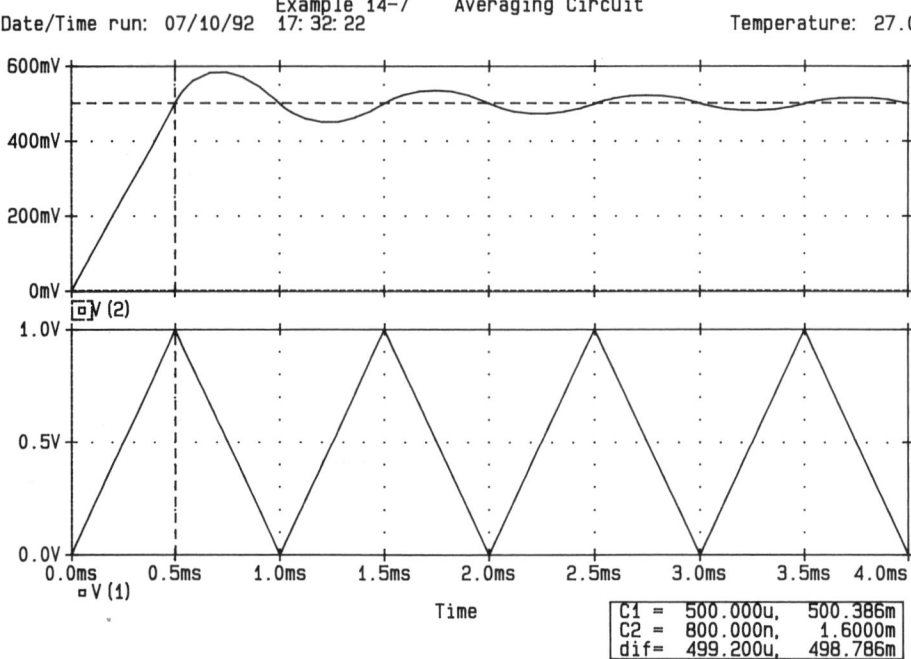

C1 =	500.000u,	500.386m
C2 =	800.000n,	1.6000m
dif=	499.200u,	498.786m

Figure 14-21 Transient response for the averaging circuit of Example 14-7.

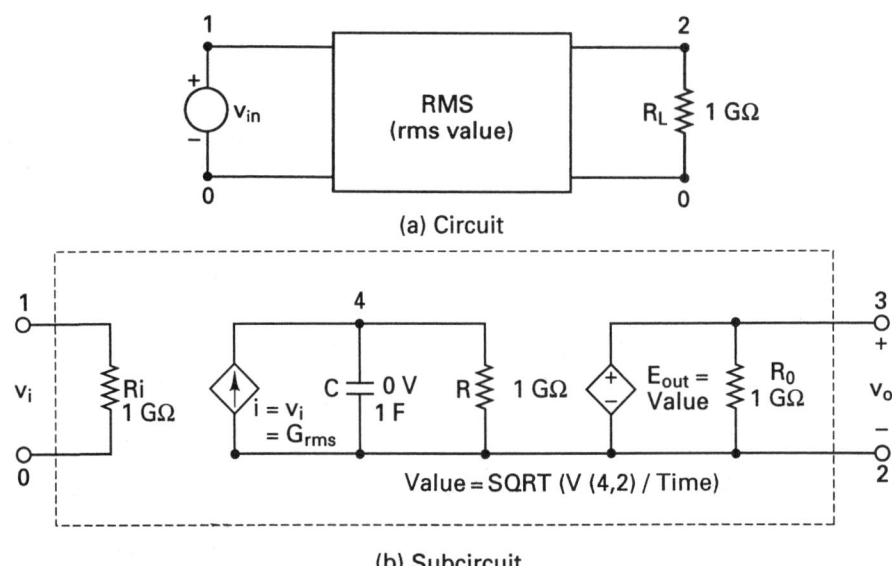

(a) Circuit

(b) Subcircuit

Figure 14-22 Rms value circuit.

■ ■

Example 14-8   RMS circuit

```
SOURCE      ■ Vin   1    0    SIN (0    1V    60HZ)
CIRCUIT     ■ ■ RL   2    0    1G
              *  Calling subcircuit RMS:
              X1    1    2    0    RMS
              *      Vi+   Vo+   Vo-    model name
              *  Subcircuit definition for RMS:
              .SUBCKT    RMS       1    3    2
              *      model name   Vi+   Vo+   Vo-
              RI    1    0    1G
              GAVR  2    4    VALUE = {V(1)*V(1)}
              C     4    2    1    IC=0V
              R     4    2    1G
              EOUT  3    2    VALUE = {SQRT(V(4,2)/TIME)}
              RO    3    2    1G
              .ENDS    RMS                  ; End of subcircuit definition
ANALYSIS    ■ ■ ■ .TRAN  10US  33.33MS  UIC  ; Use initial condition in transient analysis
              .PLOT    TRAN   V(2)    V(1)         ; Prints on the output file
              .OPTIONS ABSTOL = 1.00N  RELTOL = 0.01  VNTOL = 0.1  ITL5=0
              .PROBE                          ; Graphics post-processor
            .END
```
■ ■

The plots of the transient response for the input and output voltages are shown in Fig. 14-23. The rms value of a sine wave with 1 V peak is $1 \text{ V}/\sqrt{2} = 0.707$ V.

Example 14-9

The input signal to the integrating circuit of Fig. 14-24(a) is $v_{in} = 2 \sin(120\pi t)$. The plot of current i against V_{in}/H is shown in Fig. 14-24(b). Use PSpice to calculate the transient response for a duration of 0 to 33.33 ms in steps of 10 μs, and to plot the output voltage V_o against the input voltage V_{in}.

Solution   The input frequency is $f = 60$ Hz. The list of the circuit file is as follows:

■ ■

Example 14-9   Hysteresis loop

```
SOURCE      ■ Vin   1    0    SIN (0    2V    60HZ)
CIRCUIT     ■ ■ .PARAM   H = 2                        ; Hysteresis
              RL    3    0    1G
              RI    1    0    1G
              GTAB  0    4    TABLE {V(1)/(H/2)} =    ; Table for ratio Vin/(H/2)
              +    (-2 -1K)   (-1, 0)   (1, 0)   (2  1K)
              C     4    0    1    IC=0V
              R     4    0    1G
              EOUT  3    0    4    0    1
ANALYSIS    ■ ■ ■ .TRAN  0.1MS  16.67MS  UIC ; Use initial condition in transient analysis
              .PLOT    TRAN   V(3)    V(1)              ; Prints on the output file
              .OPTIONS ABSTOL = 1.00N  RELTOL = 0.01  VNTOL = 0.1  ITL5=0
              .PROBE                          ; Graphics post-processor
            .END
```
■ ■

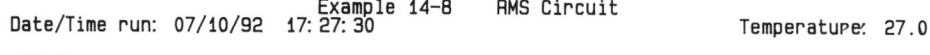

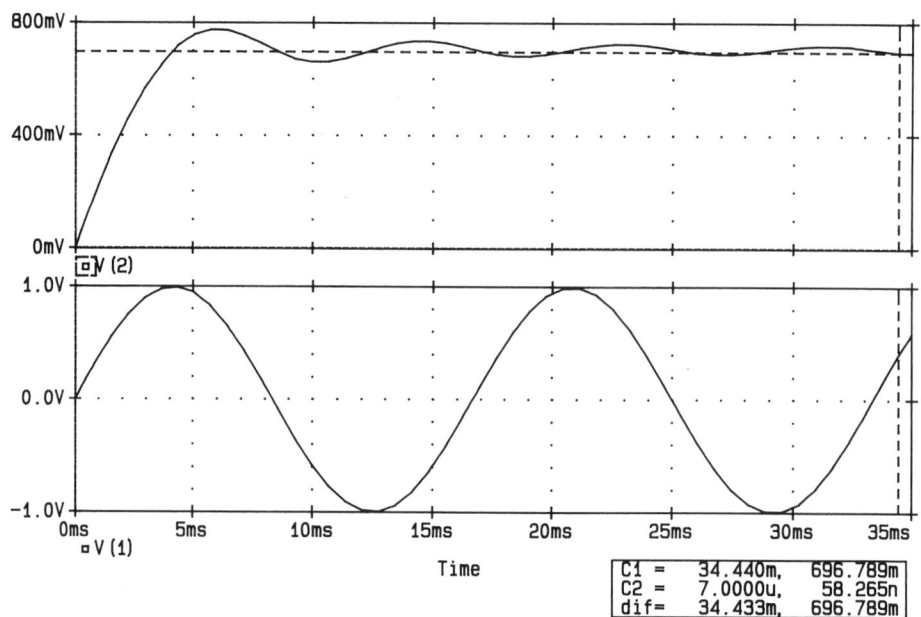

Figure 14-23   Transient response for the rms circuit of Example 14-8.

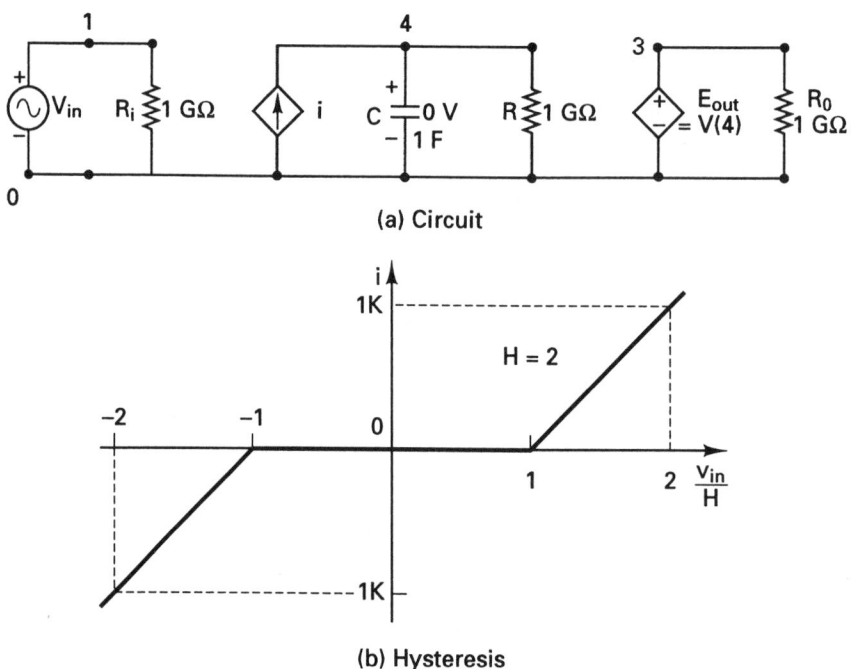

(a) Circuit

(b) Hysteresis

Figure 14-24   Hysteresis circuit.

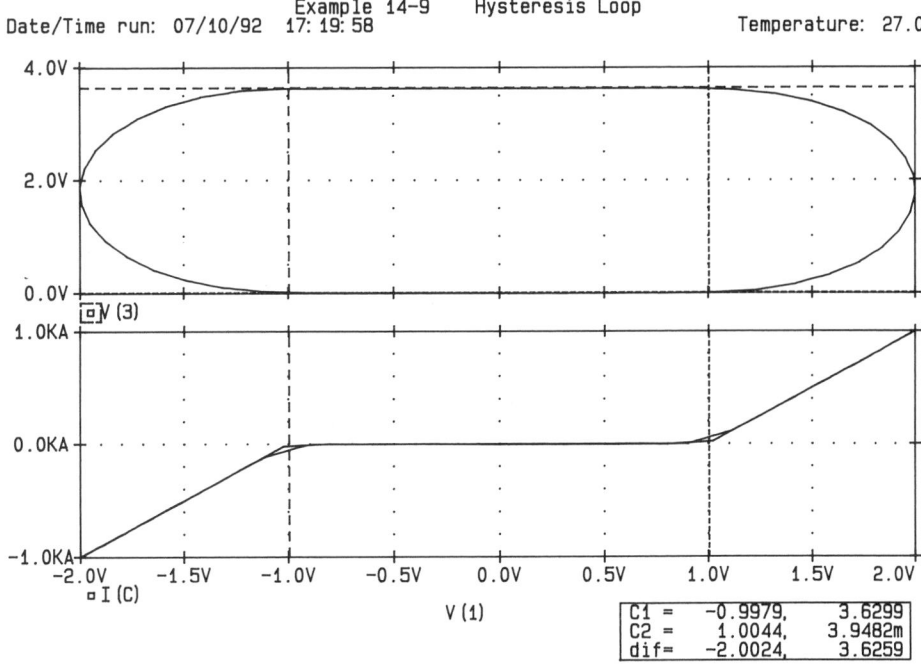

Example 14-9   Hysteresis Loop

C1 =	−0.9979,	3.6299
C2 =	1.0044,	3.9482m
dif=	−2.0024,	3.6259

Figure 14-25   Output voltage against input voltage for Example 14-9.

The plot of the output voltage V(3) against the input voltage V(1) is shown in Fig. 14-25. The output remains constant during the hysteresis band.

14-5 CLOSED-LOOP CURRENT CONTROL

Closed-loop control is used in many power electronics circuits to control the shape of a particular current (e.g., inductor motor with current control, and current source inverter). In the following example we illustrate the simulation of a current-controlled rectifier circuit such that the output current of the rectifier is half a sine wave, thereby giving a sine wave at the input side of the rectifier.

Example 14-10

A diode rectifier followed by a boost converter is shown in Fig. 14-26. The input voltage is $v_s = 170 \sin(120\pi t)$. The circuit is operated closed-loop, so that the output voltage is $V_o = 220 \pm 0.2$ V, and the input current is sinusoidal with an error of ± 0.2 A. Use PSpice to plot the transient response of output voltage v_o and the input current i_s. The model parameters for the BJT are IS=2.33E−27 and BF=13 and for the diode are IS=2.2E−15, BV=1200V, CJO=1PF, and TT=0.

Solution   The peak input voltage is $V_m = 170$ V. Assuming a diode drop of 1 V, the peak rectified voltage becomes $170 - 2 = 168$ V. The input frequency is $f = 60$ Hz. The block diagram for closed-loop control is shown in Fig. 14-27(a). The error of the output voltage after passing through a controller generates the reference current for

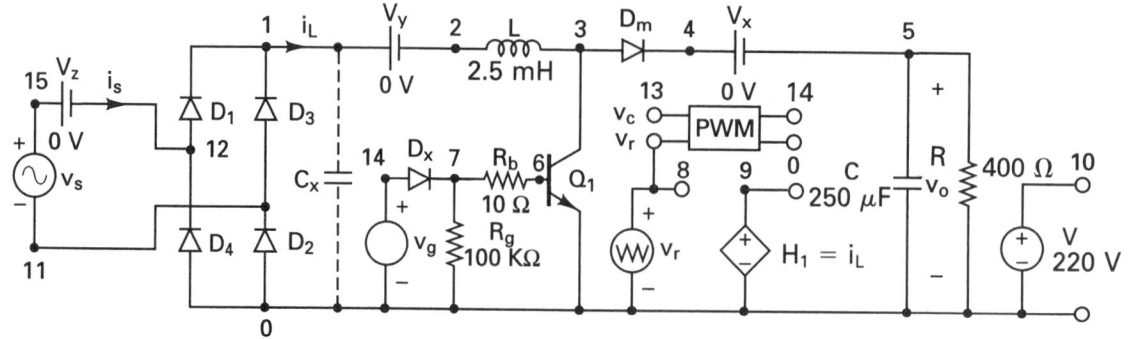

Figure 14-26 Diode rectifier with input current control.

the current controller, whose output is the carrier signal for the PWM generator. Equating rectified output power to the load power, we get

$$V_{in(dc)}I_{in(dc)} = \frac{V_o^2}{R}$$ (14-9)

Since for a rectified sine wave $V_{(average)} = 2V_{(peak)}/\pi$, Eq. (14-9) becomes

$$\frac{2V_m}{\pi}\frac{2I_m}{\pi} = \frac{V_o^2}{R}$$

which gives the multiplication constant δ as

$$\delta = \frac{I_m}{V_m} = \frac{\pi^2 V_o^2}{4RV_m^2}$$

$$= \frac{\pi^2 \times 220^2}{4 \times 400 \times 168^2} = 0.01$$ (14-10)

A value of δ higher than 0.01 would increase the sensitivity and give better transient response. But the PWM generator will be operated in overmodulation region if it is too high. Let us assume $\delta = 0.05$.

We use a proportional controller with an error band of ± 0.2. The circuit for PSpice simulation of the current controller is shown in Fig. 14-27(b). The multiplication is implemeneted with VALUE. The voltage and current controllers are implemented by TABLE. The list of the subcircuit CONTR for Fig. 14-27(b) is as follows:

```
*    Subcircuit defintion for CONTR current controller:
.SUBCKT   CONTR     2         3        4            5              6
*         model   desired   output   rec. input   input current  carrier
*         name    voltage   voltage  voltage      signal (voltage) signal
RG4     4    0    10MEG
RG      2    3    10MEG
E1      8    0    TABLE  {V(2,3)} =              ; Voltage controller
+     (-0.2, 2)  (0, 1)   (0.2, 0)
RG1     8    0    10MEG
E2      5    7    VALUE = {0.05*V(4)*V(8)}   ; Reference current
RG2     7    0    10MEG
```

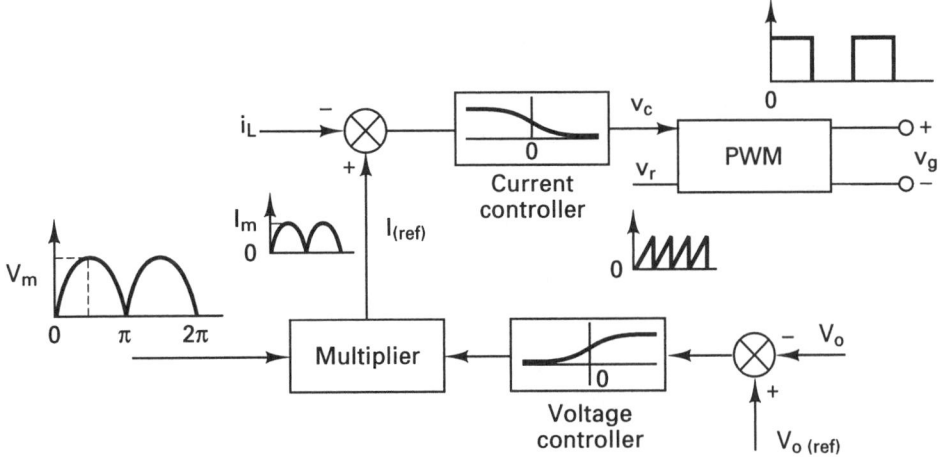

(a) Block diagram

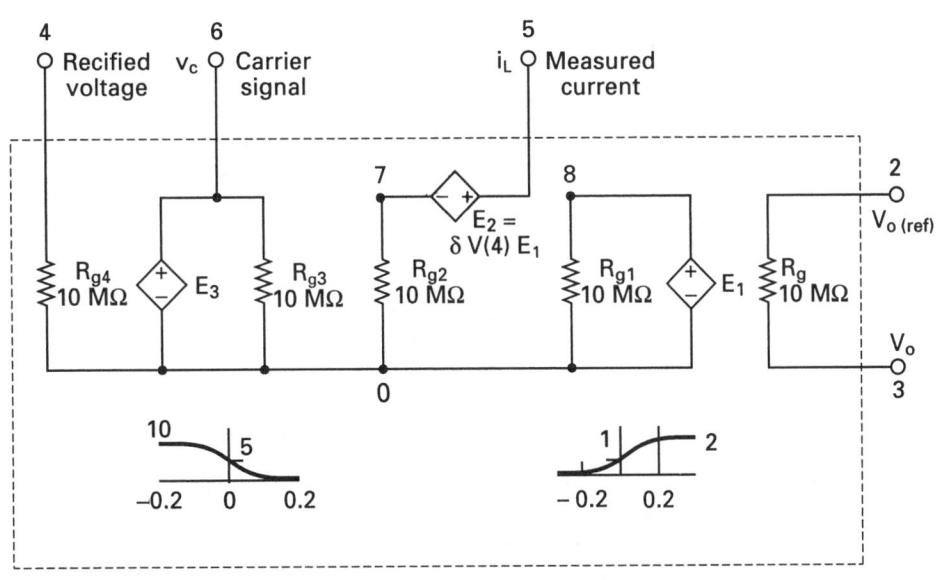

(b) Subcircuit CONTR current controller

Figure 14-27   Block diagram for closed-loop control.

```
E3      6    0     TABLE  {V(7)} =                ; Current controller
+       (−0.2, 0)   (0, 5)   (0.2, 10)
RG3     6    0     10MEG
.ENDS   CONTR
```

The subcircuit definition for PWM is described in Section 12-2. Using 20 pulses per half-cycle of the input voltage, the switching frequency of the PWM generator is $f_s = 40 \times 60 = 2.4$ kHZ, and the switching period is $T_s = 416.67 \mu s$. The list of the circuit file for Fig. 14-26 is as follows:

■ ■

Example 14-10   Diode rectifier with PWM current control

```
SOURCE     ■ VS    15   11    SIN (0V  170V  60HZ)
CIRCUIT    ■ ■ VZ  15   12    DC   0V        ; Measures supply current
             D1    12    1    DMOD           ; Rectifier diodes
             D2     0   11    DMOD
             D3    11    1    DMOD
             D4     0   12    DMOD
             CX     1    0    1PF
             VY     1    2    DC   0V        ; Voltage source to measure input current
             Vr     8    0    PULSE (-10V  0V  0  208.33US  208.33US 1NS 416.67US)
             VDR   10    0    DC   220V
             H1     9    0    VY   1.0
             Rg     7    0    100K
             DX    14    7    DM             ; Blocking diode
             RB     7    6    10             ; Transistor base resistance
             L      2    3    2.5MH
             DM     3    4    DMOD           ; Freewheeling diode
             VX     4    5    DC   0V        ; Voltage source to measure inductor current
             R      5    0    400            ; Load resistance
             C      5    0    250UF IC=0V    ; Load filter capacitor
             .MODEL  DMOD  D(IS=2.2E-15 BV=1200V CJO=1PF TT=0) ; Diode model parameters
             .MODEL  DM D(IS=2.2E-15 BV=1200V CJO=0 TT=0) ; Diode model parameters
             Q1     3    6    0    0    2N6546        ; BJT switch
             .MODEL  2N6546  NPN (IS=2.33E-27 BF=13) ;
             *  Subcircuit call for PWM control:
             XPW    8   13   14    0   PWM     ; Control voltage for transistor Q1
             *   Subcircuit call for CONTR current controller:
             XCONT  10   5    1    9   13   CONTR
             *  Subcircuit definition CONTR must be inserted.
             *  Subcircuit definition PWM must be inserted.
ANALYSIS   ■ ■ ■ .TRAN   0.1MS   35MS                    ; Transient analysis
             .PROBE                                      ; Graphics post-processor
             .FOUR   60HZ   I(VZ)                         ; Fourier analysis
             .OPTIONS ABSTOL = 1.00U RELTOL = 0.01 VNTOL = 0.1 ITL5=0 ; Convergence
           .END
```

■ ■

The plots of the output voltage V(5) and input current I(VZ) are shown in Fig. 14-28. The current controller forces the input current to be in phase with the input supply voltage and to follow a sinusoidal reference current. This improves the input power factor. A small filter can remove the high frequency components of the input current. The upper bound of the output voltage is limited by the proportional controller. It should be noted that the load voltage has not yet reached the steady-state condition. The lower limit depends on the time constant of the load circuit. It requires a careful design for determining the values of L and C, the controller parameters and the switching frequency. A proportional–integral (PI) controller,

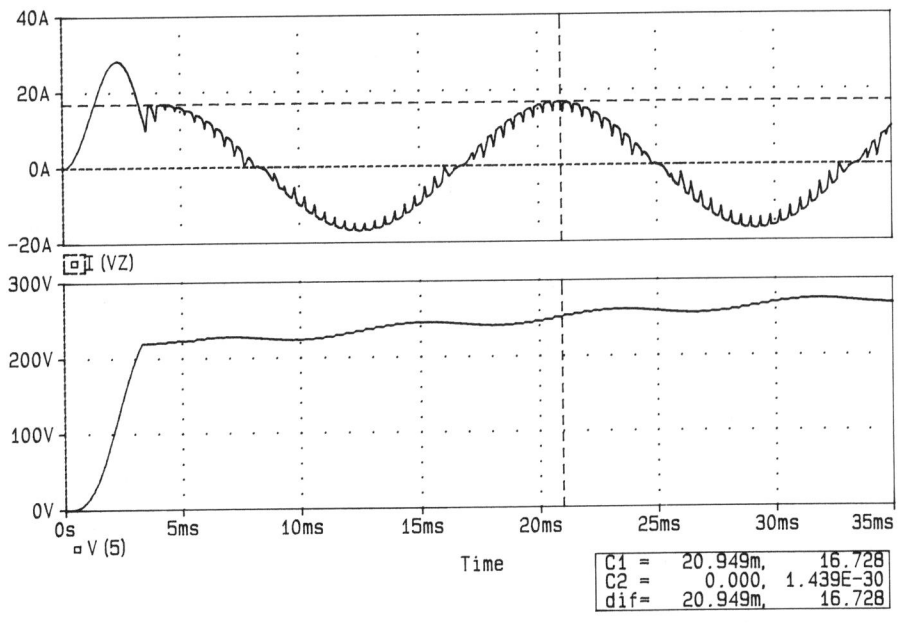

Figure 14-28 Output voltage and input current for Example 14-10.

together with a large value for the load filter capacitor, should give faster response of the output voltage.

SUGGESTED READING

1. *Linear Circuits: Operational Amplifier Macromodels.* Dallas: Texas Instruments, 1990.

2. G. Boyle, B. Cohn, D. Pederson, and J. Solomon, "Macromodeling of integrated circuit operational amplifiers," *IEEE Journal of Solid-State Circuits*, Vol. SC-9, No. 6, December 1974, pp. 353–364.

3. I. Getreu, A. Hadiwidjaja, and J. Brinch. "An integrated-circuit comparator macromodel," *IEEE Journal of Solid-State Circuits*, Vol. SC-11, No. 6, December 1976, pp. 826–833.

4. S. Progozy, "Novel applications of SPICE in engineering education," *IEEE Transactions on Education*, Vol. 32, No. 1, February 1990, pp. 35–38.

PROBLEMS

14-1. A full-wave precision rectifier is shown in Fig. P14-1. The input voltage is $v_i = 0.1 \sin(2000\pi t)$. Plot the transient response of the output voltage for a duration of 0 to 1 ms in steps of 10 μs. The op-amp μA741 can be modeled as a macromodel as

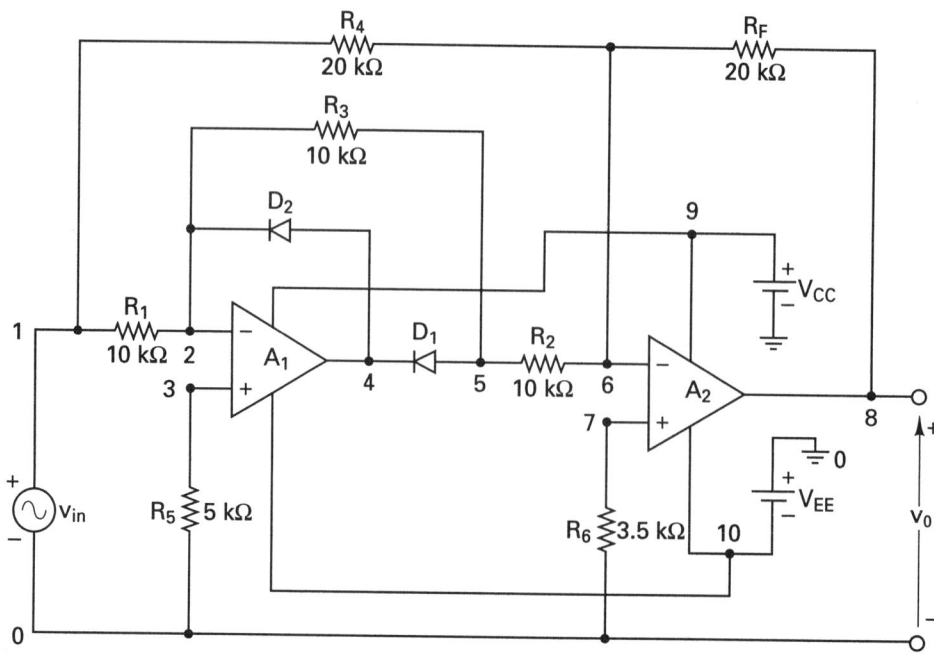

Figure P14-1

shown in Fig. 14-3. The supply voltages are $V_{CC} = 12$ V and $V_{EE} = -12$ V.

14-2. Plot the dc transfer characteristics for Fig. P14-2. The input voltage is varied from -10 to $+10$ V in steps of 0.1 V. The zener voltages are $V_{Z1} = V_{Z2} = 6.3$ V. The op-amp, which is modeled by the circuit in Fig. 14-1(a), has $R_i = 2$ MΩ, $R_o = 75$ Ω, $C_1 = 1.5619$ μF, and $R_1 = 10$ kΩ.

14-3. A feedback-control system is shown in Fig. P14-3. The reference input v_r is the step voltage of 1 V. Use PSpice to plot the transient response of the output voltage for a duration of 0 to 10 s in steps of

10 ms. The gain K is to be varied from 0.5 to 2 with an increment of 0.5. Assume that all initial conditions are zero and that $K_1 = 0.5$.

14-4. The constant μ of Van der Pol's equation is 4. Use PSpice to plot **(a)** the transient response of the output signal for a duration of 0 to 20 s in steps of 10 ms, and **(b)** the phase plane (dy/dt against y). Assume an initial disturbance of 1.

14-5. Repeat Example 14-9 for the i against V_{in}/H characteristic shown in Fig. P14-5.

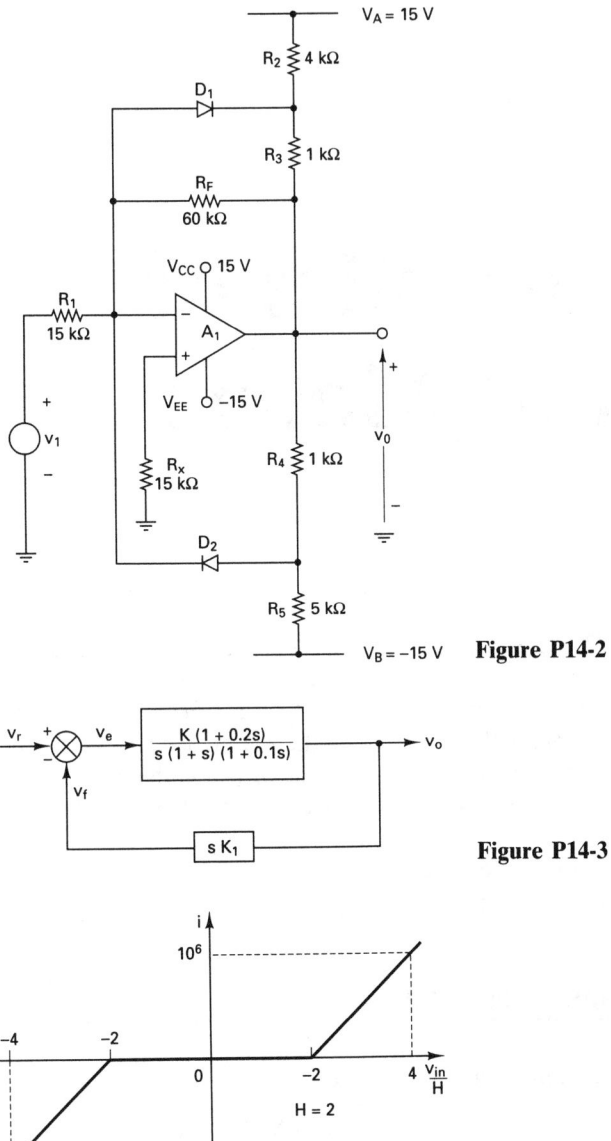

Figure P14-2

Figure P14-3

Figure P14-5

Chapter 15

■■■■■■■■■

Characteristics of Electrical Motors

15-1 INTRODUCTION

The PSpice simulation of power converters can be combined with the equivalent circuit of electrical machines to obtain their control characteristics. The machines can be represented in SPICE by a linear or a nonlinear magnetic circuit or as a function or table form. We use linear circuit models to obtain:

Dc motor characteristics
Induction motor characteristics

15-2 DC MOTOR CHARACTERISTICS

The motor back emf is given by

$$E_g = K\omega I_f$$

and the developed motor torque

$$T_d = KI_a I_f$$

can be represented by polynomial sources. The torque T_d is related to load torque T_L and motor speed ω by

$$T_d = J\frac{d\omega}{dt} + B\omega + T_L$$

$$T_d - T_L - B\omega = J\frac{d\omega}{dt}$$

Thus, integrating the net torque will give the motor speed ω, which after further integration gives the shaft position θ.

Example 15-1

The armature of a separately excited dc motor is controlled by a dc chopper operating at a frequency of $f_c = 1$ kHz and a duty cycle of $k = 0.8$. The dc supply voltage to the armature is $V_s = 220$ V. The field current is also controlled by a dc chopper operating at a frequency of $f_s = 1$ kHz and a duty cycle of $\delta = 0.5$. The dc supply voltage to the field is $V_f = 280$ V. The motor parameters are:

Armature resistance, $R_m = 0.1$ Ω

Armature inductance, $L_m = 10$ mH

Field resistance, $R_f = 10$ Ω

Field inductance, $L_f = 20$ mH

Back-emf constant, $K = 0.91$

Viscous torque constant, $B = 0.3$

Motor inertia, $J = 1$

Load torque, $T_L = 50, 100,$ and 150 N·m

Use PSpice to plot the transient response of the armature and field currents, the torque developed, and the motor speed for a duration of 0 to 4 ms in steps of 10 μs.

Solution The armature and field circuits for PSPice simulation are shown in Fig. 15-1(a) and (b), respectively. The net torque, which is obtained by subtracting the viscous (T_B) and load torque (T_L) from the torque developed (T_d), is integrated to obtain the motor speed as shown in Fig. 15-1(c). The motor speed is integrated to obtain the shaft position as shown in Fig. 15-1(d). The list of the circuit file is as follows:

■ ■

Example 15-1 Dc separately excited motor with variable load torques

```
SOURCE  ■ VS    1    0    DC    220V                  ; Armature supply
          Vg1   6    0    PULSE (0    20V   0   1NS   1NS   0.8MS   1MS)
          Rg1   6    0    10MEG
          VF    7    0    DC    280V                  ; Field supply
          Vg2   11   0    PULSE (0    20V   0   1NS   1NS   0.5MS   1MS)
          Rg2   11   0    10MEG
CIRCUIT ■ ■ .PARAM  VISCOUS = 0.3                     ; Viscous constant
          .PARAM  J = 1                               ; Motor inertia
          .PARAM  TL = 100                            ; Load torque
          .STEP   PARAM  TL   50   150    50          ; Load torque varied
          S1    1    2    6    0    SMOD              ; Voltage-controlled switch
          .MODEL  SMOD  VSWITCH (RON=0.01 ROFF=10E+6 VON=10V VOFF=5V)
          D1    0    2    DMOD
          .MODEL  DMOD  D(IS=2.2E-15 BV=1200V CJO=0 TT=0)  ; Diode model parameters
          RM    2    3    0.1
          LM    3    4    10MH
          VX    4    5    DC    0V                     ; Senses the armature current
          E1    5    0    VALUE = {0.91*V(15)*I(VY)}
          RF    8    9    10
```

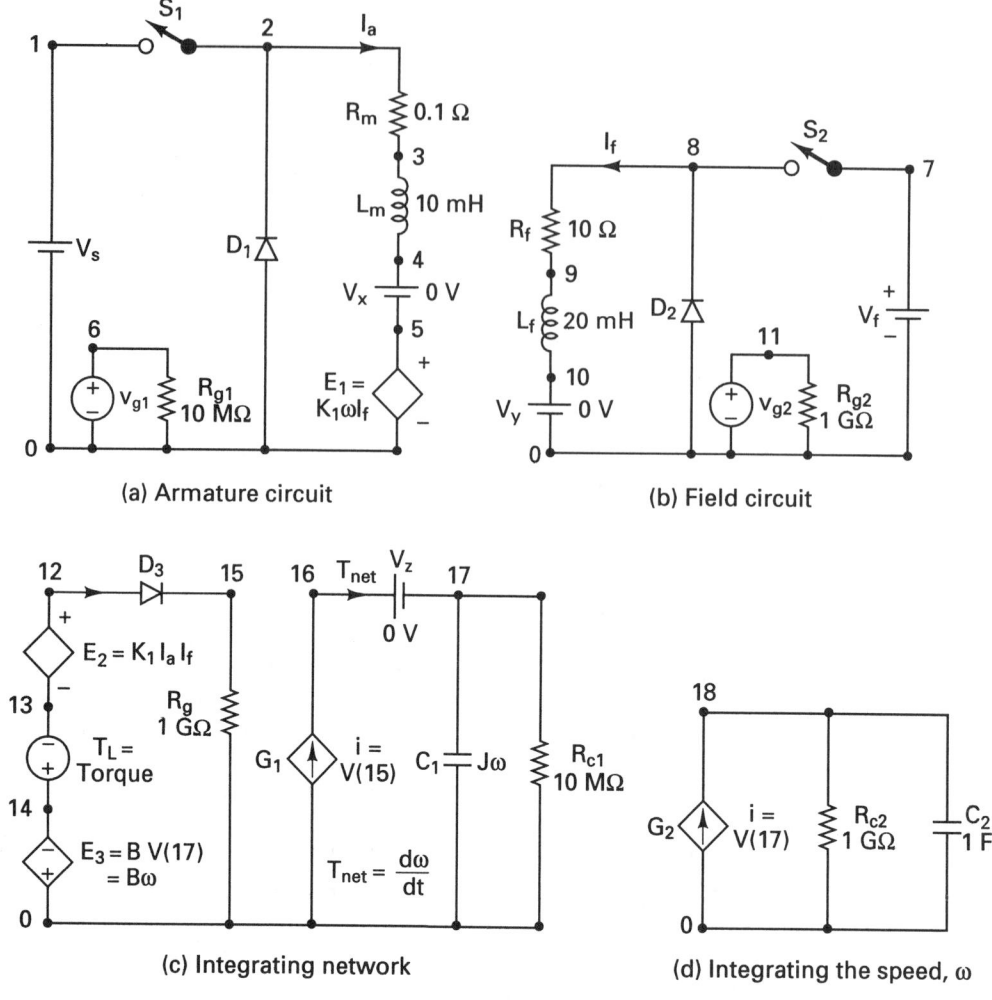

Figure 15-1 Chopper-controlled dc motor for PSpice simulation.

```
LF     9    10    20MH
VY     10   0     DC     0V                       ; Senses the field current
S2     7    8     11     0       SMOD             ; Voltage-controlled switch
D2     0    8     DMOD
E2     12   13    VALUE = {0.91*I(VX)*I(VY)}      ; Torque developed
VL     14   13    {TL}                            ; Load torque
E3     0    14    VALUE = {VISCOUS*V(17)}         ; Viscous torque
D3     12   15    DMOD
Rg     15   0     1G
G1     0    16    15     0       1                ; Net torque
VZ     16   17    DC     0V                       ; Measures the net torque
C1     17   0     {J}    IC=0V                    ; Load inertia
Rc1    17   0     1G
```

```
            G2    0    18    17    0    1              ; Velocity to position
            C2   18     0     1   IC=0V
            Rc2  18     0    1G
ANALYSIS  ■ ■ ■  .TRAN 10US 4MS UIC  ; Transient analysis with initial condition
            .PLOT  TRAN  V(3)  V(1)            ; Prints on the output file
            .OPTIONS ABSTOL = 1.00N  RELTOL = 0.01  VNTOL = 0.1  ITL5=50000
            .PROBE                      ; Graphics post-processor
.END
■ ■ ■ ■ ■ ■ ■ ■ ■ ■ ■ ■ ■ ■ ■ ■ ■ ■ ■ ■ ■ ■ ■ ■ ■ ■ ■ ■ ■ ■ ■ ■ ■ ■ ■ ■
```

The plots of the transient response for the armature I(VX) and field currents I(VY) are shown in Fig. 15-2. The plots for the torque developed V(12,13) and the

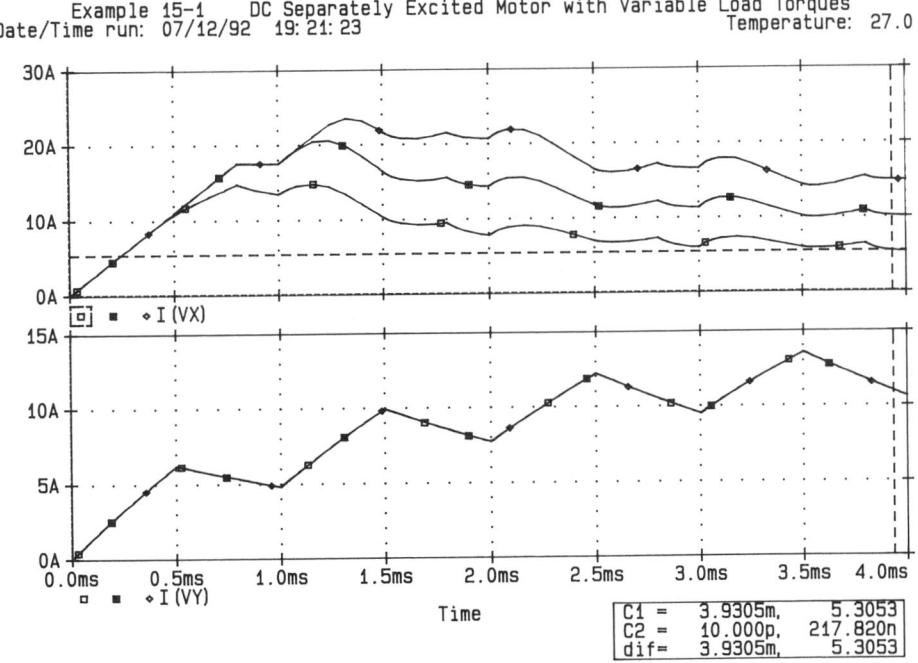

Figure 15-2 Plots of armature and field currents for Example 15-1.

motor speed V(17) are shown in Fig. 15-3. The field current has not reached steady-state conditions. The armature current reaches a peak before settling down. The torque is pulsating due to pulsating armature and field currents.

Example 15-2

Use PSpice to plot the transient response of the motor speed in Example 15-1 if the load torque T_L is subjected to a step change as shown in Fig. 15-4.

Solution The list of the circuit file is as follows:

■ ■

Example 15-2   Dc separately excited motor with step load torque change

```
SOURCE  ■ VS    1    0    DC       220V           ; Armature supply
          Vg1   6    0    PULSE (0   20V   0   1NS   1NS   0.8MS   1MS)
```

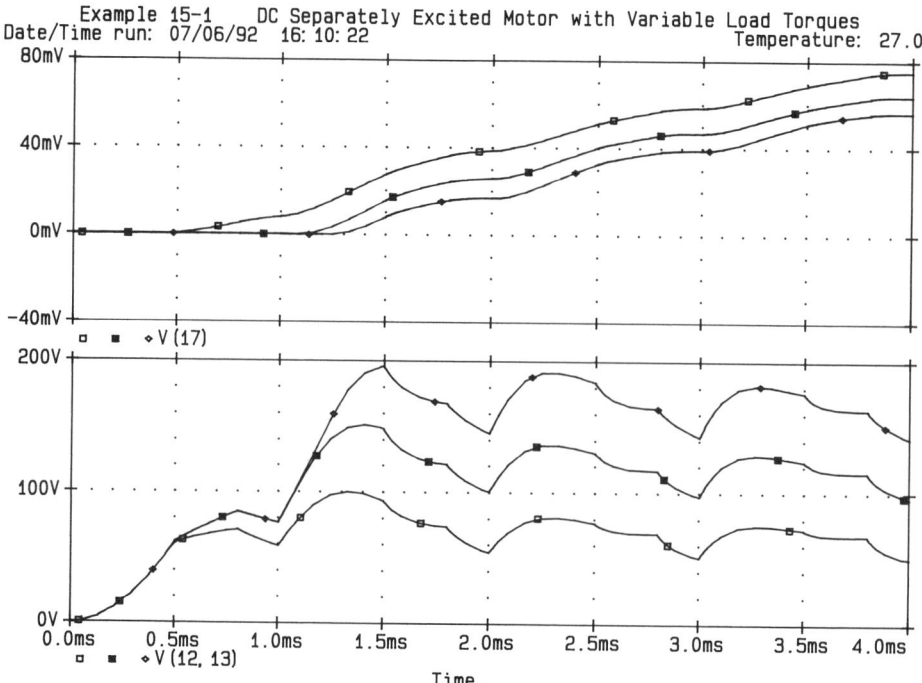

Figure 15-3 Developed torque and motor speed for Example 15-1.

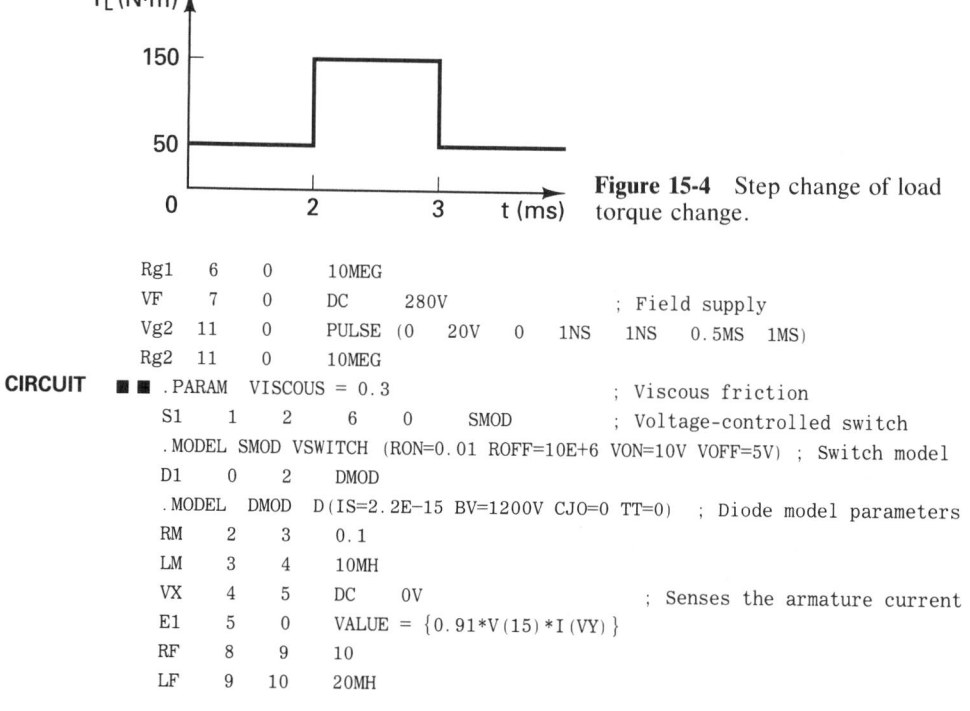

Figure 15-4 Step change of load torque change.

```
        Rg1    6    0     10MEG
        VF     7    0     DC     280V              ; Field supply
        Vg2    11   0     PULSE (0   20V   0   1NS   1NS   0.5MS   1MS)
        Rg2    11   0     10MEG
CIRCUIT ■ ■ .PARAM  VISCOUS = 0.3                  ; Viscous friction
        S1     1    2     6     0     SMOD          ; Voltage-controlled switch
        .MODEL  SMOD  VSWITCH (RON=0.01 ROFF=10E+6 VON=10V VOFF=5V) ; Switch model
        D1     0    2     DMOD
        .MODEL  DMOD  D(IS=2.2E-15 BV=1200V CJO=0 TT=0)  ; Diode model parameters
        RM     2    3     0.1
        LM     3    4     10MH
        VX     4    5     DC     0V                 ; Senses the armature current
        E1     5    0     VALUE = {0.91*V(15)*I(VY)}
        RF     8    9     10
        LF     9    10    20MH
```

```
          VY    10    0    DC    0V                      ; Senses the field current
          S2    7     8    11    0    SMOD               ; Voltage-controlled switch
          D2    0     8    DMOD
          E2    12    13   VALUE = {0.91*I(VX)*I(VY)}    ; Torque developed
          VL    14    13   PULSE (50 150 2MS 1NS 1NS 1MS 20MS)  ; Load torque
          E3    0     14   VALUE = {VISCOUS*V(17)}       ; Viscous torque
          D3    12    15   DMOD
          Rg    15    0    1G
          G1    0     16   15    0    1                  ; Net torque
          VZ    16    17   DC    0V                      ; Measures the net torque
          C1    17    0    1    IC=0V
          Rc1   17    0    1G
          G2    0     18   17    0    1                  ; Velocity to position
          C2    18    0    1    IC=0V
          Rc2   18    0    1G
ANALYSIS  ■ ■ ■ .TRAN  10US  4MS   UIC  ; Transient analysis with initial condition
          .PLOT    TRAN    V(3)   V(1)           ; Prints on the output file
          .OPTIONS ABSTOL = 1.00N  RELTOL = 0.01  VNTOL = 0.1  ITL5=50000
          .PROBE                               ; Graphics post-processor
.END
■ ■ ■ ■ ■ ■ ■ ■ ■ ■ ■ ■ ■ ■ ■ ■ ■ ■ ■ ■ ■ ■ ■ ■ ■ ■ ■ ■ ■ ■ ■ ■ ■ ■ ■ ■ ■ ■ ■ ■ ■ ■ ■ ■ ■
```

The plots of the transient response of the load torque V(14,13) and the motor speed V(17) are shown in Fig. 15-5. As expected, an increase in load torque slows down the speed rise.

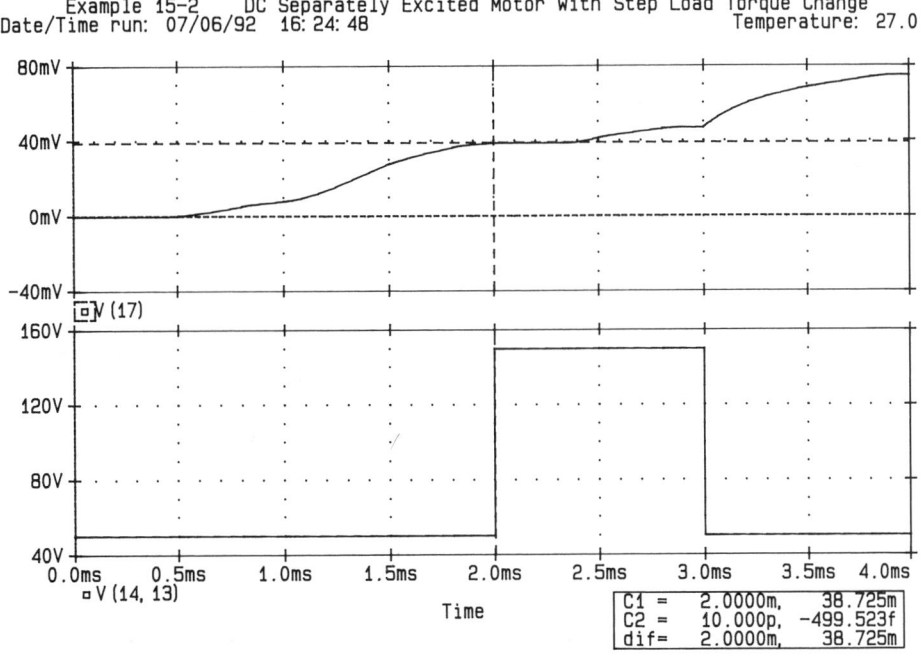

Figure 15-5 Plots of the load torque and the motor speed for Example 15-2.

PSpice can simulate the equivalent circuit of an induction motor and generator to determine their control characteristics. Parameters such as supply frequency, rotor resistance, and rotor slip can be varied to find the effects of control variables on the performance of motor drives. An inverter that controls the motor voltage, current, and frequency can be added to the motor circuit to simulate an inverter-fed induction motor drive.

Example 15-3

The equivalent circuit of an induction motor is shown in Fig. 15-6. Use PSpice to plot the torque developed against frequency for slip $s = 0.1, 0.25, 0.5$, and 0.75. The supply frequency is to be varied from 0.1 to 100 Hz. The motor parameters are:

Stator resistance, $R_s = 0.42\ \Omega$

Stator inductance, $L_s = 2.18$ mH

Rotor resistance, $R_r = 0.42\ \Omega$ (referred to stator)

Rotor inductance, $L_r = 2.18$ mH (referred to stator)

Magnetizing inductance, $L_m = 58.36$ mH (referred to stator)

Solution The effective resistance due to slip is

$$R_{\text{slip}} = R_r \frac{1 - s}{s}$$

By varying the slip, R_{slip}, the torque developed, T_L, can be varied. The list of the circuit file is as follows:

■ ■

Example 15-3 Torque–speed characteristic of induction motor

```
SOURCE   ■ VS    1    0    AC    170V              ; Input voltage of  170V
CIRCUIT  ■ ■ .PARAM  SLIP = 0.05                   ; Slip
           .PARAM  RRES = 0.42                     ; Rotor resistance
           .PARAM  RSLIP = {RRES*(1-slip)/slip}
```

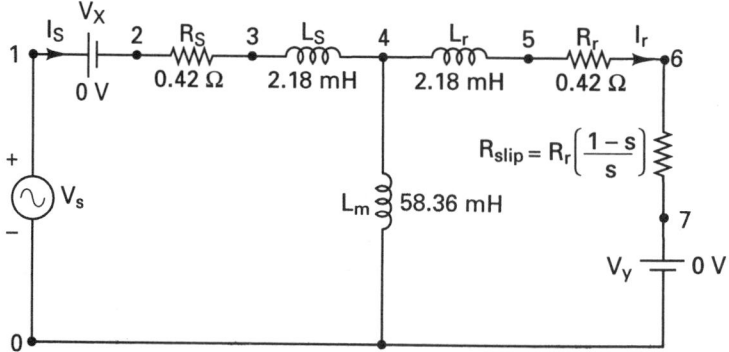

Figure 15-6 Equivalent circuit for inductor motor.

```
.STEP PARAM SLIP LIST  0.1  0.25  0.5  0.75  ; Slip values
VX    1    2    DC    0V                       ; Senses the stator current
RS    2    3    {RRES}
LS    3    4    2.18MH
LM    4    0    58.36MH
LR    4    5    2.18MH
RR    5    6    {RRES}
RX    6    7    {RSLIP}
VY    7    0    DC    0V                       ; Senses the rotor current
.AC    DEC    100    0.1HZ    100HZ             ; Ac analysis
.OPTIONS ABSTOL = 1.00N  RELTOL = 0.01  VNTOL = 0.1  ITL5=50000
.PROBE                                          ; Graphics post-processor
.END
```

The plots of the torque developed, $T_d = V(5) \times I(VY)/2$, against frequency for various slips are shown in Fig. 15-7. The torque developed increases as the slip is reduced. For a fixed slip there is a region of constant torque.

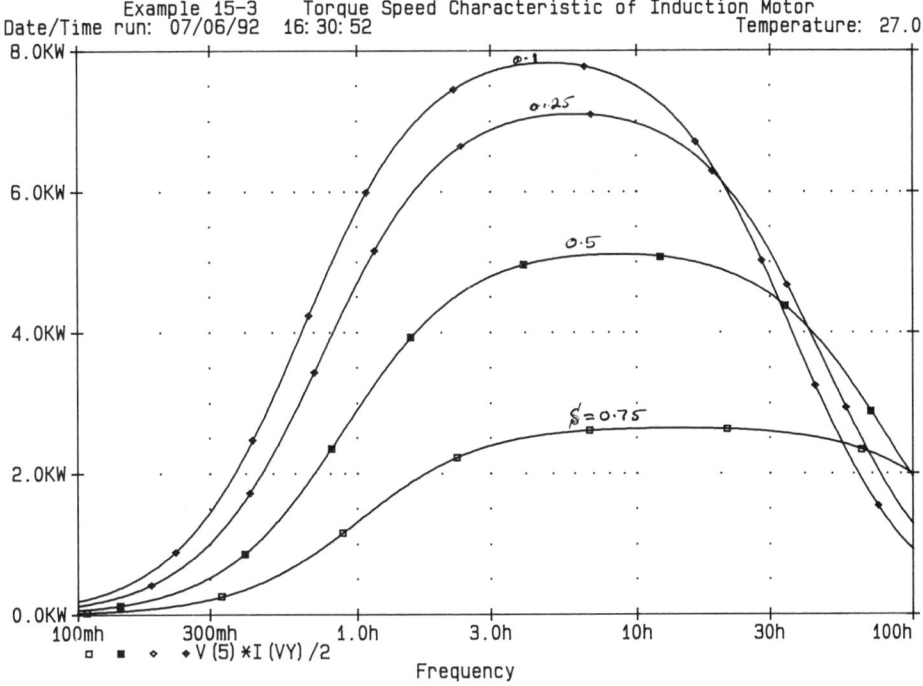

Figure 15-7 Plots of developed torque against frequency for Example 15-3.

Example 15-4

Repeat Example 15-3 for rotor resistance $R_r = 0.1\ \Omega$, $0.2\ \Omega$, $0.3\ \Omega$, and $0.42\ \Omega$. The slip is kept fixed at $s = 0.1$.

Solution The list of the circuit file is as follows:

■ ■

Example 15-4   Torque–speed characteristic with variable rotor resistance

| SOURCE | ■ VS | 1 | 0 | AC | 170V | | ; Input voltage of 170V |

```
SOURCE   ■ VS      1    0    AC    170V                    ; Input voltage of 170V
CIRCUIT  ■ ■ .PARAM  SLIP = 0.1                            ; Slip
            .PARAM  RRES = 0.42                            ; Rotor resistance
            .PARAM  RSLIP = {RRES*(1-slip)/slip}
            .STEP   PARAM  RRES  LIST  0.1  0.2  0.3  0.42 ; List values
            VX      1    2    DC    0V                     ; Senses the stator current
            RS      2    3    {RRES}
            LS      3    4    2.18MH
            LM      4    0    58.36MH
            LR      4    5    2.18MH
            RR      5    6    {RRES}
            RX      6    7    {RSLIP}
            VY      7    0    DC    0V                      ; Senses the rotor current
ANALYSIS  ■ ■ ■ .AC   DEC  100  0.1HZ  100HZ               ; Ac analysis
            .OPTIONS ABSTOL = 1.00N  RELTOL = 0.01  VNTOL = 0.1  ITL5=50000
            .PROBE                                          ; Graphics post-processor
          .END
```

■ ■

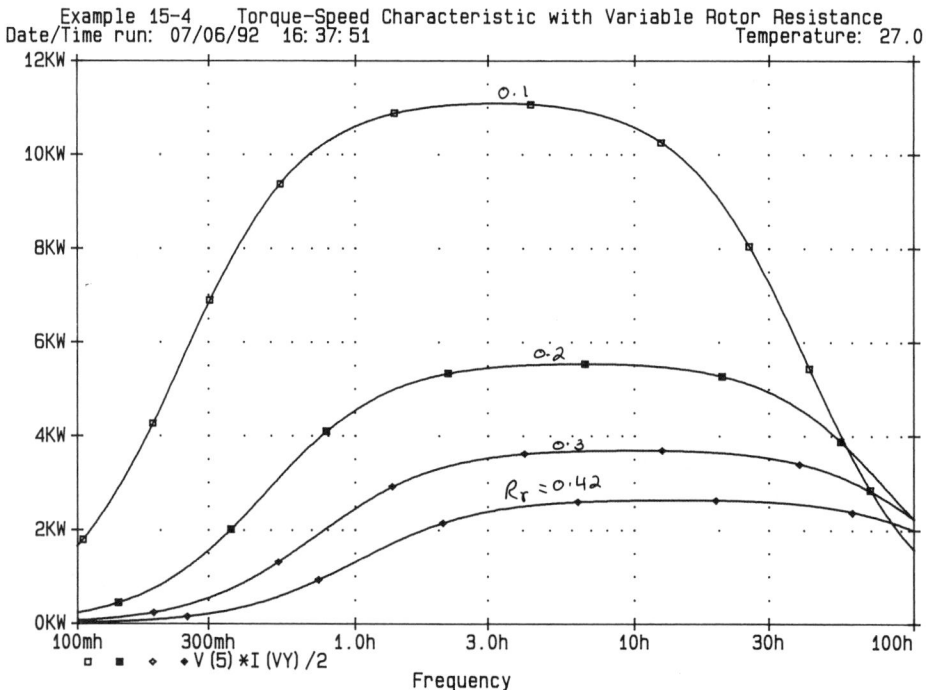

Figure 15-8 Plots of developed torque against frequency for Example 15-4.

The plots of the torque developed, $T_d = V(5) \times I(VY)/2$, against frequency for various slips are shown in Fig. 15-8. The torque increases as the rotor resistance is reduced.

SUGGESTED READING

1. J. F. Lindsay and M. H. Rashid, *Electromechanics and Electrical Machinery*. Englewood Cliffs, N.J.: Prentice Hall, 1986.
2. Y. C. Liang and V. J. Gosbel, ''DC machine models for SPICE2 simulation,'' *IEEE Transactions on Power Electronics*, Vol. 5, No. 1, January 1990, pp. 16–20.
3. M. H. Rashid, *Power Electronics: Circuits, Devices, and Applications*, 2nd ed. Englewood Cliffs, N.J.: Prentice Hall, 1993.

PROBLEMS

15-1. The dc chopper of Fig. 15-1 is operating at a frequency of $f_c = 2$ kHz and a duty cycle of $k = 0.75$. The dc supply voltage to the armature is $V_s = 220$ V. The field current is also controlled by a dc chopper operating at a frequency of $f_s = 2$ kHz and a duty cycle of $\delta = 0.75$. The dc supply voltage to the field is $V_f = 220$ V. The motor parameters are;

Armature resistance, $R_m = 1\ \Omega$

Armature inductance, $L_m = 5$ mH

Field resistance, $R_f = 10\ \Omega$

Field inductance, $L_f = 10$ mH

Back-emf constant, $K = 0.91$

Viscous torque constant, $B = 0.4$

Motor inertia, $J = 0.8$

Load torque, $T_L = 10, 100, 200$ N·m

Use PSpice to plot the transient response of **(a)** the armature and field currents, **(b)** the torque developed, and **(c)** the motor speed for a duration of 0 to 4 ms in steps of 10 μs.

15-2. Use PSpice to plot the transient response of motor speed in Problem 15-1 if the load torque is subjected to a step change as shown in Fig. P15-2.

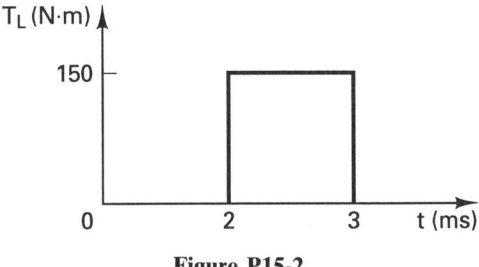

Figure P15-2

15-3. The parameters of the induction motor equivalent circuit as shown in Fig. 15-6 are:

Stator resistance, $R_s = 1.01\ \Omega$

Stator inductance, $L_s = 3.4$ mH

Rotor resistance, R_r
$= 0.69\ \Omega$ (referred to stator)

Rotor inductance, L_r
$= 5.15$ mH (referred to stator)

Magnetizing inductance, L_m
$= 115.4$ mH (referred to stator)

Use PSpice to plot the torque developed against frequency for slip $s = 0.1, 0.20, 0.4,$ and 0.6. The supply frequency is to be varied from 0.1 to 200 Hz.

15-4. Repeat Problem 15-3 for rotor resistances $R_r = 0.1, 0.2, 0.5,$ and $0.69\ \Omega$. The slip is kept fixed at $s = 0.15$.

Chapter 16

■■■■■■■■■

Difficulties

16-1 INTRODUCTION

An input file may not run for various reasons, and it is necessary to know what to do when the program does not work. To run a program successfully requires knowledge of what would not work, why not, and how to fix the problem. Although there could be many reasons why a program does not work, in this chapter we cover the problems commonly encountered and their solutions. The problems could be due to one or more of the following:

Large circuits
Running multiple circuits
Large outputs
Long transient runs
Convergence
Analysis accuracy
Negative component values
Power-switching circuits
Floating nodes
Nodes with fewer than two connections
Voltage source and inductor loops

16-2 LARGE CIRCUITS

The entire description of an input file must fit into RAM during the analysis. However, none of the results of any analysis are stored into RAM. All results (including intermediate results for the .PRINT and .PLOT statements) go to the output file or one of the temporary files. Therefore, whether the run would fit into RAM depends on how big the input file is.

The size of an input file can be found by using the ACCT option in the .OPTIONS statement and looking at the MEMUSE number printed at the end of runs that ran successfully. MEMUSE is the peak memory use of that circuit. If the circuit file does not fit, the possible remedies are:

1. To break it up into pieces and run the pieces separately.
2. To reduce the amount of memory taken up by other resident software (e.g., DOS, utilities). The total memory available can be checked by the DOS command CHKDSK.
3. To buy more memory (up to 640 kilobytes, the most that PC-DOS recognizes).

16-3 RUNNING MULTIPLE CIRCUITS

A set of circuits may be run as a single job by putting all the circuits into one input file. Each circuit begins with a title statement and ends with an .END command, as usual. PSPICE1.EXE will read through all the circuits in the input file and process each one in sequence. The output file will contain the outputs from each circuit in the same order as they appear in the input file. This technique is most suitable for running a set of large circuits overnight, especially with SPICE or the professional version of PSpice. However, Probe can be used in this situation, because only the results of the last circuit will be available for graphical output by Probe.

16-4 LARGE OUTPUTS

A large output file will be generated if an input file is run with several circuits, or for several temperatures, or with the sensitivity analysis. This will not be a problem with a hard disk. For a PC with floppy disks, the diskette may be filled with the output file. The best solution for this is to:

1. Direct the output to the printer instead of a file, or
2. Direct the output to an empty diskette instead of the one containing PSPICE1.EXE, by assigning the PSpice programs to drive A: and the input and output files to drive B:. The command to run a circuit file would be

```
A:PSPICE  B:EX2-1.CIR  B:EX2-1.OUT
```

It is recommended that SPICE run on a hard disk.

Long transient analysis runs can be avoided by choosing the appropriate limit options. The limits that affect the transient analysis are:

1. Number of print steps in a run, LIMPTS
2. Number of total iterations in a run, ITL5
3. Number of data points that Probe can handle

The number of print steps in a run is limited to the value of the LIMPTS option. It has a default value of 0 (meaning no limit) but can be set to a positive value as high as 32,000 (e.g., .OPTIONS LIMPTS=6000). The number of print steps is simply the final analysis time divided by the print interval time (plus one). The size of the output file that is generated by PSpice can be limited in case of errors by the LIMPTS option.

The total number of iterations in a run is limited to the value of the ITL5 option. It has a default value of 5000 but can be set as high as 2×10^9 (e.g., .OPTIONS ITL5=8000). The limit can be turned off by setting ITL5 = 0. This is the same as setting ITL5 to infinity and is often more convenient than setting it to a positive number. It is advisable to set ITL5 = 0.

Probe limits the data points to 16,000. This limit can be overcome by using the third parameter on the .TRAN statement to suppress part of the output at the beginning of the run. For a transient analysis from 0 to 10 ms in steps of 10 μs and printing output from 8 to 10 ms, the command would be

```
.TRAN    10US    10MS    8MS
```

Note. The limit options from Tables 6-1 and 6-2 can be typed into the circuit file as an .OPTIONS command. Alternatively, the limit options can be set from the *change options menu*, in which case PSpice writes these options into the circuit file automatically.

16-6 CONVERGENCE

PSpice uses iterative algorithms. These algorithms start with a set of node voltages, and each iteration calculates a new set, which is expected to be closer to a solution of Kirchhoff's voltage and current laws. That is, an initial guess is used and the successive iterations are expected to converge to the solution. Convergence problems may occur in:

Dc sweep
Bias-point calculation
Transient analysis

16-6.1 DC Sweep

If the iterations do not converge onto a solution, the analysis fails. The dc sweep skips the remaining points in the sweep. The most common cause of failure of the dc sweep analysis is an attempt to analyze a circuit with regenerative feedback, such as Schmitt triggers. The dc sweep is not appropriate for calculating the hysteresis of such circuits, because it is required to jump discontinuously from one solution to another at the crossover point.

To obtain hysteresis characteristics, it is advisable to use transient analysis with a piecewise linear (PWL) voltage source to generate a very slowly rising ramp. There is no CPU-time penalty for this, because PSpice will adjust the internal time step to be large away from the crossover point and small in that region. A very slow ramp assures that the switching time of the circuit will not affect hysteresis levels. This is similar to changing the input voltage slowly until the circuit switches. With a PWL source in transient analysis, the hysteresis characteristics due to upward and downward switching can be calculated.

Example 16-1

An emitter-coupled Schmitt trigger circuit is shown in Fig. 16-1(a). Plot the hysteresis characteristics of the circuit from the results of the transient analysis. The input voltage, which is varied slowly from 1.5 to 3.5 V and from 3.5 to 1.5 V, is as shown in Fig. 16-1(b). The model parameters of the transistors are IS=2.33E−27, BF=13, CJE=1PF, CJC=607.3PF, and TF=26.5NS. Print the job statistical summary of the circuit.

Solution The input voltage is varied very slowly from 1.5 to 3.5 V and from 3.5 to 1.5 V as shown in Fig. 16-1(b). The list of the circuit file is as follows:

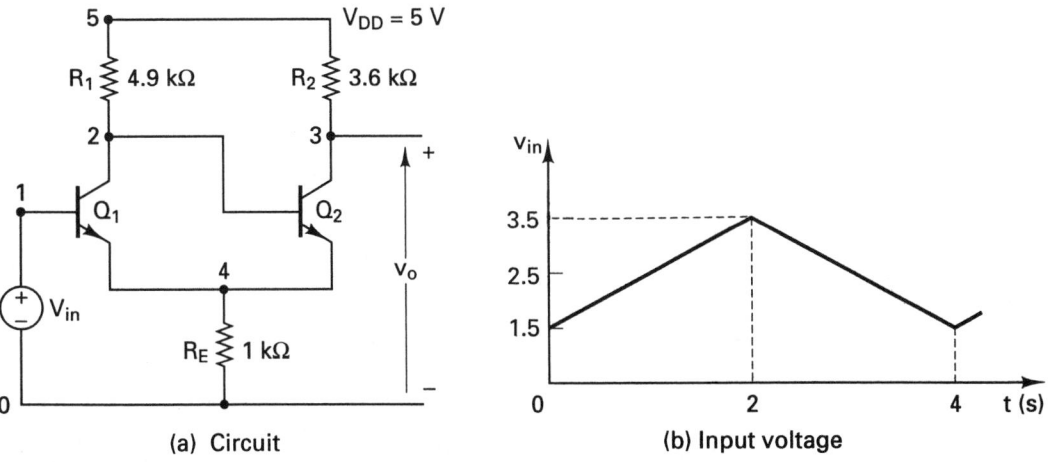

Figure 16-1 Schmitt trigger circuit.

■ ■

Example 16-1   Emitter-coupled trigger circuit

SOURCE	■ VDD	5	0	DC	5V		; Dc supply voltage of 5 V
	VIN	1	0	PWL (0	1.5V 2 3.5V 4 1.5V)		; PWL waveform
CIRCUIT	■ ■ R1	5	2	4.9K			
	R2	5	3	3.6K			
	RE	4	0	1K			
	Q1	2	1	4	4	2N6546	; Transistor Q1
	Q2	3	2	4	4	2N6546	; Transistor Q2

.MODEL 2N6546 NPN(IS=2.33E-27 BF=13 CJE=1PF CJC=607.3PF TF=26.5NS)

.OPTIONS   ACCT   ; Printing the accounts summary

ANALYSIS ■ ■ ■ .TRAN 0.01 4 ; Transient analysis from 0 to 4 s in steps of 0.01 s

.PROBE   ; Graphics post-processor

.END

■ ■

The hysteresis characteristic for Example 16-1 is shown in Fig. 16-2. The job statistical summary, which is obtained from the output file, is as follows:

```
****      JOB STATISTICS SUMMARY
   NUNODS   NCNODS   NUMNOD   NUMEL   DIODES    BJTS    JFETS   MFETS GASFETS
      6        6        6        7       0        2        0       0        0
  NDIGITAL   NSTOP    NTTAR    NTTBR   NTTOV    IFILL     IOPS   PERSPA
      0        8       23       23      54        0       36    64.063
  NUMTTP    NUMRTP   NUMNIT   DIGTP   DIGEVT   DIGEVL     MEMUSE
    210       40      896       0       0        0       9914
```

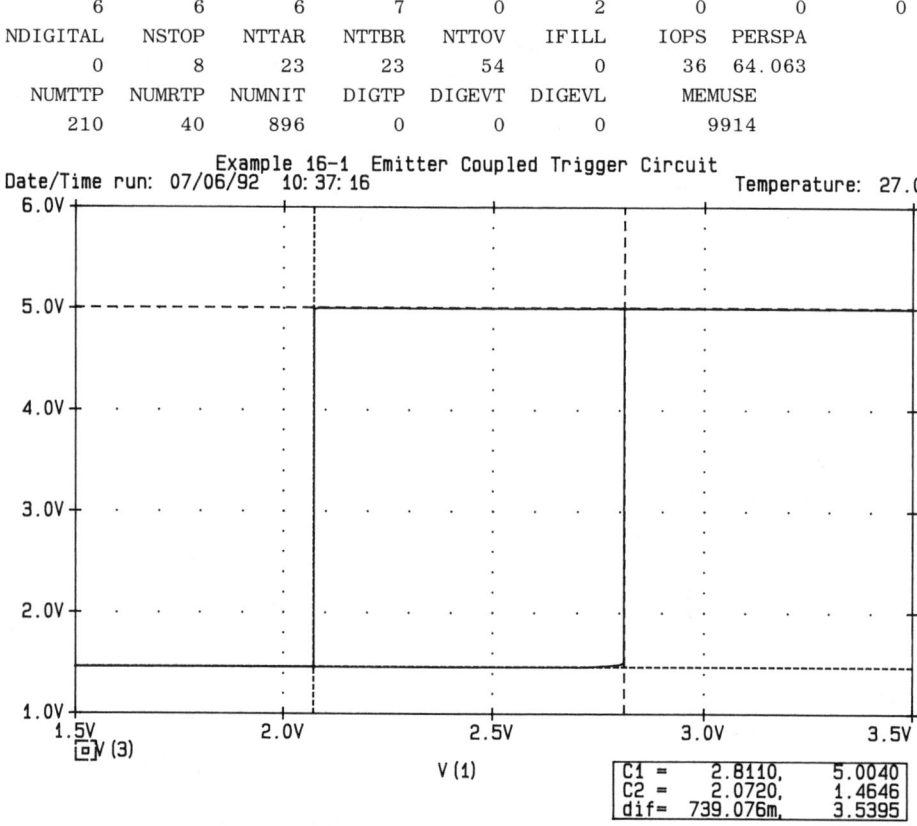

Figure 16-2 Hysteresis characteristics for Example 16-1.

	SECONDS	ITERATIONS
MATRIX SOLUTION	.77	5
MATRIX LOAD	4.42	
READIN	.50	
SETUP	.11	
DC SWEEP	0.00	0
BIAS POINT	.99	92
AC and NOISE	0.00	0
TRANSIENT ANALYSIS	9.56	896
OUTPUT	0.00	
TOTAL JOB TIME	10.54	

16-6.2 Bias-Point Calculation

Failure of the bias-point calculation prevents other analyses (e.g., ac analysis, sensitivity, etc.). The problems in calculating the bias point can be minimized by the .NODESET statement [e.g., .NODESET V(1)=0V]. By giving PSpice "hints" in the form of initial guesses for node voltages, it starts out that much closer to the solution. A little judgment must be used in assigning appropriate node voltages.

It is rare to have a convergence problem in the bias-point calculation. This is because PSpice contains an algorithm for scaling the power supplies automatically if it is having trouble finding a solution. This algorithm first tries to find a bias point with the power supplies at full scale. If there is no convergence, the power supplies are cut back to one-fourth strength and the program tries again. If there is still no convergence, the supplies are cut by another factor of 4, to one-sixteenth strength; and so on. At power supplies of 0 V, the circuit definitely has a solution with all nodes at 0 V and the program will find a solution for some supply value scaled far enough back. It then uses that solution to help it work its way back up to a solution with the power supplies at full strength. If this algorithm is in effect, a message such as

```
Power supplies cut back to 25%
```

(or some other percentage) appears on the screen while the program calculates the bias point.

16-6.3 Transient Analysis

In the case of failure due to convergence, the transient analysis skips the remaining time. The few remedies that are available for transient analysis, are:

1. To change the relative accuracy RELTOL from 0.001 to 0.01
2. To set the iteration limits at any point during transient analysis by option ITL4. Setting ITL4 = 50 (by the statement .OPTIONS ITL4=50) will allow 50 iterations at each point. As a result of more iteration points, a longer simulation time will be required. It is *not* recommended for circuits that do not have a convergence problem in transient analysis.

The accuracy of PSpice's results is controlled by the parameters RELTOL, VNTOL, ABSTOL, and CHGTOL in the .OPTIONS statement. The most important of these is RELTOL, which controls the relative accuracy of all the voltage and currents that are calculated. The default value of RELTOL is 0.001 (0.1%).

VNTOL, ABSTOL, and CHGTOL set the best accuracy for the voltages, currents, and capacitor charges/inductor fluxes, respectively. If a voltage changes its sign and it gets close to zero, RELTOL will force PSpice to calculate more and more accurate values of that voltage because 0.1% of its value becomes a tighter and tighter tolerance. This would prevent PSpice from ever letting the voltage cross zero. To prevent this problem, VNTOL can limit the accuracy of all voltages to a finite value, and the default value is 1 μV. Similarly, ABSTOL and CHGTOL can limit the currents and charges (or fluxes), respectively.

The default values for the error tolerances in PSpice are the same as in the University of California–Berkeley's SPICE2. However, they differ from that of the commercial HSpice program as shown in Table 16-1. RELTOL = .001 (0.1%) is more accurate than that necessary for many applications. The speed can be increased by setting RELTOL = 0.01 (1%), and this would increase the average speed-up by a factor of 1.5. In most power electronics circuits, the default values can be changed without affecting the results significantly.

TABLE 16-1 TOLERANCES

	PSpice	SPICE2	HSpice
RELTOL	0.001	0.001	0.01
VNTOL	1 μV	1 μV	50 μV
ABSTOL	1 μA	1 μA	1 nA

Note. The limit options from Table 6-1 can be typed into the circuit file as an .OPTIONS command. Alternatively, the limit options can be set from the *change options menu*, in which case PSpice writes these options into the circuit file automatically.

16-8 NEGATIVE COMPONENT VALUES

PSpice allows negative values for resistors, capacitors, and inductors. It should calculate a bias point or dc sweep for such a circuit. The .AC and .NOISE analyses can handle negative components. In the case of resistors, their noise contribution comes from their absolute values, and the components are not allowed to generate negative noise. However, negative components, especially negative capacitors and inductors, may cause instabilities in time, and the transient analysis may fail for a circuit with negative components.

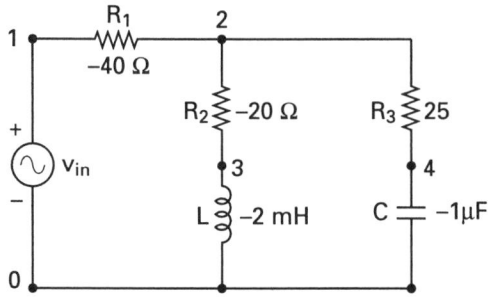

Figure 16-3 Circuit with negative components.

Example 16-2

A circuit with negative components is shown in Fig. 16-3. The input voltage is V_{in} = 120 V (peak), 60 Hz. Use PSpice to calculate the currents I(R1), I(R2), and I(R3) and the voltage V(2).

Solution The list of the circuit file is as follows:

■ ■

Example 16-2   Circuit with negative components

SOURCE	■ VIN	1	0	AC	120V	; Ac input voltage of 120 V
CIRCUIT	■ ■ R1	1	2	−40		; Negative resistances
	R2	2	3	−20		
	R3	2	4	25		
	L	3	0	−2MH		; Negative inductance
	C	4	0	−1UF		; Negative capacitance
ANALYSIS	■ ■ ■ .AC	LIN	1	60HZ	60HZ	; Ac analysis
	.PRINT	AC	IM(R1)	IM(R2)	IM(R3)	VM(2)
	.END					

■ ■

The results of PSpice simulation are as follows:

```
****      AC ANALYSIS                        TEMPERATURE =   27.000 DEG C
  FREQ          IM(R1)        IP(R1)      VM(2)         VP(2)
6.000E+01     2.000E+00     2.000E+00   1.509E−02     4.003E+01
              JOB CONCLUDED
              TOTAL JOB TIME            1.26
```

16-9 POWER SWITCHING CIRCUITS

The SPICE program was developed to simulate integrated circuits containing many small, fast transistors. Because of the "integrated-circuit emphasis," the default values of the overall parameters are not optimal for simulating power circuits. Convergence problems can be minimized by paying special attention to:

Model parameters of diodes and transistors
Error tolerances
Snubbing resistor
Quasi-steady-state conditions

Diodes and transistors. The default values for all parasitic resistances and capacitances in .MODEL statements are zero. If the parameters RS and CJO are not specified in a .MODEL for a device, the device will have no ohmic resistance and no junction capacitance. With RS = 0, the circuit may have nothing to limit the forward current through the device, and the current can easily become large enough to cause numerical problems. With CJO = 0 (and TT = 0), the device will have zero switching time and the transient analysis may find itself trying to make a transition in zero time. This will cause PSpice to make the internal time step smaller and smaller until it gives up and reports a transient convergence problem.

Error tolerances. The main error tolerance is RELTOL (default = 0.001 = .1%). This is not affected by power circuits. VNTOL is used for setting the most accurate voltage (default = 1 μV), and ABSTOL is used for setting the most accurate current (default = 1 pA). The dynamic range of PSpice is about 12 orders of magnitude. In a circuit with currents in the kiloampere range, ABSTOL = 1 pA will exceed this range and may cause a convergence problem. For power circuits it is often necessary to adjust ABSTOL higher than its default value of 1 pA. The recommended settings for VNTOL and ABSTOL are about nine orders of magnitude smaller than the typical voltages and currents in the circuit. Almost all power circuits should work with the settings:

ABSTOL = 1 μA	For a circuit with currents in the kiloampere range
ABSTOL = 1 mA	For a circuit with currents in the megaampere range
VNTOL = 1 μV	For a circuit with voltages in the kilovolt range

Snubbing resistor. In circuits containing inductors, there may be spurious ringing between the inductors and parasitic capacitances. Let us consider the diode circuit of Fig. 16-4 with L = 1 mH. The parasitic capacitance of the bridge can ring against the inductor with a very high frequency, on the order of megahertz. This ringing is the result of parasitic capacitance only, *not* the actual behavior of the circuit. During transient simulation, PSpice will take unnecessary small internal time steps and cause a convergence problem. The simplest solution is to add a snubbing resistor R_{snub} as shown by dashed lines in Fig. 16-4. The value

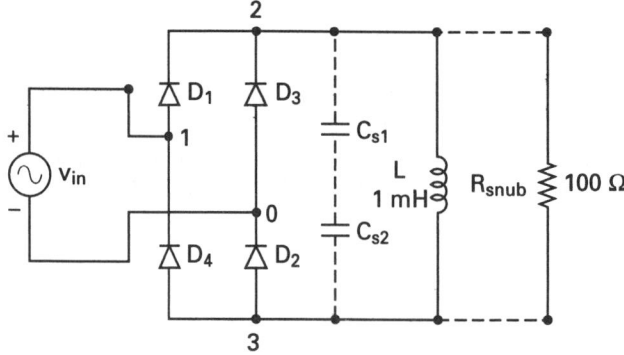

Figure 16-4 Snubbing resistor.

of R_{snub} should be chosen to match the impedance of the inductor at the corner frequency of the circuit. At low frequencies, the impedance of L is low and R_{snub} has little effect on the circuit's behavior. At high frequencies, the impedance of L is high, and R_{snub} prevents it from supporting the ringing. The action of R_{snub} is similar to the physical mechanism, primarily eddy current losses, which limit the frequency response of an inductor.

If components in series with an inductor switch off while current is still flowing in the inductor, the di/dt can be high, causing large spikes and convergence problems. A snubbing resistor can keep such spikes to a large but tractable size and thereby eliminate such convergence problems.

Quasi-steady-state condition. Running transient analysis on power switching circuits can lead to long run times. PSpice must keep the internal time step short compared to the switching period, but the circuit's response extends generally over many switching cycles. This problem can be solved by transforming the switching circuit into an equivalent circuit without switching. The equivalent circuit represents a sort of "quasi steady state" of the actual circuit and can accurately model the actual circuit's response as long as the inputs do not change too fast. This is illustrated in Example 16-3.

Example 16-3

A single-phase bridge-resonant inverter is shown in Fig. 16-5. The transistors and diodes can be considered as switches whose on-state resistance is 10 mΩ and the on-state voltage is 0.2 V. Plot the transient response of the capacitor voltage and the current through the load from 0 to 2 ms in steps of 10 μs. The output frequency of the inverter is f_o = 4 kHz.

Solution When transistors Q_1 and Q_2 are turned on, the voltage applied to the load will be V_s, and the resonant oscillation will continue for the entire resonant period, first through Q_1 and Q_2 and then through diodes D_1 and D_2. When transistors Q_3 and Q_4 are turned on, the load voltage will be $-V_s$, and the oscillation will continue for another entire period, first through Q_3 and Q_4 and then through diodes D_3 and D_4. The resonant period of the circuit is calculated approximately as

$$\omega_r = \left(\frac{1}{LC} - \frac{R^2}{4L^2}\right)^{1/2}$$

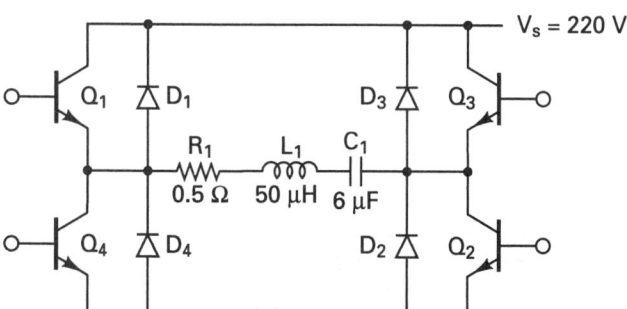

Figure 16-5 Single-phase bridge-resonant inverter.

For $L = L_1 = 50\ \mu H$, $C = C_1 = 6\ \mu F$, and $R = R_1 + R_{1(sat)} + R_{2(sat)} = 0.5 + 0.2 + 0.2 = 0.54\ \Omega$, $\omega_r = 57572.2$ rad/s, and $f_r = \omega_r/2\pi = 9162.9$ Hz. The resonant period is $T_r = 1/f_r = 1/9162.9 = 109.1\ \mu s$. The period of the output voltage is $T_o = 1/f_o = 1/4000 = 250\ \mu s$.

The switching action of the inverter can be represented by two voltage-controlled switches as shown in Fig. 16-6(a). The switches are controlled by voltages as shown in Fig. 16-6(b). The on-time of switches, which should be approximately equal to the resonant period of the output voltage, is assumed to be 112 μs. Switch S_2 is delayed by 115 μs to take account of overlap. The model parameters of the switches are RON=0.01, ROFF=10E+6, VON=0.001, and VOFF=0.0.

The list of the circuit file is as follows:

■ ■

Example 16-3   Full-bridge resonant inverter

SOURCE ■ *   The controlling voltage for switch S1:

    V1   1   0   PULSE (0   220V 0 1US 1US 110US   250US)

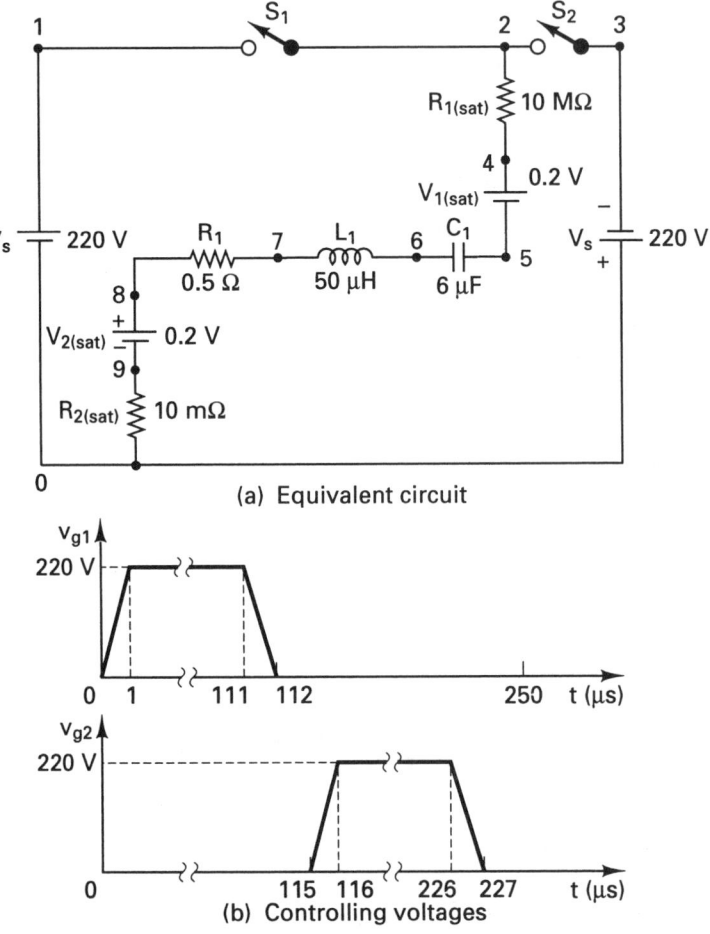

Figure 16-6   Equivalent circuit for Fig. 16-5.

```
             *   The controlling voltage for switch S2 with a delay time of 115 μs:
             V2   3    0    PULSE (0 −220V 115US 1US 1US 110US 250US)
CIRCUIT   ■ ■ S1   1    2    1    0    SMOD  ; Voltage-controlled switches
             S2   2    3    0    3    SMOD
             *   Switch model parameters for SMOD:
             .MODEL   SMOD   VSWITCH  (RON=0.01  ROFF=10E+10E+6  VON=0.001 VOFF=0.0)
             RSAT1   2    4    10M
             VSAT1   4    5    DC   0.2V
             RSAT2   9    0    10M
             VSAT2   8    9    DC   0.2V
             *  Assuming an initial capacitor voltage of −250V to reduce settling time:
             C1   5    6    6UF   IC=−250V
             L1   6    7    50UH
             R1   7    8    0.5
ANALYSIS  ■ ■ ■ *  Transient analysis with UIC condition:
             .TRAN   1US   500US   UIC  ; Transient analysis with UIC condition
             .PROBE                     ; Graphics post-processor
      .END
```

The transient response for Example 16-3 is shown in Fig. 16-7.

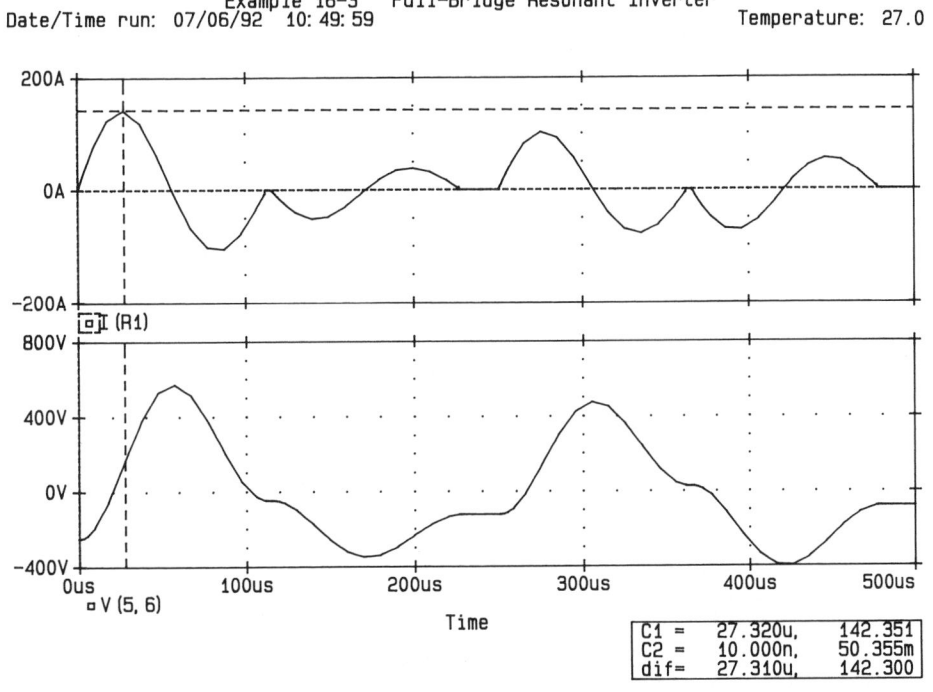

Figure 16-7 Transient responses for Example 16-3.

PSpice requires that there should be no floating nodes. If there is any, PSpice will indicate a read-in error in the screen, and the output file will contain a message, which is similar to

```
ERROR: Node 15 is floating
```

This means that there is no dc path from node 15 to ground. A dc path is a path through resistors, inductors, diodes, and transistors. This is a very common problem, and it can occur in many circuits, as shown in Fig. 16-8.

Node 4 in Fig. 16-8(a) is floating and does not have a dc path. This problem can be avoided by connecting node 4 to node 0 as shown by dashed lines (or by

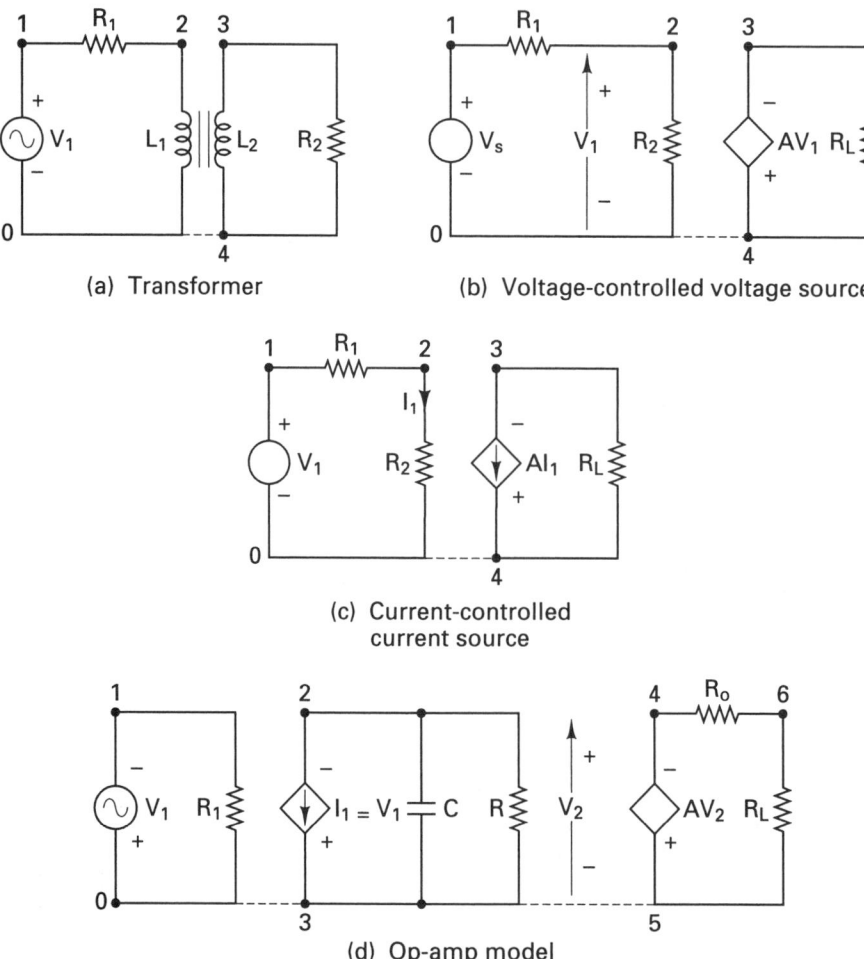

(a) Transformer

(b) Voltage-controlled voltage source

(c) Current-controlled current source

(d) Op-amp model

Figure 16-8 Typical circuits with floating nodes.

connecting node 3 to node 2). A similar situation can occur in voltage-controlled and current-controlled sources as shown in Fig. 16-8(b) and (c). The model of op-amps as shown in Fig. 16-8(d) has many floating nodes, which should be connected to provide dc paths to ground. For examples, nodes 0, 3, and 5 could be connected together or, alternatively, nodes 1, 2, and 4 may be joined together.

The two sides of a capacitor have no dc path between them. If there are many capacitors in a circuit as shown in Fig. 16-9, nodes 3 and 5 do not have dc

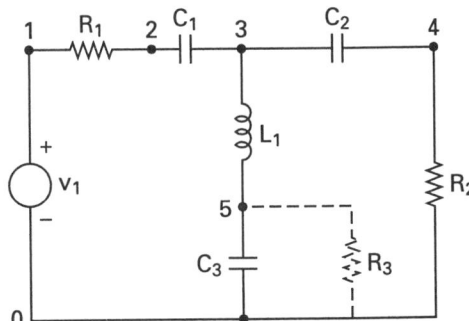

Figure 16-9 Typical circuit without dc path.

paths. Dc paths can be provided by connecting a very large resistance R_3 (say, 100 MΩ) across capacitor C_3, as shown by dashed lines.

Example 16-4

A passive filter is shown in Fig. 16-10. The output is taken from node 9. Plot the magnitude and phase of the output voltage, separately against the frequency. The frequency should be varied from 100 Hz to 10 kHz in steps of one decade and 10 points per decade.

Solution The nodes between C_1 and C_3, C_3 and C_5, C_5 and C_7 do not have dc paths to the ground. Therefore, the circuit cannot be analyzed without connecting resis-

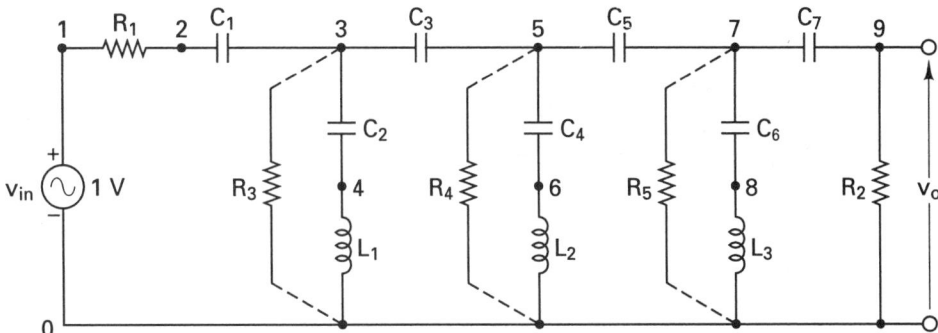

$R_1 = 10$ kΩ, $R_2 = 10$ kΩ, $R_3 = R_4 = R_5 = 200$ MΩ
$C_1 = 7$ nF, $C_2 = 70$ nF, $C_3 = 6$ nF, $C_4 = 22$ nF, $C_5 = 7.5$ nF
$C_6 = 12$ nF, $C_7 = 10.5$ nF, $L_1 = 1.5$ mH
$L_2 = 1.75$ mH, $L_3 = 2.5$ mH

Figure 16-10 Passive filter.

tors R_3, R_4, and R_5 as shown in Fig. 16-10 by dashed lines. If the values of these resistance are very high, say 200 MΩ, their influence on the ac analysis would be negligible. The list of the circuit file is as follows:

■■■

Example 16-4 Passive filter

SOURCE	■ VIN	1	0	AC	1V	; Input voltage is 1 V peak

CIRCUIT ■■ R1   1   2   10K

R2   9   0   10K

*   Resistances R3, R4, and R5 are connected to provide dc paths:

R3   3   0   200MEG

R4   5   0   200MEG

R5   7   0   200MEG

C1   2   3   7NF

C2   3   4   70NF

C3   3   5   6NF

C4   5   6   22NF

C5   5   7   7.5NF

C6   7   8   12NF

C7   7   9   10.5NF

L1   4   0   1.5MH

L2   6   0   1.75MH

L3   8   0   2.5MH

ANALYSIS ■■■ * Ac analysis for 100 Hz to 10 kHz with a decade increment and

* 10 points per decade:

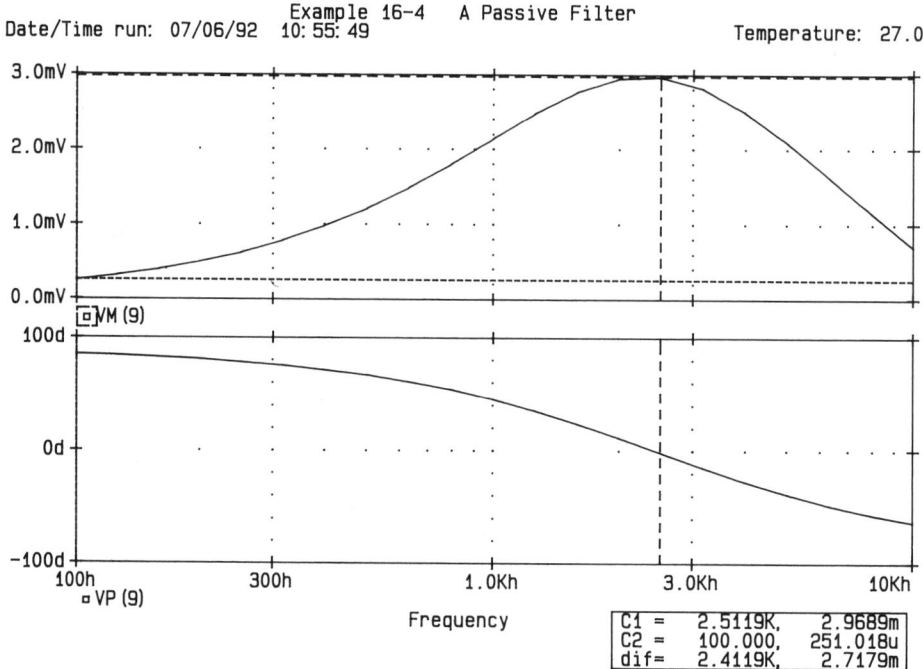

Example 16-4 A Passive Filter

Date/Time run: 07/06/92 10:55:49   Temperature: 27.0

C1 =	2.5119K,	2.9689m
C2 =	100.000,	251.018u
dif=	2.4119K,	2.7179m

Figure 16-11 Frequency response for Example 16-4.

```
.AC     DEC   10    100    10KHZ
* Plot the results of ac analysis for the magnitude of voltage
* at node 9.
.PLOT   AC    VM(9)   VP(9)              ; Plots on output file
.PLOT   AG    VP(9)                      ; Plots on output file
.PROBE                                   ; Graphics post-processor
.END
```

The frequency response for Example 16-4 is shown in Fig. 16-11.

16-11 NODES WITH FEWER THAN TWO CONNECTIONS

PSpice requires that every node be connected to at least two other nodes. Otherwise, PSpice will give an error message, which is similar to

```
ERROR: Fewer than two connections at node 10
```

This means that node 10 must have at least another connection. A typical situation is shown in Fig. 16-12(a), where node 3 has only one connection. This problem can be solved by short-circuiting resistance R_2 as shown by dashed lines.

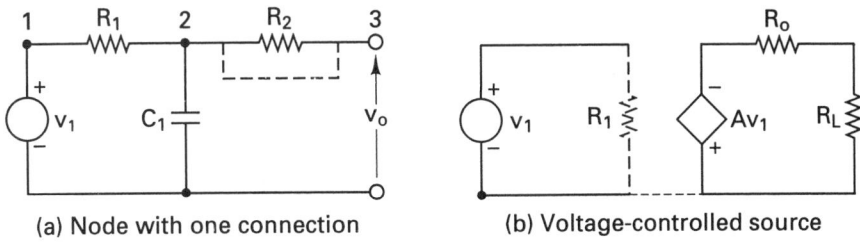

(a) Node with one connection   (b) Voltage-controlled source

Figure 16-12   Typical circuits with less than two connections at a node.

An error message may be indicated in the output file for a circuit with voltage-controlled sources as shown in Fig. 16-12(b). The input to the voltage-controlled source will not be considered to have connections during the check by PSpice. This is because the input draws no current and it has infinite impedance. A very high resistance (say, 10 GΩ) may be connected from the input to the ground as shown by dashed lines.

16-12 VOLTAGE SOURCE AND INDUCTOR LOOPS

PSpice requires that there be no loops with zero resistance. Otherwise, PSpice will indicate a read-in error on the screen, and the output file will contain a message similar to

```
ERROR:  Voltage loop involving V5
```

This means that the circuit has a loop of zero-resistance components, one of which is V5. The zero-resistance components in PSpice are independent voltage sources (V), inductors (L), voltage-controlled voltage sources (E), and current-controlled voltage sources (H). Typical circuits with such loops are shown in Fig. 16-13.

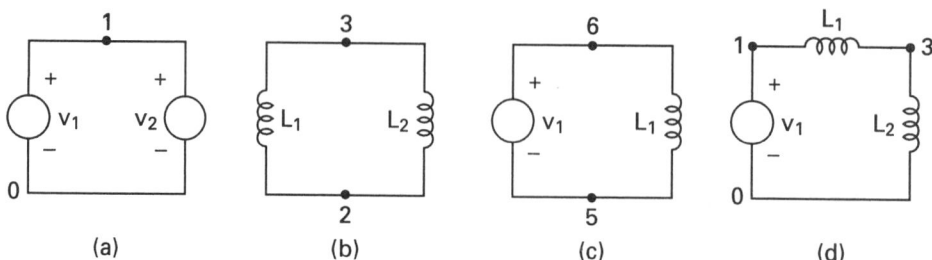

Figure 16-13 Typical circuits with zero-resistance loops.

It does not matter whether the values of the voltage sources are zero or not. Having a voltage source of E in a zero resistance, the program will need to divide E by 0. But having $E = 0$ V, the program will need to divide 0 by 0, which is also impossible. It is therefore the presence of a zero-resistance loop that is the problem, not the values of the voltage sources. A simple solution is to add a series resistance to at least one component in the loop. The resistor's value should be small enough so that it does not disturb the operation of the circuit. However, the resistor's value should *not* be less than 1 $\mu\Omega$.

16-13 RUNNING PSpice FILE ON SPICE

PSpice will give essentially the same results as SPICE-2G from the University of California–Berkeley (referred to as SPICE). There could be some small differences, especially for values crossing zero due to the corrections made for convergence problems. The semiconductor device models are the same as in SPICE.

There are a number of features of PSpice that are not available in SPICE. These are:

1. Extended syntax for output variables (e.g., in .PRINT and .PLOT). SPICE allows only voltages of the form $V(x)$ or $V(x,y)$ and currents through voltage sources. Group delay is not available.

2. Extra devices:
   Gallium arsenide model
   Nonlinear magnetic (transformer) model
   Voltage- and current-controlled switch models

3. Optional models for resistors, capacitors, and inductors. The temperature coefficients for capacitors and inductors and exponential temperature coefficients for resistors.

4. The model parameters RG, RDS, L, W, and WD are not available in the MOSFET's .MODEL statement in SPICE.

5. Extensions to the dc sweep. SPICE restricts the sweep variable to be the value of an independent current or voltage source. SPICE does not allow sweeping of model parameters or temperature.

6. The .LIB and .INCLUDE statements.

7. SPICE requires the input (.CIR) file to be uppercase.

16-14 RUNNING SPICE FILE ON PSpice

PSpice will run any circuit, which SPICE-2G from the University of California–Berkeley (referred to as SPICE) will run with these exceptions:

1. Circuits that use .DISTO (small-signal distortion) analysis, which has errors in Berkeley SPICE. Also, the special distortion output variables (HD2, DIM3, etc.) are not available. Instead of the .DISTO analysis, we recommend running a transient analysis and looking at the output spectrum with the Fourier transform mode of Probe. This technique shows the distortion (spectral) products for both small-signal and large-signal distortion.

2. The IN = option on the .WIDTH statement is not available. PSpice always reads the entire input file regardless of how long the input lines are.

3. Temperature coefficients for resistors must be put into a .MODEL statement instead of on the resistor statement. Similarly, the voltage coefficients for capacitors and the current coefficients for inductors are used in the .MODEL statements.

SUGGESTED READING

1. *PSpice Manual.* Irvine, Calif.: MicroSim Corporation, 1992.

2. W. Blume, "Computer circuit simulation," *Byte*, Vol. 11, No. 7. July 1986, p. 165.

PROBLEMS

16-1. For the inverter circuit in Fig. P16-1, plot the hysteresis characteristics. The input voltage is varied slowly from -5 to $+5$V and from $+5$ to -5V. The model parameters of the PMOS are VTO$=-2.5$, KP$=4.5$E-3, CBD$=5$PF, CBS$=2$PF, CGSO$=1$PF, CGDO$=1$PF, and CGBO$=1$PF.

16-2. For the circuit in Fig. P16-2, plot the hysteresis characteristics from the results of the transient analysis. The input voltage is varied slowly from -4 to $+4$V and from $+4$ to -4V. The op-amp can be modeled as a macromodel, as shown in Fig. 14-3. The description of the macromodel is listed in library file EVAL.LIB. The supply voltages are $V_{CC} = 12$ V and $V_{EE} = -12$ V.

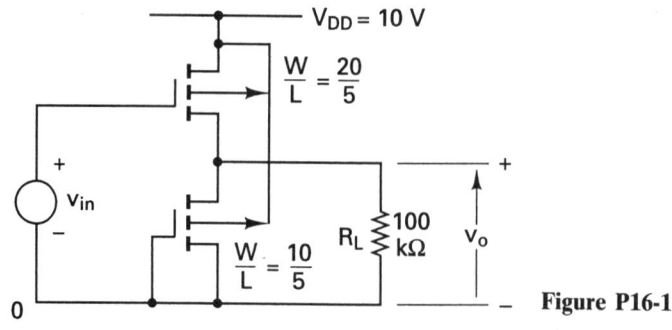

Figure P16-1

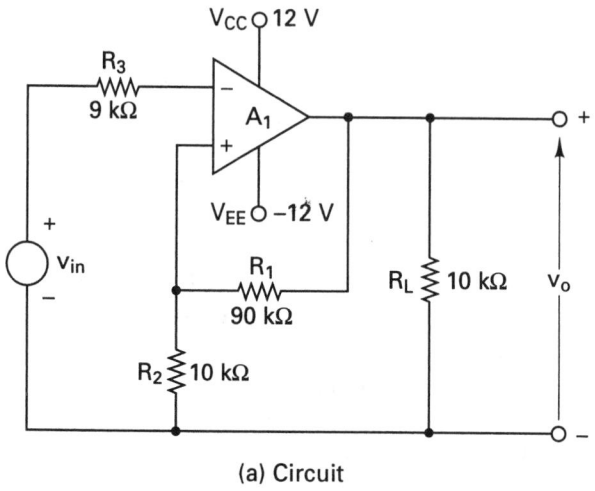

(a) Circuit

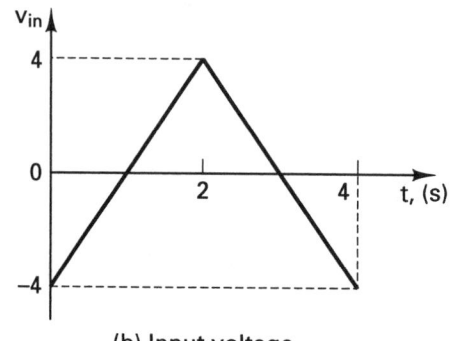

(b) Input voltage

Figure P16-2

Appendix: Running PSpice on PCs

PSpice programs are available in high-density (1.2-megabyte) diskettes or normal (360-kilobyte) diskettes. The first step is to have the directory listing of the files on the program diskettes. The next step is to print and read the README.DOC file. It contains a brief description of the type of display and hard copy that are allowed by PSpice and Probe. It also contains the system requirement and instructions on running PSpice programs.

PSpice will run on any IBM-PC or the Macintosh II or compatible computer. The student version of PSpice does not require the coprocessor for running Probe. The display could be on monochrome or color graphics monitors. There is no requirement for special features of printers. The types of printers and display can be set by editing the PROBE.DEV file or the setup menu. The simulation of a circuit requires:

Installing PSpice in PCs
Creating input files
Run command
DOS (disk operating system) commands

A-1 INSTALLING PSpice IN PCs

The steps to be followed to install a PSpice program in PCs with a hard drive are:

1. Create the directory PSPICE by typing

```
MD PSPICE   or   MD\PSPICE
```

2. Copy all the PSpice files from the diskettes into the hard drive by typing

   COPY a:*.* (and then hit return)

3. Change your current directory to PSPICE by typing

   CD PSPICE or CD\PSPICE

4. Type

   PS

5. Now the PSpice menu should appear in your monitor screen.
6. The file menu should be highlighted; if not, highlight by pressing the arrow keys.
7. Hit the return key.
8. Move to the display/printer setup submenu and hit the return key.
9. Choose the type of display and printer.

Display (e.g., IBMVGA)	(Press F4 for options/choices)
Port (e.g., PRN for printer)	(Press F4 for options/choices)
Printer (e.g., EPSON printer)	(Press F4 for options/choices)

10. Press ESC to get out.
11. Move to the current file menu and hit the return key.
12. Type the name of your new circuit file or the location and name of your existing circuit file. Hit the return key.
13. You should be in editing mode, and type the circuit descriptions in order to create (or edit) your circuit file.
14. Move to the analysis menu and run the circuit simulation.
15. Move to the Probe menu and run Probe for a graphical display of output plots on the monitor.
16. Use the browse output menu for a look through your output file.

Notes

1. Move to a menu by arrow keys and hit the return key to work on the menu.
2. Always press ESC to get out of a menu.

A-2 CREATING INPUT FILES

PSpice has a built-in text editor. The input files can also be created by text editors. The text editor that is always available is EDLIN. It comes with DOS and is described in the DOS user's guide. There are other editors, such as Program Editor (from WordPerfect Corporation). Word-processing programs such as WordStar, WordStar 2000, and Word) may also be used to create the input file. The word processor normally creates a file that is not a text file. It contains embedded characters to determine margins, paragraph boundaries, pages, and so on. However, most word processors have a command or mode to create a text file without these control characters. For example, WordStar 200 creates text files with the UNIFORM format.

A-3 RUN COMMAND

PSpice can be run from the menu, but PSpice can also be run by typing

```
PSPICE  ⟨input file⟩  ⟨output file⟩
```

By default, the input file has the extension of .CIR, and the output file has the extension of .OUT. The name of the output file defaults to the name of the input file. If the input file, EX2-1.CIR, is on the default drive, the following commands are equivalent:

```
PSPICE  EX2-1
PSPICE  EX2-1.CIR
PSPICE  EX2-1.CIR  EX2-1
PSPICE  EX2-1.CIR  EX2-1.OUT
```

The output file can be assigned to the printer that is connected to the PC by

```
PSPICE  EX1  PRN
```

The commands that will instruct PSpice as to the location of the input file and the program files will depend on the type of disk drives. There are two types of disks: with a fixed (or hard) disk and without a fixed (or floppy) disk.

With a fixed disk. Running PSpice with a fixed disk is straightforward. The PSPICE.BAT file must be in the default drive. Running PSpice will call the PSPICE.BAT file, which in turn will call the PSPICE1.EXE file and then the PROBE.EXE file, if required. PSPICE1.EXE creates one temporary file for storing intermediate results and deletes this temporary file when it finishes. If the circuit file EX2-1.CIR is on a diskette on drive A:, the command for running PSpice is

```
PSPICE A:EX2-1.CIR  A:EX2-1.OUT
```

Without a fixed disk. The input file and the PSPICE.BAT file must be on diskette 1 in drive A:. The PSPICE.BAT file is searched by DOS for programs and commands. Running PSpice will cause PSPICE.BAT to call PSPICE1.EXE from drive B:. PSPICE1.EXE creates one temporary file for storing intermediate results, and automatically deletes the temporary file after writing to the output file when it finishes. The command for running the input file EX2-1.CIR is

```
PSPICE A:EX2-1.CIR
```

A-4 DOS COMMANDS

The DOS commands that are frequently used are:

To format a brand new diskette on drive A:,

```
FORMAT A:
```

To list the directory of a diskette on drive A:,

```
DIR  A:
```

To delete the file EX2-1.CIR on drive A:,

```
Delete A:EX2-1.CIR  (or  Erase A:EX2-1.CIR)
```

TO copy the file EX2-1.CIR on drive A: to the file EX2-2.CIR on drive B:,

```
COPY  A:EX2-1.CIR  B:EX2-2.CIR
```

To copy all the files on diskette in drive A: to diskette on drive B:,

```
COPY  A:*.*  B:
```

To type the contents of file EX2-1.OUT on drive A:,

```
TYPE  A:EX2-1.OUT
```

To print the contents of the file EX2-1.CIR on drive A: to the printer, first activate the printer by pressing Ctrl (Control) and Prtsc (Print Screen) keys together and then type

```
TYPE A:EX2-1.CIR
```

The printer can be deactivated by pressing the Ctrl (control) and Prtsc (print screen) keys again.

Reference Table I

Circuit Elements and Sources	1st Letter	Model Type-Name
Bipolar Junction Transistor	Q	NPN/PNP
Capacitor	C	CAP
Current-Controlled Current Source	F	
Current-Controlled Switch	W	ISWITCH
Current-Controlled Voltage Source	H	
Diode	D	D
Exponential Source		EXP
GaAs MES Field-Effect-Transistor	B	GASFET
Independent Current Source	I	
Independent Voltage Source	V	
Inductor	L	IND/CORE
Junction Field-Effect-Transistor	J	NJF/PJF
MOS Field-Effect-Transistor	M	NMOS/PMOS
Mutual Inductors (Transformer)	K	
Piece-Wise Linear Source		PWL
Polynomial Source		POLY(n)
Pulse Source		PULSE
Resistor	R	RES
Single-Frequency Frequency-Modulation Source		SFFM
Sinusoidal Source		SIN
Transmission Line	T	
Voltage-Controlled Current Source	G	
Voltage-Controlled Switch	S	VSWITCH
Voltage-Controlled Voltage Source	E	

Reference Table II

Analysis and Functions	Commands
AC/Frequency Analysis	.AC
DC Operating Point	.OP
DC Sweep	.DC
End of Subcircuit	.ENDS
Fourier Analysis	.FOUR
Frequency Response Transfer Function	FREQ
Function Definition	.FUNC
Global Nodes	.GLOBAL
Graphical Post-Processor	.PROBE
Include File	.INC
Initial Conditions	.IC
Library File	.LIB
Model Definition	.MODEL
Node Setting	.NODESET
Noise Analysis	.NOISE
Options	.OPTIONS
Parameter Definition	PARAM
Parameter Variation	.PARAM
Parametric Analysis	.STEP
Plot Output	.PLOT
Print Output	.PRINT
Sensitivity Analysis	.SENS
Subcircuit Definition	.SUBCKT
Table	TABLE
Temperature	.TEMP
Transfer Function	.TF
Transient Analysis	.TRAN
Value	VALUE
Width	.WIDTH

■■■■■■■■
Bibliography

Allen, Phillip E., *CMOS Analog Circuit Design*. New York: Holt, Rinehart and Winston, 1987.

Antognetti, Paolo, and Guiseppe Massobrio, *Semiconductor Device Modeling with SPICE*. New York: McGraw-Hill, 1988.

Banzhaf, Walter, *Computer-Aided Circuit Analysis Using SPICE*. Englewood Cliffs, N.J.: Prentice Hall, 1989.

Bugnola, Dimitri S., *Computer Programs for Electronic Analysis and Design*. Reston, Va.: Reston Publishing Company, 1983.

Chattergy, Rahul, *SPICEY Circuits: Elements of Computer-Aided Analysis*. Boca Raton, Fla.: CRC Press, 1992.

Chua, Leon O., and Pen-Min Lin, *Computer-Aided Analysis of Electronic Circuits: Algorithms and Computational Techniques*. Englewood Cliffs, N.J.: Prentice Hall, 1975.

Ghandi, S. K., *Semiconductor Power Devices*. New York: Wiley, 1977.

Gray, Paul R., and Robert G. Meyer, *Analysis and Design of Analog Integrated Circuits*. New York: Wiley, 1984.

Grove, A. S., *Physics and Technology of Semiconductor Devices*. New York: Wiley, 1967.

Hodges, D. A., and H. G. Jackson, *Analysis and Design of Digital Integrated Circuits*. New York: McGraw-Hill, 1988.

McCalla, William J., *Fundamentals of Computer-Aided Circuit Simulation*. Norwell, Mass.: Kluwer Academic, 1988.

MicroSim Corporation, *PSpice Manual*. Irvine, Calif.: MicroSim Corporation, 1992.

Nagel, Laurence W., *SPICE2: A Computer Program to Simulate Semiconductor Circuits*, Memorandum ERL-M520, May 1975, Electronics Research Laboratory, University of California, Berkeley.

Nashelsky, Louis, and Robert Boylestad, *BASIC for Electronics and Computer Technology*. Englewood Cliffs, N.J.: Prentice Hall, 1988.

Rashid, M. H., *Power Electronics: Circuits, Devices, and Applications*, 2nd ed. Englewood Cliffs, N.J.: Prentice Hall, 1993.

Rashid, M. H., *SPICE for Circuits and Electronics Using PSpice*. Englewood Cliffs, N.J.: Prentice Hall, 1990.

Spence, Robert, and John P. Burgess, *Circuit Analysis by Computer: From Algorithms to Package*. London: Prentice Hall International (UK), 1986.

Tuinenga, Paul W., *SPICE: A Guide to Circuit Simulation and Analysis Using PSpice*, 2nd ed. Englewood Cliffs, N.J.: Prentice Hall, 1992.

van der Ziel, Aldert, *Noise in Solid State Devices*. New York: Wiley, 1986.

Index

U

UIC (Use Initial Conditions), 110

V

VALUE, 45
Values:
  element, 6
    scaling, 7
Variables:
  ac analysis, 27
  dc sweep, 23
  noise analysis, 29
  output, 10, 23
  defining, 23
  transient analysis, 23

VNTOL, 368
Voltage-controlled current source, 43
Voltage-controlled switch, 69
Voltage-controlled voltage source, 41
Voltage source:
  current-controlled, 44
  independent, 37
  voltage-controlled, 41

W

.WIDTH statement, 92

X

X (subcircuit call) device, 78

SPICE FOR CIRCUITS AND ELECTRONICS USING PSpice®
Muhammad H. Rashid

Hardware requirements: IBM PC or compatible, 640K minimum memory, a hard drive, 1.2mb diskette drive, MS DOS 3.0 or later. (Floating point coprocessor optional.)

Please send the item(s) checked below. PAYMENT ENCLOSED (check or money order only). The Publisher will pay all shipping and handling charges.

____ PSpice® Student Version Disks (two 5¼" disks) IBM PC compatible. $7.50 each set. (73476-4) Release 5.0

____ PSpice® Student Version Disks (two 3½" disks) IBM PC compatible. $9.00 each set. (73475-6) Release 5.0

____ PSpice® Student Version Disks (three 3½" disks) MAC II compatible. $15.50 each set. (73474-9) Release 5.0

NAME _____

DEPT. _____

SCHOOL _____

CITY _____ STATE _____ ZIP _____

NOTE: PROFESSIONAL/REFERENCE BOOKS ARE TAX DEDUCTIBLE.
Prices subject to change without notice. Please add sales tax for your area.

Tear out this card and fill in all necessary information. Then enclose this card with your check or money order only in an envelope and mail to:

Book Distribution Center
PRENTICE HALL
Route 59 at Brook Hill Drive
West Nyack, NY
10995

SPICE FOR CIRCUITS AND ELECTRONICS USING PSpice®
Muhammad H. Rashid

Hardware requirements: IBM PC or compatible, 640K minimum memory, a hard drive, 1.2mb diskette drive, MS DOS 3.0 or later. (Floating point coprocessor optional.)

Please send the item(s) checked below. PAYMENT ENCLOSED (check or money order only). The Publisher will pay all shipping and handling charges.

____ PSpice® Student Version Disks (two 5¼" disks) IBM PC compatible. $7.50 each set. (73476-4) Release 5.0

____ PSpice® Student Version Disks (two 3½" disks) IBM PC compatible. $9.00 each set. (73475-6) Release 5.0

____ PSpice® Student Version Disks (three 3½" disks) MAC II compatible. $15.50 each set. (73474-9) Release 5.0

NAME _____

DEPT. _____

SCHOOL _____

CITY _____ STATE _____ ZIP _____

NOTE: PROFESSIONAL/REFERENCE BOOKS ARE TAX DEDUCTIBLE.
Prices subject to change without notice. Please add sales tax for your area.